Lecture Notes in Computer Science 13880

More information about this series at https://link.springer.com/bookseries/558

Michael Hanus · Daniela Inclezan (Eds.)

Practical Aspects of Declarative Languages

25th International Symposium, PADL 2023
Boston, MA, USA, January 16–17, 2023
Proceedings

Editors
Michael Hanus
University of Kiel
Kiel, Germany

Daniela Inclezan
Miami University
Oxford, OH, USA

ISSN 0302-9743 ISSN 1611-3349 (electronic)
Lecture Notes in Computer Science
ISBN 978-3-031-24840-5 ISBN 978-3-031-24841-2 (eBook)
https://doi.org/10.1007/978-3-031-24841-2

This Springer imprint is published by the registered company Springer Nature Switzerland AG
The registered company address is: Gewerbestrasse 11, 6330 Cham, Switzerland

Preface

This volume contains the papers presented at the 25th International Symposium on Practical Aspects of Declarative Languages (PADL 2023). The symposium was co-located with the 50th ACM SIGPLAN Symposium on Principles of Programming Languages (POPL 2023). The symposium took place in Boston, Massachusetts, USA, during January 16–17, 2023.

PADL is a well-established forum for researchers and practitioners to present original work emphasizing novel applications and implementation techniques for all forms of declarative programming, including programming with sets, functions, logics, and constraints. PADL 2023 especially welcomed new ideas and approaches related to applications, design and implementation of declarative languages going beyond the scope of the past PADL symposia, for example, advanced database languages and contract languages, as well as verification and theorem proving methods that rely on declarative languages.

Originally established as a workshop (PADL 1999 in San Antonio, Texas), the PADL series developed into a regular annual symposium; other previous editions took place in Boston, Massachusetts (2000), Las Vegas, Nevada (2001), Portland, Oregon (2002), New Orleans, Louisiana (2003), Dallas, Texas (2004), Long Beach, California (2005), Charleston, South Carolina (2006), Nice, France (2007), San Francisco, California (2008), Savannah, Georgia (2009), Madrid, Spain (2010), Austin, Texas (2012), Rome, Italy (2013), San Diego, California (2014), Portland, Oregon (2015), St. Petersburg, Florida (2016), Paris, France (2017), Los Angeles, California (2018), Lisbon, Portugal (2019), New Orleans, Louisiana (2020), online (2021), and Philadelphia, Pennsylvania (2022).

This year, the call for papers resulted in 42 abstract submissions from which 36 were finally submitted as full papers. The Program Committee accepted 15 technical papers and four application papers. Each submission was reviewed by at least three Program Committee members in a single blind process and went through a five-day online discussion period by the Program Committee before a final decision was made. The selection was based only on the merit of each submission and regardless of scheduling or space constraints.

The program also included two invited talks:

- Robert Bruce Findler (Northwestern University, Evanston, USA), "Modern Macros"
- Chris Martens (Northeastern University, Boston, USA), "Towards Declarative Content Generation for Creativity Support Tools"

We would like to express our thanks to the Association of Logic Programming (ALP) and the Association for Computing Machinery (ACM) for their continuous support of the symposium, and Springer for their longstanding, successful cooperation with the PADL series. We are very grateful to the 23 members of the PADL 2023

Program Committee and external reviewers for their invaluable work. Many thanks to Marco Gavanelli, the ALP Conference Coordinator. The chairs of POPL 2023 were also of great help in steering the organizational details of the event.

We are happy to note that the conference paper evaluation was successfully managed with the help of EasyChair.

January 2023

Michael Hanus
Daniela Inclezan

Organization

Program Chairs

Michael Hanus	Kiel University, Germany
Daniela Inclezan	Miami University, USA

Program Committee

Andreas Abel	Gothenburg University, Sweden
Annette Bieniusa	TU Kaiserslautern, Germany
Joachim Breitner	Epic Games, Germany
William Byrd	University of Alabama at Birmingham, USA
Pedro Cabalar	University of Corunna, Spain
Francesco Calimeri	University of Calabria, Italy
Stefania Costantini	Università dell'Aquila, Italy
Esra Erdem	Sabanci University, Turkey
Martin Gebser	University of Klagenfurt, Austria
Robert Glück	University of Copenhagen, Denmark
Gopal Gupta	University of Texas at Dallas, USA
Tomi Janhunen	Tampere University, Finland
Patricia Johann	Appalachian State University, USA
Yukiyoshi Kameyama	University of Tsukuba, Japan
Ekaterina Komendantskaya	Heriot-Watt University, UK
Simona Perri	University of Calabria, Italy
Enrico Pontelli	New Mexico State University, USA
Tom Schrijvers	Katholieke Universiteit Leuven, Belgium
Paul Tarau	University of North Texas, USA
Peter Thiemann	University of Freiburg, Germany
Peter Van Roy	Université catholique de Louvain, Belgium
Janis Voigtländer	University of Duisburg-Essen, Germany
Ningning Xie	University of Cambridge, UK

Additional Reviewers

Kinjal Basu
Aysu Bogatarkan
Roger Bosman
Francesco Cauteruccio
Mohammed El-Kholany
Serdar Erbatur
Masood Feyzbakhsh Rankooh
Müge Fidan
Yusuf Izmirlioglu
Benjamin Kovács
Cinzia Marte
Elena Mastria
Francesco Pacenza
Kristian Reale

Christoph Reichenbach	Sarat Chandra Varanasi
Elmer Salazar	Anssi Yli-Jyrä
Pierre Tassel	Jessica Zangari
Birthe van den Berg	Ester Zumpano

Abstracts of Invited Talks

Modern Macros

Robert Bruce Findler

Findler Northwestern University, Evanston, IL 60208, USA
robby@northwestern.edu
https://users.cs.northwestern.edu/robby/

Racket's approach to macros is the latest point in an evolution that started in 1963 with Lisp's macros. Building on those ideas, Racket's macros have evolved so far that, to a modern macro programmer, macros are more helpfully understood as extending and manipulating the compiler's front end than as a mechanism for textual substitution or syntactic abstraction.

Having a malleable compiler front end naturally enables succinct implementations of many domain-specific and embedded languages. A look at the Racket ecosystem reveals a wealth of examples. Scribble, a language for writing texts uses a LaTeX inspired syntax and has been used to write thousands of pages of documentation, dozens of research papers, and at least two books. Redex, a language for writing and testing operational semantics, has helped numerous researchers debug their semantics and explore their ideas. Racket's sister-language, Typed Racket, boasts numerous type-level innovations and full-fledged interoperability with untyped code. Beside these large efforts, Racket's macros also have enabled extensions on the medium scale as well, being the basis for its pattern matcher, class system, contract system, family of for loops, and more. On the small scale, project-specific macros are common in Racket codebases, as Racket programmers can lift the language of discourse from general programming-language constructs to project-specific concerns, aiding program comprehension and shrinking codebase size.

In this talk, I'll discuss the essential aspects of Racket's macro system design, showing how they enable languageoriented programming and provide an intellectual foundation for understanding modern macros. These aspects all center on the idea of automatically managing scope and taking advantage of its automatic management.

Going beyond implementing languages, the data structures supporting automatic scope management have proven central to DrRacket (the Racket IDE), specifically its rename refactoring and its ability to navigate codebases via uses and definitions of identifiers. Recently, Racketeers have begun to explore how more aspects of Racket's macro system can support sophisticated IDE tooling for programming languages in the

Racket ecosystem. I will try to paint a picture of where we hope to go with that work as well.

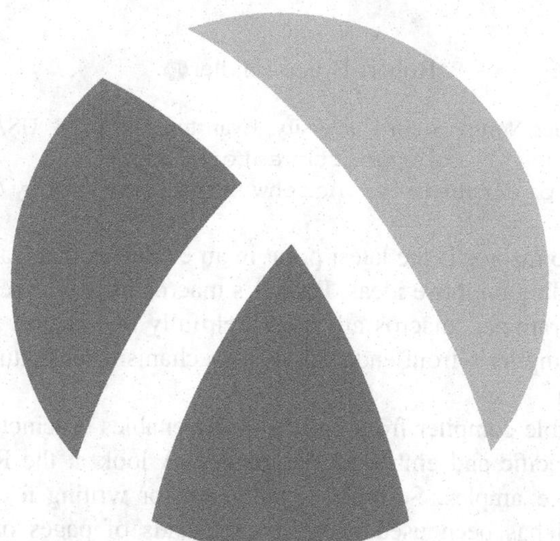

Towards Declarative Content Generation for Creativity Support Tools

Chris Martens

Northeastern University
c.martens@northeastern.edu
https://www.convivial.tools

Abstract. Procedural content generation (PCG) represents a class of computational techniques in the context of games, art, graphics, and simulation, for algorithmically producing novel digital artifacts more typically handcrafted by artists and designers, such as planetary terrain, texture assets, level map layouts, and game quests and stories. There is growing interest in PCG for its potential to enable creativity support tools (CST): software that can assist and proactively collaborate with human practitioners to iterate on generative algorithms towards a creative design goal. While declarative programming tools, especially logic programming and solvers, offer many appealing benefits for PCG, there remain barriers to their adoption. In this paper, we focus on the specific challenge of *inspecting execution*: the important role in PCG of being able to visualize a generative algorithm's intermediate states and see them as meaningful.

Keywords: Logic Programming · Procedural Content Generation · Creativity Support Tools

1 Procedural Terrain Generation

Suppose you want to fill a 2D square grid of cells with a map for a fictional setting. This motive arises in games, fantasy roleplaying games, and worldbuilding for storytelling; hobby sites for this purpose abound on the internet. Fantasy map construction is an example of a more general interest in creativity support tools (CSTs): software that can assist and proactively collaborate with human practitioners to iterate on generative algorithms towards a creative design goal.

We can formalize a simplified version of the map generation problem as one of *terrain generation*: given a grid size and a range of possible heights, fill in each cell with a valid height such that the result represents a reasonable depiction of terrestrial terrain. A common elaboration of a "reasonable" notion of smoothness: e.g., a cell's height should be at most 1 (or some other maximum) unit of difference from its neighbors' heights. Other criteria for correctness or metrics for desirability may exist, but often in a PCG task, these are not known in advance: we iterate with the system to discover what examples meet our stated criteria, then refine those criteria as we go.

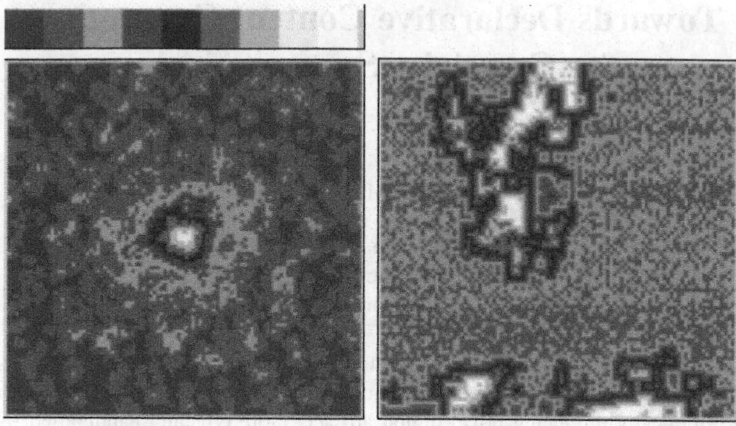

Fig. 1. Terrain height maps generated from the CA approach (left), initialized with a center-biased random map, and by the ASP code in Figure 2 (right), along with a constraint that the minimum and maximum heights in the map differ by at least 3. The bar at the top shows the range of heights from 0–8 as their corresponding colors.

A classic approach to solving this problem is with a cellular automaton (CA). A CA can simulate "erosion" of terrain through local rules, by transferring the height of a cell to its neighbors whenever its height exceeds them by more than the maximum. Let us call this the "procedural" (content generation) approach.

A more declarative approach would be to use Answer Set Programming (ASP) to directly capture the constraint of interest, an approach previously identified for PCG [3]. In ASP, logic programs, extended with a kind of disjunction and strong negation, represent a specification, and an answer set solver finds all stable models (maximal sets of ground predicates) that satisfy the specification. Figure 2 shows the most relevant part of an ASP program to generate terrain. Let us call this the "declarative" (content generation) approach.

Figure 1 shows two example maps resulting from each of these two approaches. Although these maps were motivated by the same initial specification—the idea of a maximum height difference between neighboring cells—their realizations differ significantly on the basis of the programming style used. One compelling direction for future research is how to formally relate these two programs (or their solution spaces): can we prove that the CA code satisfies the constraint in the ASP? Is one solution set a subset of the other? Could we maintain a check on these relationships as we edit one or the other program?

2 Inspecting Generative Processes

PCG systems in practice are evaluated along a number of criteria, some of which are sometimes in tension with each other, such as controllability, speed, solution quality, and solution diversity [1]. Declarative approaches like ASP are appealing due to their

```
#const max_height = 8.
possible_height(0..max_height).

1 { height(X,Y, H) : possible_height(H) } 1 :- cell(X,Y).

nbr_height(H, N) :- possible_height(H), possible_height(N),
  H-N <= 1, N-H <= 1.

:- height(X1, Y1, H1), height(X2, Y2, H2),
    nbr((X1, Y1), (X2, Y2)), not nbr_height(H1, H2).
```

Fig. 2. A snippet of an answer set program that populates a grid with heights such that adjacent heights differ by at most 1.

especially strong affordances for controllability (which Smith refers to as "sculpting the design space" [2] and solution quality. However, a characteristic that has particular importance for creativity support tools is the *ability to inspect a program's intermediate states as it runs*. With a simulation-based approach, such as CA, we can see each iteration of the map as the process unfolds as though watching erosion animate in real time. Likewise, evolutionary algorithms, another popular iterative approach, allow inspection of each "generation" of individuals as they are evolved. Exposing each step of iteration enables mental modeling of the process and therefore a more collaborative relationship between the programmer and the creative process. It also exposes potential for inserting a "human in the loop," as in simulation-based gameplay.

Declarative programming's famous slogan that programs represent "what, not how" things are computed seems to embrace this drawback as a feature. However, it may be possible and interesting to instead take inspectability as a challenge: what would it mean to stop a logic program at each step of its execution and reveal its internal state as a partial (either unsound or incomplete) solution? Could intermediate proof states correspond to regions of the possibility space that the PCG programmer might want to pause and explore? How can we better enable the kind of iterative refinement that makes programming high-dimensional visual, spatial, and temporal data appealing in the first place?

References

1. Shaker, N., Togelius, J., Nelson, M.J.: Procedural Content Generation in Games. Springer, Cham (2016). https://doi.org/10.1007/978-3-319-42716-4
2. Smith, A.M., Butler, E., Popovic, Z.: Quantifying over play: constraining undesirable solutions in puzzle design. In: FDG. pp. 221–228 (2013)
3. Smith, A.M., Mateas, M.: Answer set programming for procedural content generation: a design space approach. IEEE Trans. Comput. Intell. AI Games **3**(3), 187–200 (2011)

Contents

Functional Programming

RICE: An Optimizing Curry Compiler

Steven Libby(✉)

University of Portland, Portland, OR 97203, USA
libbys@up.edu

Abstract. This paper introduces the RICE compiler, an optimizing compiler for the functional logic language Curry, written entirely in Curry. We show the advantages of using Curry as an implementation language by creating a robust optimizer that is easy to extend. Furthermore, we show the benefits of optimizing Curry programs directly, as opposed to translating to another high level language. When we compiled programs in RICE, they ran anywhere from 3 to 1000 times faster than programs compiled using the current state of the art compilers.

1 Introduction

Functional logic programming is a very powerful technique for expressing complicated ideas in a simple form. Curry implements these ideas with a clean, easy to read syntax, which is similar to Haskell [41], a well known functional programming language. Like Haskell, Curry is also lazy. Curry extends Haskell with two concepts from logic programming, non-determinism, and free (or logic) variables. Non-determinism can be implemented with overlapping rules, such as the choice operator

$x \; ? \; y = x$
$x \; ? \; y = y$

A free variable is a variable that is not in the scope of the current function. The value of a free variable is not defined by the user, but it may be constrained.

These features are very useful for solving constraint problems. Consider the problem of scheduling a test for a large class of students. Since the class is so large, the students cannot take the test at the same time. To solve this problem, we allow each student to choose all times that they are available to take the test. After they have selected their times, we partition the students into groups, where each group corresponds to a testing time, and the size each group is less than a given capacity. This is a solvable problem in any language, but the solution in Curry, as shown in Fig. 1, is both concise and easily understood.

While readability and ease of development are important in a language, one complaint leveled against Curry is that current implementations are too slow. We do not intend to debate the merit of these claims. Instead, we propose a new compiler with a focus on optimization.

There are currently two mature Curry compilers, PAKCS [21] and KiCS2 [16]. PAKCS compiles Curry to Prolog in an effort to leverage Prolog's non-determinism and free variables. KiCS2 compiles Curry to Haskell in an effort to

M. Hanus and D. Inclezan (Eds.): PADL 2023, LNCS 13880, pp. 3–19, 2023.
https://doi.org/10.1007/978-3-031-24841-2_1

```
type Time = Int
type Name = String
type Student = (Name, [Time])
type Test = [Name]

schedule :: [Student] → [Test]
schedule students = buildSchedule students
  where buildSchedule = foldr takeTests []

takeTest :: Student → [Test] → [Test]
takeTest (name, times) tests = foldr1 (?) (map (testAt name tests) times)

testAt :: Name → [Test] → Time → [Test]
testAt name tests k
  | length test < capacity = ts1 ++ (name : test) ++ ts2
    where (ts1, test : ts2) = splitAt k tests
```

Fig. 1. A program to schedule students for a test with multiple times. Each student consists of a name, and a list of times they are available to take the test. We non-deterministically pick a test for each student according to their availability. We then check that the number of students scheduled for that test is below some capacity. If so, we schedule the student.

leverage Haskell's higher-order functions and optimizing compiler. Both compilers have their advantages. PAKCS tends to perform better on non-deterministic expressions with free variables, whereas KiCS2 tends to perform much better on deterministic expressions. Unfortunately neither of these compilers perform well in both circumstances.

Sprite [10], an experimental compiler, aims to fix these inefficiencies. The strategy is to compile to a virtual assembly language known as LLVM. So far, Sprite has shown promising improvements over both PAKCS and KiCS2 in performance, but it is not readily available for testing at the time of this writing.

Similarly MCC [40] also worked to improve performance by compiling to C. While MCC often ran faster than both PAKCS or KiCS2, it could perform very slowly on common Curry examples. It is also no longer in active development.

One major disadvantage of all four compilers is that they all attempt to pass off optimization to another compiler. PAKCS attempts to have Prolog optimize the non-deterministic code; KiCS2 attempts to use Haskell to optimize deterministic code; Sprite attempts to use LLVM to optimize the low level code; and MCC did not optimize its code. Unfortunately none of these approaches work very well. While some implementations of Prolog can optimize non-deterministic expressions, they have no concept of higher-order functions, so there are many optimizations that cannot be applied. KiCS2 is in a similar situation. In order to incorporate non-deterministic computations in Haskell, a significant amount of code must be threaded through each computation. This means that any non-deterministic expression cannot be optimized in KiCS2. Finally, since LLVM does not know about either higher-order functions or non-determinism, it loses many easy opportunities for optimization.

Another option for efficient execution is partial evaluation. Recently, many researchers [44,45] have developed a strong theory of partial evaluation for functional logic programs. While these results are promising, partial evaluation is not currently automatic in Curry. Guidance is required from the programmer to run the optimization. Furthermore, the partial evaluation fails to optimize several common programs.

We propose a new framework for compiling Curry programs which we call the Reduction Inspired Compiler Environment (RICE) [39]. We have used this framework to create a Curry compiler by the same name. RICE compiles Curry programs directly to C code and applies optimizations to the Curry programs themselves. This allows us to apply optimizations to non-deterministic functions. We also introduce the GAS (General Algebra System) libaray writing compiler optimizations. This is an easily extendable optimization framework, and is the workhorse behind our compiler. The main contribution of this paper is the discussion of the optimizations made to Curry programs, as well as the decisions we made in designing the run-time system.

In Sect. 2 we discuss Curry programs and their compilation. In Sect. 3 we present the GAS libary, and show how to use it to implement optimizations. Next, in Sect. 4, we show our results compared to current Curry compilers. Finally, in Sect. 5, we conclude and discuss future work.

2 The Curry Run-Time System

In this section we give a description of the decisions we made while implementing the run-time system for Curry. We discuss two novel contributions for improving the performance of this system. First, we outline an optimization for Backtracking, which allows us to avoid backtracking deterministic expressions. Second, we allow nested case expressions in compiled code, as opposed to PAKCS and KiCS2 which split nested cases into separate functions.

Curry is a functional logic programming language, which grew out of the efforts to combine the functional and logic paradigms [17,19,25,27,29,37,47,48]. A Curry program forms a Limited Overlapping Inductively Sequential (LOIS) Graph Rewrite System (GRS) [20]. This means that the rules for each Curry function can be grouped together in a structure called a definitional tree [4]. In practice, Curry programs are compiled to an intermediate representation known as FlatCurry [2].

FlatCurry encodes the evaluation strategy [8] in nested case expressions, and looks similar to Haskell's Core representation. It has a few notable differences. Unsurprisingly there is the inclusion of non-determinism via the ? operator. We also include the `let free` clause for introducing free variables, and the $\bot$ expression to represent a non-deterministic branch of a computation that has failed. There are also no variable patterns in a case. If the source program contains a variable pattern, then the pattern is converted to a choice of a case expression, and the expression of the variable pattern. The final difference is that this is a combinator language. There are no lambda expressions, and functions cannot be defined in a let clause. The definition of FlatCurry can be seen in Fig. 2.

$f \Rightarrow f\ \overline{v} = e$	Function Definition
$e \Rightarrow v$	Variable
$\mid\ l$	Literal
$\mid\ e_1\ ?\ e_2$	Choice
$\mid\ \bot$	Failed Computation
$\mid\ f\ \overline{e}$	Function Application
$\mid\ C\ \overline{e}$	Constructor Application
$\mid$ `let` $\overline{v = e}$ `in` e	Variable Declaration
$\mid$ `let` $\overline{v}$ `free in` e	Free Variable Declaration
$\mid$ `case` e `of` $\overline{p \rightarrow e}$	Case Expression
$p \Rightarrow C\ \overline{v}$	Constructor Pattern
$\mid\ l$	Literal Pattern

Fig. 2. Syntax definition for FlatCurry This is largely the same as previous presentations [2,9], but we have elected to add a $\bot$ expression to represent exempt branches [4]. The notation $\overline{e}$ refers to a variable length list $e_1\ e_2 \ldots e_n$.

It would be pointless to optimize Curry programs if we generated slow code. To avoid this, when there was a choice in how we compiled FlatCurry, we chose the option that we believed had the best run-time performance. In order to compile FlatCurry to efficient code, we chose a similar compilation scheme to Sprite [10]. Expressions are represented as labeled directed graphs, and case expressions are compiled to a tagged dispatch loop. For deterministic expressions, evaluation is similar to lazy functional language implementations, such as the G-machine or STG-machine [14,31].

For non-deterministic expressions, there are several areas where we deviated from previous compilers to produce more efficient code. In order to more closely match the rewriting semantics of Curry [2], PAKCS and KiCS2 transform FlatCurry programs so that expressions can only contain a single case expression. KiCS2 calls this normalization [15]. RICE and Sprite both allow nested case expressions because it is more efficient, and support function inlining. However, it nested cases also complicate the implementation of non-determinism.

There are several different ways to support non-determinism in Curry. Backtracking, as in Prolog, was implemented in PAKCS and MCC. It is the simplest method and is usually efficient, but it can become very inefficient if a computation has to backtrack a large deterministic expression, such as fib 100 ? 2. After producing a result, PAKCS would need to undo all of the computations of fib 100, just to produce the answer 2.

KiCS2 and Sprite both support Pull Tabbing [3]. When encountering non-determinism Pull Tabbing will move the determinism up one level. As an example, the expression f (0 ? 1) would be rewritten as f 0 ? f 1. This is called a pull tab because the choice moves up the expression like a zipper tab moves up the zipper when you pull it. However, this pull tab step is not valid in Curry if expressions are shared [5]. To account for this, each node in the expression graph needs to track information about each pull tab. Pull Tabbing does not have the drawback of backtracking deterministic expressions, but it requires much more overhead to implement.

Finally, there is Bubbling [6], which is similar to Pull Tabbing. Instead of moving the non-deterministic choice up a single level in the expression graph, the choice is moved to a dominator [38] of the expression. A dominator is a node that is common to every path from the root of the expression to that node. Unlike Pull Tabbing, Bubbling is a correct transformation on Curry programs [7], so we do not need to store any information about bubbling steps at run-time. The down side is that computing dominators can be costly, however there is hope for a more efficient solution with not too much run-time overhead [11]. Currently there are no implementations of Bubbling, but it looks promising.

We chose to use Backtracking for RICE. Our backtracking stack holds reductions, which we represent as a pair of expressions. When pushing a rewrite onto the backtracking stack, we push the node that is being reduced along with a copy of the original value of that node. When backtracking we pop nodes off the stack and overwrite the reduced node with its backup copy.

Backtracking requires less overhead than pull tabbing, and to avoid the issue of backtracking deterministic computations, we avoid storing any deterministic expressions on the backtracking stack. This is justified by looking at non-determinism as a run-time property. Each node in our expression graph contains a flag to mark whether the node's evaluation depended on a non-deterministic expression. If it did, then it is pushed on the backtracking stack when it is reduced; otherwise it was a deterministic expression, so backtracking would have no effect on it.

RICE also needs to address the problem of nested case expressions. If we allow both let expressions and nested case expressions with no restrictions, then FlatCurry programs no longer correspond to rewrite systems. Consider the following function.

```
weird = let x = False ? True
        in case x of
             True → True
```

If we were to reduce *weird*, then we create a node for x which is reduced to *False* and put on the backtracking stack, then *weird* is reduced to $\bot$, and is put on the backtracking stack. When we backtrack, it is unclear how we should undo the reduction of $\bot$. If we just replace $\bot$ with *weird*, then when we evaluate *weird* again, it will repeat this entire process, causing an infinite loop.

Instead, we want to undo *weird* up to a point in the computation where x has already been created. KiCS2 solved this problem by disallowing nested case expressions, so our code would be decomposed to the following:

$$weird = let\ x = False\ ?\ True\ \texttt{in}\ weird_1\ x$$
$$weird_1\ x = \texttt{case}\ x\ \texttt{of}$$
$$True \rightarrow True$$

We did not want to lose the efficiency of nested case expressions, but we still need a point to backtrack to, such as $weird_1$. We can get the best of both worlds by duplicating the code of weird.

$$weird = \texttt{let}\ x = False\ ?\ True$$
$$\texttt{in case}\ x\ \texttt{of}$$
$$True \rightarrow True$$

$$weird_1\ x = \texttt{case}\ x\ \texttt{of}$$
$$True \rightarrow True$$

When we push *weird* onto the backtracking stack, instead of making a copy of *weird*, we push $weird_1\ x$. Now when backtracking, our original node is overwritten with $weird_1\ x$, and evaluation can continue similar to KiCS2.

One reasonable objection to this scheme is the code size. We are generating much more code then KiCS2 does. This situation is not as bad as it seems at first. We can see that the size of each newly created function is the same size as a branch in a case of the original function. Adding all of these newly created functions together give a total size of no greater than twice that of the original function.

There are several more small run-time enhancements we added to RICE, but we feel these are the two most substantial improvements over previous compilers. In the next section we discuss our optimization framework, as well as what optimizations we included in RICE.

3 GAS and Optimizations

One of our primary contributions in this compiler was a library to simplify the development and testing of new optimizations and transformations on Curry programs by supplying rewrite rules. We call this library the General Algebra System (GAS), and it is the primary driver of the compiler. While it was constructed for the RICE compiler specifically, GAS performs FlatCurry to FlatCurry transformations, so it can be incorporated into other tools for program transformation.

GAS performs FlatCurry to FlatCurry transformations by supplying simple rewrite rules. This idea is inspired from corresponding ideas in the GHC compiler [32,35,36]. We have extended this idea in two notable ways. First, the rewrite rules are just Curry functions, they are not a DSL implemented on top of Curry. Second, the transformation does not need to traverse the entire tree. A transformation is a function of the type *Expr* → *Expr*, where *Expr* is the type of FlatCurry expression in the compiler. The function will pattern match on a piece of FlatCurry syntax, and return the resulting replacement. For example, consider the let-floating transformation.

$f\ (\texttt{let}\ x = e_1\ \texttt{in}\ e_2) \Rightarrow \texttt{let}\ x = e_1\ \texttt{in}\ f\ e_2$

In our implementation this is simply:

$float\ (Comb\ ct\ f\ [Let\ [(x, e1)]\ e2) = Let\ [(x, e1)]\ (Comb\ ct\ f\ [e2])$

The *Let* and *Comb* constructors represent a let expression and function application expression respectively.

If we want to apply this transformation, we can use the function *simplify*, which takes a transformation f and an expression e, and traverses e bottom up applying f at every location where it could apply. This continues until there are no more spots where the expression can be transformed. In theory we could traverse e in any order, but we've found that a bottom up traversal gives the best performance. This is already a well understood technique [32,35]. We extend it in three ways.

First, since our transformations are allowed to be non-deterministic, composing two transformations t_1 and t_2 is as simple as $t_1\ ?\ t_2$. We developed a number of optimizations for the compiler, and they all matched a local pattern they were optimizing. In other compilers, to combine these optimizations we would either need to create a monolithic function that traverses the syntax tree and includes all of the optimizations, or we would need to write several small tree traversals that are run in sequence. Either way, we are duplicating a large amount of code. Instead we were able to combine all of our optimizations quite easily. Here ft is a function table, which contains information needed for performing β-reduction.

$$\begin{aligned} optimizations\ ft = {} & caseCancel\ ?\ deforest\ ?\ deadCode\ ? \\ & constantFold\ ?\ condFold\ ?\ unapply\ ? \\ & flatten\ ?\ alias\ ?\ letFloat\ ?\ caseVar \\ & inline\ ?\ caseInCase\ ?\ reduce\ ft \end{aligned}$$

Second, we can extend a transformation simply by adding another parameter. For example, the Case-in-case [35] transformation may need a unique variable name, so we add a fresh variable name as a parameter.

$caseInCase :: VarIndex \rightarrow Expr \rightarrow Expr$

While this does require us to change the type of our transformations to $VarIndex \rightarrow Expr \rightarrow (Expr, VarIndex)$. However, it's easy to update any previous optimizations to the correct type with a simple wrapper functions.

$addParam\ opt\ \_\ e = (opt\ e, 0)$

Finally, we extended this system to report which transformations are performed. This is done by adding the transformation name to the return value of an expression. This allows us to print out every transformation that was performed on a Curry function during compilation with minimal impact to the compiler writer.

One important question remains for us to answer: since transformations are non-deterministic functions, then how do we limit the search space and end up with a single optimized expression? This is the job of the *simplify* function. When it chooses a spot to apply an optimization, it will search using *oneValue* function from the *FindAll* library. This way at each step, a single expression is returned regardless of how non-deterministic our transformation is.

We used this system to implement a number of optimizations in our compiler including fundamental optimizations like case canceling, Inlining, constant folding, dead code elimination, fold/build deforestation [26], and several others. However, the implementation of some of these are non-trivial. Specifically, it can be difficult to gauge the correctness of these optimizations. For example, when inlining functions in Haskell, β-reduction [13,34] is always valid but not always a good idea [34]. This is not the case for Curry. Consider the following expression:

$double\ x = x + x$
$main = double\ (0\ ?\ 1)$

It would be incorrect to reduce *main* to $(0\ ?\ 1) + (0\ ?\ 1)$. This is similar to the problem of call-time choice semantics vs. run-time choice semantics [28]. It turns out that this problem can be solved. If we convert our FlatCurry programs to A-Normal Form (ANF) [24], that is, we only allow function application to variables or literals, then β-reduction is valid again. After converting to ANF, our main function becomes:

$main = \texttt{let}\ y = 0\ ?\ 1$
$\qquad\quad \texttt{in}\ double\ y$

This can be reduced to:

$main = \texttt{let}\ y = 0\ ?\ 1$
$\qquad\quad \texttt{in}\ y + y$

However, some optimizations such as Common Subexpression Elimination [13], cannot be made valid in Curry. This is because CSE introduces sharing where none existed previously. To see this, we can look at the same example as our inlining transformation. If the programmer really intended $(0\ ?\ 1) + (0\ ?\ 1)$, then optimizing this via CSE would create the expression $let\ x = 0\ ?\ 1\ in\ x + x$. This would take an expression that would normally produce the set of answers $\{0, 1, 1, 2\}$, and changing it to an expression that only produces $\{0, 1\}$. Therefore, common subexpression cannot be an optimization.

We should note that this is not unique to Curry. Problems with inlining and CSE are common in strict languages like ML, but they're also important to consider for lazy functional languages like Haskell. While the transformations are semantically correct, they may increase the run-time or memory usage of the compiler. Furthermore ANF is not necessary to preserve the semantics of Curry, but it is sufficient.

While all of the optimizations and transformations we have discussed so far are easily described as transformation on Curry source programs, there are a few optimizations where the implementation is more complicated. Consider the case of evaluating an expression as the scrutinee of a case expression. Marlow et al. [33] point out that it would be silly to create a thunk for *case* (*null xs*) *of* ... when we are only going to use the results of the scrutinee in one expression. However, we cannot avoid this thunk creation so easily in Curry. It could be the case that *null xs* is a non-deterministic expression. Therefore it is important that we keep its value around. We may need to backtrack to it later.

The solution to this problem is not as straightforward as our solution to inlining. In fact, it involves cooperation between the optimizer and the run-time system. The optimizer marks each variable that only appears in a case expression. Then at run-time we store those marked nodes in the static area of the program rather than allocating memory on the heap. In fact, we only need a single, statically allocated node to hold any case expression. We call this optimization case shortcutting, because we are avoiding the construction of the scrutinee of a case expression. The need for this optimization seems to suggest that the self evaluating model for closures from the STG-machine [31] would be a bad fit for Curry.

Another optimization that requires cooperation between the optimizer and the run-time is the implementation of Unboxing, as described by Peyton Jones et al. [46]. However, there is no interaction between non-determinism, free variables, and unboxing. In fact, this implementation of unboxing simplified the implementation of free variables for primitive types. Instead of creating primitive operations where the arguments could be unevaluated expressions, or even non-deterministic expressions, we can simply force evaluation with a case expression. Consider the expression *plusInt* :: *Int* → *Int* → *Int* from the standard Prelude. This is the function that implements addition on *Int* values. It is defined externally for KiCS2, and in PAKCS it fully evaluates its arguments, then calls an external implementation. This puts a strain on the run-time system because dealing with non-deterministic results, failures, and primitive operations all need to be handled by run-time code; this can be error prone and difficult to maintain.

If we treat unboxed values as first class citizens, we can rewrite the *plusInt* function. The integer type is now a box that holds a single value called *Int#*. This value can only be a primitive C `int` value. Now, we can write our *plusInt* as a Curry function that calls a primitive *prim_plusInt*. This does not seem like a big gain over PAKCS, but our *prim_plusInt* is actually just translated to the + in C code. Furthermore, this implementation works well with β-reduction and case canceling. We are usually able to eliminate at least one case, and often both cases. The implementation for *plusInt* is given below.

```
data Int = Int Int#

plusInt :: Int → Int → Int
plusInt x y = case x of
                  Int x' → case y of
                              Int y' → Int (prim_plusInt x' y')

prim_plusInt :: Int# → Int# → Int#
prim_plusInt = external
```

We have described several optimizations to Curry programs, and the Curry run-time system, but this would be pointless if they did not offer any improvement. In the next section we compare RICE to KiCS2, MCC, PAKCS, and GHC to see how effective this compiler is. We also compare unoptimized RICE code to optimized RICE code in order to see the effect of the optimizations.

4 Results

To test the efficacy of our compiler, we used several programs from the KiCS2 compiler benchmarks [16]. We have removed some tests, specifically those related to functional patterns, since they are an extension to Curry, and not the focus of our work. RICE does support this extension, but we currently do not have any optimizations that target functional patterns. We hope to add some in the future. We have also added a few tests to demonstrate the effect of deforestation because the original test suite did not include any examples where it could be applied.

In order to characterize the effectiveness of our optimizations, we are interested in two measurements. First, we want to show that optimized programs consume less memory. Second, we want to show that the execution time of the programs is improved. The first goal is easy to achieve. We augment the run-time system with a counter that we increment every time we allocate memory. When the program is finished running, we print out the number of memory allocations.

The second goal of measuring execution time is much more difficult. There are many factors which can affect the execution time of a program. To help alleviate these problems, we took the approach outlined by Mytkowicz et al. [43]. All programs were run multiple times, and compiled in multiple environments for each compiler. We took the lowest execution time. We believe these results are as unbiased as we can hope for; however, it is important to remember that our results may vary across machines and environments. For most of our results, the RICE compiler is a clear winner.

For brevity, we give a short description of the tests that were run:

- fib.curry computes Fibonacci numbers
- fibNondet.curry computes Fibonacci numbers with a non-deterministic input
- tak.curry computes a long, mutually recursive function with a numeric argument
- cent.curry computes all expressions containing the numbers 1 to 5 that evaluate to 100
- half.curry computes half of a number n by finding a solution to $x + x =:= n$
- ndTest.curry computes a non-deterministic function with many results
- perm.curry non-deterministically produces all permutations of a list
- queensPerm solves the n-queens problem by permuting a list and checking if it is a valid solution
- primesort.curry non-deterministically sorts a list of very large prime numbers
- sort.curry sorts a list by finding a sorted permutation
- last.curry computes the last element in a list using free variables

- schedule.curry is the scheduling program from the introduction
- queensDet.curry computes solutions to the n-queens problem using a backtracking solution and list comprehension
- reverseBuiltin.curry reverses a list without using functions or data types defined in the standard Prelude
- reverseFoldr.curry reverses a list using a reverse function written as a fold
- reversePrim.curry reverses a list using the built-in reverse function and primitive numbers
- sumSquares.curry computes $sum \circ map\ sqaure \circ filter\ odd \circ enumFromTo\ 1$
- buildFold.curry computes a long chain of list processing functions
- primes.curry computes a list of primes
- sumPrimes.curry computes $sum \circ filter\ isPrime \circ enum\ 2$

The results of running the tests are given in Table 1 and Table 2. In Table 1 we compare RICE to PAKCS version 2.3.0 [23], MCC version 0.9.12 [1], and KiCS2 version 2.3.0 [22]. All results are normalized to RICE. We optimized these compilers as much as possible to get the best results. For example, KiCS2 executed much quicker when run in the primitive depth first search mode. We increased the input size for tak, buildFold, and sumPrimes in order to get a better comparison with these compilers. However, we were not able to run the buildFold or reverseBuiltin tests for the PAKCS compiler. They were both killed by the Operating System before they could complete. We timed every program with KiCS2 [16], PAKCS [21], and the MCC [40] compiler. Unfortunately we were not able to get an accurate result on how much memory any of these compilers allocated, so we were unable to compare our memory results.

In Table 2 We compare unoptimized programs to their optimized counterparts in RICE. All results are normalized to the optimized version.

We also show how our compiler compares against GHC in Table 3. Because most examples include non-determinism or free variables, we are unable to run those. We run our optimized code against unoptimized GHC and optimized GHC.

There are many interesting results in Tables 1 and 2 that we feel are worth pointing out. First, it should be noted that the MCC compiler performed very well, not only against both KiCS2 and PAKCS, but it also against RICE. In most examples, it was competitive with the unoptimized code and ahead of it in several tests. It even outperformed the optimized version in the cent example. We are currently unsure of why this happened, but we have two theories. First, the code generation and run-time system of MCC may just be more efficient than RICE. While we worked hard to make the run-time system as efficient as possible, it was not the focus of this compiler. MCC also translated the code to Continuation Passing Style [13] before generating target code. This may be responsible for the faster execution times. Our second theory is that MCC supports an older version of Curry that does not include type classes. The current implementation of typeclasses in FlatCurry involves passing a dictionary of all of the functions implemented by a type class as a hidden parameter to each function that uses the type class. MCC may have performed better simply by not having to deal with the overhead of unpacking that dictionary.

Table 1. Comparison of execution time for PAKCS, KiCS2, MCC, and RICE. All times are normalized to RICE.

	PAKCS	KiCS2	MCC	RICE
fib	2,945	16	7	1
fibNondet	2,945	839	8	1
tak	7,306	14	19	1
cent	152	62	0.65	1
half	1,891	49	3	1
ndtest	491	18	2	1
perm	73	6	2	1
queensPerm	5,171	27	1	1
primesort	9,879	3	7	1
sort	923	35	1	1
last	∞	42	1	1
schedule	5,824	20	2	1
queensDet	4,573	5	5	1
reverseBuiltin	∞	3	2	1
reverseFoldr	13,107	8	4	1
reversePrim	1,398	9	3	1
sumSquare	140	10	22	1
buildFold	∞	24	9	1
primes	10,453	51	12	1
sumPrimes	2,762	3	4	1

Generally RICE compares very favorably with all of the current compilers, only losing out to MCC on the cent example. We focus on the KiCS2 compiler because that was the best performing compiler that is still in active development. With this comparison, RICE performs very well, showing anywhere form a 2x to 50x execution speed-up on all programs. Even comparing against MCC, we typically see a 2x speed-up. The only exceptions are cent and programs that cannot be optimized, such as perm. We also see a very impressive speedup on fibNondet compared to KiCS2. However, this is a known issue with the evaluation of non-deterministic expressions with functions with non-linear rules. We believe that this is important to note, because these programs are common in Curry. This is the reason that we could not use KiCS2 to develop RICE.

When looking at memory usage, we start to see why some examples are so much faster. In the cases of fib, fibNondet, tak, and buildFold, we have eliminated all, or nearly all, heap allocations. This would be a great result on its own, but it gets even better when we compare it to GHC. Compiling the same fib algorithm on GHC produced code that ran about three times as fast as our optimized RICE code, and when we turned off Optimizations for GHC we ran faster by a

Table 2. Results for execution time between the RICE compiler with and without optimizations. Time values are normalized to *unoptimized*, so they are the ratio, of the execution time over *unoptimized*'s execution time.

	Unoptimized time	Optimized time	Unoptimized memory	Optimized memory
fib	7.69	1	1,907,000.00	1
fibNondet	7.69	1	381,400.00	1
tak	14.29	1	94,785,000.00	1
cent	2.33	1	1.24	1
half	1.70	1	1.00	1
ndtest	1.96	1	0.84	1
perm	1.32	1	1.00	1
queensPerm	5.56	1	6.59	1
primesort	2.70	1	1.66	1
sort	2.63	1	1.70	1
last	1.59	1	1.63	1
schedule	1.25	1	1.35	1
queensDet	12.50	1	10.34	1
reverseBuiltin	1.79	1	1.00	1
reverseFoldr	3.03	1	2.20	1
reversePrim	3.03	1	2.00	1
sumSquare	6.25	1	4.17	1
buildFold	12.50	1	6,666,666.50	1
primes	3.13	1	1.67	1
sumPrimes	5.88	1	4,601.05	1

Table 3. Comparison of RICE and GHC on deterministic programs.

	RICE	GHC unoptimized	GHC optimized
fib	1.00	4.60	0.32
tak	1.00	4.07	0.35
queensDet	1.00	0.94	0.07
reverseFoldr	1.00	2.33	0.66
buildFold	1.00	2.77	0.31
primes	1.00	0.36	0.20
sumPrimes	1.00	1.17	0.15

factor of 4 as seen in Table 3. It is not surprising to us that our code ran slower than GHC. The run-time system is likely much faster than ours, and there are several optimizations in GHC that we have not implemented. In fact, we would be shocked if it managed to keep up. It is surprising and encouraging that we were competitive at all. It suggests that Curry is not inherently slower than Haskell. We believe that a more mature Curry compiler could run as fast as GHC for deterministic functions. This would give us the benefits of Curry, such as non-determinism and free variables, without sacrificing the speed of modern functional languages.

5 Conclusion and Future Work

We were quite happy with the results of this project, although we readily admit that some of these results are not surprising. The functional community has been aware of many of these optimizations for decades. However, even optimizations that were easy to implement in a functional language often had to be carefully considered in Curry. One example that comes to mind is fold/build deforestation. This is a well known optimization in Haskell with an elegant proof relying on logical relations and free theorems. Unfortunately free theorems are not automatically valid in Curry [18].

We believe that RICE is an important step in Functional Logic programming. We have shown that there are significant performance benefits to optimizing Curry code itself, and not just translating it to another high level language. Because many of the optimizations are performed on FlatCurry and could be easily incorporated into other compilers, we hope that these results are beneficial for any compiler writers in Curry,

Currently RICE compiles Curry to C, but this was mostly chosen because of familiarity with C. We hope to migrate RICE over to LLVM soon. We would also like to develop the theory of Functional Logic optimizations more fully. First, it would be nice to have a more complete theory of inlining that included lambda expressions. We chose to convert all expressions to A-Normal Form in RICE, and while this is sufficient to preserve the semantics of Curry, it is likely unnecessary. Several optimizations were hampered by the fact that we were working in a combinator based language. We believe that inlining lambda expressions would allow more optimizations to fire.

We would also like to work on optimizations that incorporate non-determinism directly. We believe that bubbling might provide a useful theory for moving choice expressions around safely. It may even be possible through inlining, β-reduction and bubbling to create a single definition tree out of our optimization function that combined 13 smaller optimizations. This would certainly be more efficient.

Finally, we would like to work on the efficiency of the GAS library itself. Currently there is no system to make sure the best optimization is chosen. It just applies the first optimization that will fit at a particular point. It may be worthwhile to change out implementation to using set functions, and developing a strategy for picking the best optimization, rather than picking an arbitrary transformation with *oneValue*.

We would also like to be able to bootstrap RICE. Currently it is compiled with PAKCS, and we believe we would get much better performance if we could compile RICE in itself. However, RICE does not support set functions yet.

We have presented the RICE Optimizing Curry compiler. The compiler was primarily built to test the effectiveness of various optimizations on Curry programs. While testing these optimizations, we have also built an efficient evaluation method for backtracking Curry programs, as well as a general system for describing and implementing optimizations. The compiler itself is written in Curry.

This system is built on previous work from the functional language community, and the Haskell community in particular. The work by Appel and Peyton-Jones about functional compiler construction formed the basis for the design of the compiler [12,13,30]. The work on general optimizations [35], inlining [34], unboxing [46], deforestation [26], and the STG-machine [31,33] were all instrumental in the creation of this compiler.

While there has been some work on optimizations for functional-logic programs, there does not seem to be a general theory of optimization. Peemöller and Ramos et al. [44,45] have developed a theory of partial evaluation for Curry programs, and Moreno [42] has worked on the Fold/Unfold transformation from Logic programming.

We hope that our work can help bridge the gap to traditional compiler optimizations. The benefits of optimizing Curry code directly are clearly shown here. There is still more work to do to improve performance, but we believe that RICE shows a promising direction for Curry compilers.

References

1. MCC 0.9.12-dev: The Munster Curry Compiler, 27 July 2015. http://danae.uni-muenster.de/curry/
2. Albert, E., Hanus, M., Huch, F., Oliver, J., Vidal, G.: Operational semantics for declarative multi-paradigm languages. J. Symb. Comput. **40**(1), 795–829 (2005)
3. Alqaddoumi, A., Antoy, S., Fischer, S., Reck, F.: The pull-tab transformation. In: Third International Workshop on Graph Computation Models, Enschede, The Netherlands (2010)
4. Antoy, S.: Definitional trees. In: Kirchner, H., Levi, G. (eds.) ALP 1992. LNCS, vol. 632, pp. 143–157. Springer, Heidelberg (1992). https://doi.org/10.1007/BFb0013825
5. Antoy, S.: On the correctness of pull-tabbing. Theory Prac. Log. Program. (TPLP) **11**(4–5), 713–730 (2011)
6. Antoy, S., Brown, D., Chiang, S.: Lazy context cloning for non-deterministic graph rewriting. In: Proceedings of the 3rd International Workshop on Term Graph Rewriting, Termgraph 2006, Vienna, Austria, pp. 61–70 (2006)
7. Antoy, S., Brown, D.W., Chiang, S.-H.: On the correctness of bubbling. In: Pfenning, F. (ed.) RTA 2006. LNCS, vol. 4098, pp. 35–49. Springer, Heidelberg (2006). https://doi.org/10.1007/11805618_4

8. Antoy, S., Echahed, R., Hanus, M.: A needed narrowing strategy. J. ACM **47**(4), 776–822 (2000)
9. Antoy, S., Hanus, M., Jost, A., Libby, S.: Icurry. CoRR, abs/1908.11101 (2019)
10. Antoy, S., Jost, A.: A new functional-logic compiler for curry: sprite. CoRR, abs/1608.04016 (2016)
11. Antoy, S., Libby, S.: Making Bubbling Practical. CoRR **abs/1808.07990**, arXiv:1808.07990 (2018)
12. Appel, A.: Modern Compiler Implementation in ML: Basic Techniques. Cambridge University Press, New York (1997)
13. Appel, A.: Compiling with Continuations. Cambridge University Press, New York (2007)
14. Augustsson, L.: Compiling lazy functional languages. Ph.D. thesis (1978)
15. Brassel, B.: Implementing functional logic programs by translation into purely functional programs. Ph.D. thesis, Christian-Albrechts-Universität zu Kiel (2011)
16. Braßel, B., Hanus, M., Peemöller, B., Reck, F.: KiCS2: a new compiler from curry to haskell. In: Kuchen, H. (ed.) WFLP 2011. LNCS, vol. 6816, pp. 1–18. Springer, Heidelberg (2011). https://doi.org/10.1007/978-3-642-22531-4_1
17. Castiñeiras, I.. Correas, J., Estévez-Martín, S., Sáenz-Pérez, F.: TOY: a CFLP language and system. The Association for Logic Programming (2012)
18. Christiansen, J., Seidel, D., Voigtländer, J.: Free theorems for functional logic programs. In: Proceedings of the 4th ACM SIGPLAN Workshop on Programming Languages Meets Program Verification, PLPV 2010, New York, NY, USA, pp. 39–48. Association for Computing Machinery (2010)
19. Demoen, B., de la Banda, M.G., Harvey, W., Marriott, K., Stuckey, P.: An overview of HAL. In: Jaffar, J. (ed.) CP 1999. LNCS, vol. 1713, pp. 174–188. Springer, Heidelberg (1999). https://doi.org/10.1007/978-3-540-48085-3_13
20. Echahed, R., Janodet, J.C.: On constructor-based graph rewriting systems. Technical report 985-I, IMAG (1997). ftp://ftp.imag.fr/pub/labo-LEIBNIZ/OLD-archives/PMP/c-graph-rewriting.ps.gz
21. Hanus, M. (ed.): PAKCS 1.14.3: The Portland Aachen Kiel Curry System (2017). http://www.informatik.uni-kiel.de/pakcs
22. Hanus, M. (ed.): KiCS2 2.3.0: Compiling Curry to Haskell (2020). https://www-ps.informatik.uni-kiel.de/kics2/download.html#docker
23. Hanus, M. (ed.): PAKCS 2.3.0: The Portland Aachen Kiel Curry System (2020). https://www.informatik.uni-kiel.de/pakcs/download.html
24. Flanagan, C., Sabry, A., Duba, B.F., Felleisen, M.: The essence of compiling with continuations. In: Proceedings of the ACM SIGPLAN 1993 Conference on Programming Language Design and Implementation, PLDI 1993, pp. 237–247 (1993)
25. Fribourg, L.: Slog: A logic programming language interpreter based on clausal superposition and rewriting. In: SLP (1985)
26. Gill, A., Launchbury, J., Peyton Jones, S.L.: A short cut to deforestation. In: Proceedings of the Conference on Functional Programming Languages and Computer Architecture, FPCA 1993, , New York, NY, USA, pp. 223–232. ACM (1993)
27. Giovannetti, E., Levi, G., Moiso, C., Palamidess, C.: Kernel-LEAF: a logic plus functional language. J. Comput. Syst. Sci. **42**(2), 139–185 (1991)
28. González-Moreno, J.C., Hortalá-González, M.T., López-Fraguas, F.J., Rodríguez-Artalejo, M.: A rewriting logic for declarative programming. In: Nielson, H.R. (ed.) ESOP 1996. LNCS, vol. 1058, pp. 156–172. Springer, Heidelberg (1996). https://doi.org/10.1007/3-540-61055-3_35

29. Hermenegildo, M.V., Bueno, F., Carro, M., López, P., Morales, J.F., Puebla, G.: An overview of the ciao multiparadigm language and program development environment and its design philosophy. In: Degano, P., De Nicola, R., Meseguer, J. (eds.) Concurrency, Graphs and Models. LNCS, vol. 5065, pp. 209–237. Springer, Heidelberg (2008). https://doi.org/10.1007/978-3-540-68679-8_14
30. Peyton Jones, S.L.: The Implementation of Functional Programming Languages. Prentice-Hall International Series in Computer Science, Prentice-Hall Inc, Upper Saddle River (1987)
31. Peyton Jones, S.L., Salkild, J.: The spineless tagless G-machine. In: Proceedings of the Fourth International Conference on Functional Programming Languages and Computer Architecture, FPCA 1989, pp. 184–201. ACM (1989)
32. Jones, S.L.P.: Compiling haskell by program transformation: a report from the trenches. In: Nielson, H.R. (ed.) ESOP 1996. LNCS, vol. 1058, pp. 18–44. Springer, Heidelberg (1996). https://doi.org/10.1007/3-540-61055-3_27
33. Peyton Jones, S.: How to make a fast curry: push/enter vs eval/apply. In: International Conference on Functional Programming, pp. 4–15 (2004)
34. Peyton Jones, S., Marlow, S.: Secrets of the glasgow haskell compiler inliner. J. Funct. Program. **12**(5), 393–434 (2002)
35. Peyton Jones, S., Santos, A.: A transformation-based optimiser for haskell. Sci. Comput. Program. **32**(1), 3–47 (1997)
36. Peyton Jones, S., Tolmach, A., Hoare, T.: Playing by the rules: rewriting as a practical optimisation technique in GHC. In: Haskell 2001 (2001)
37. Kuchen, H., Loogen, R., Moreno-Navarro, J.J., Rodríguez-Artalejo, M.: The functional logic language BABEL and its implementation on a graph machine. N. Gener. Comput. **14**(4), 391–427 (1996)
38. Lengauer, T., Tarjan, R.: A fast algorithm for finding dominators in a flowgraph. ACM Trans. Program. Lang. Syst. **1**(1), 121–141 (1979)
39. Libby, S.: RICE: an optimizing curry compiler (2022). https://github.com/slibby05/rice
40. Lux, W., Kuchen, H.: An efficient abstract machine for curry. In: Beiersdörfer, K., Engels, G., Schäfer, W. (eds.) Informatik 1999. Informatik aktuell, pp. 390–399. Springer, Heidelberg (1999). https://doi.org/10.1007/978-3-662-01069-3_58
41. Marlow, S., et al.: Haskell 2010 language report (2010). Available online http://www.haskell.org/. Accessed May 2011
42. Moreno, G.: Transformation rules and strategies for functional-logic programs (2002)
43. Mytkowicz, T., Diwan, A., Hauswirth, M., Sweeney, P.F.: Producing wrong data without doing anything obviously wrong! SIGPLAN Not. **44**(3), 265–276 (2009)
44. Peemöller, B.: Normalization and partial evaluation of functional logic programs. Ph.D. thesis (2016)
45. Ramos, J.G., Silva, J., Vidal, G.: An offline partial evaluator for curry programs. In: Proceedings of the 2005 ACM SIGPLAN Workshop on Curry and Functional Logic Programming, WCFLP 2005, New York, NY, USA, pp. 49–53. ACM (2005)
46. Jones, S.L.P., Launchbury, J.: Unboxed values as first class citizens in a non-strict functional language. In: Hughes, J. (ed.) FPCA 1991. LNCS, vol. 523, pp. 636–666. Springer, Heidelberg (1991). https://doi.org/10.1007/3540543961_30
47. Scheidhauer, R.: Design, implementierung und evaluierung einer virtuellen maschine für Oz. Ph.D. thesis, Universität des Saarlandes, Fachbereich Informatik, Saarbrücken, Germany (1998)
48. Somogyi, Z., Henderson, F.: The design and implementation of Mercury. In: Joint International Conference and Symposium on Logic Programming (1996)

Program Synthesis Using Example Propagation

Niek Mulleners[1(✉)], Johan Jeuring[1], and Bastiaan Heeren[2]

[1] Utrecht University, Utrecht, The Netherlands
{n.mulleners,j.t.jeuring}@uu.nl
[2] Open University of The Netherlands, Heerlen, The Netherlands
bastiaan.heeren@ou.nl

Abstract. We present Scrybe, an example-based synthesis tool for a statically-typed functional programming language, which combines top-down deductive reasoning in the style of λ^2 with Smyth-style live bidirectional evaluation. During synthesis, example constraints are propagated through sketches to prune and guide the search. This enables Scrybe to make more effective use of functions provided in the context. To evaluate our tool, it is run on the combined, largely disjoint, benchmarks of λ^2 and Myth. Scrybe is able to synthesize most of the combined benchmark tasks.

Keywords: Program synthesis · Constraint propagation · Input-Output examples · Functional programming

1 Introduction

Type-and-example-driven program synthesis is the process of automatically generating a program that adheres to a type and a set of input-output examples. The general idea is that the space of type-correct programs is enumerated, evaluating each program against the input-output examples until a program is found that does not result in a counterexample. Recent work in this field has aimed to make the enumeration of programs more efficient, using various pruning techniques and other optimizations. Hoogle+ [5] and Hectare [7] explore efficient data structures to represent the search space. Smith and Albarghouthi [12] describe how synthesis procedures can be adapted to only consider programs in normal form. MagicHaskeller [6] and RESL [11] filter out programs that evaluate to the same result. Instead of only using input-output examples for the verification of generated programs, Myth [4,10], Smyth [8], and λ^2 [3] use input-output examples during pruning, by eagerly checking incomplete programs for counterexamples using constraint propagation.

Constraint Propagation. Top-down synthesis incrementally builds up a sketch, a program that may contain holes (denoted by ●). Holes may be annotated with constraints, e.g. type constraints. During synthesis, holes are filled

M. Hanus and D. Inclezan (Eds.): PADL 2023, LNCS 13880, pp. 20–36, 2023.
https://doi.org/10.1007/978-3-031-24841-2_2

with new sketches (possibly containing more holes) until no holes are left. For example, for type-directed synthesis, let us start from a single hole $\bullet_0$ annotated with a type constraint:

$$\bullet_0 :: \mathit{List\ Nat} \to \mathit{List\ Nat}$$

We may fill $\bullet_0$ using the function map :: $(a \to b) \to \mathit{List}\ a \to \mathit{List}\ b$, which applies a function to the elements of a list. This introduces a new hole $\bullet_1$, with a new type constraint:

$$\bullet_0 :: \mathit{List\ Nat} \to \mathit{List\ Nat} \quad \xrightarrow{\mathsf{map}} \quad \mathsf{map}\ (\bullet_1 :: \mathit{Nat} \to \mathit{Nat})$$

We say that the constraint on $\bullet_0$ is propagated through map to the hole $\bullet_1$. Note that type information is preserved: the type constraint on $\bullet_0$ is satisfied exactly if the type constraint on $\bullet_1$ is satisfied. We say that the hole filling $\boxed{\bullet_0 \mapsto \mathsf{map}\ \bullet_1}$ refines the sketch with regards to its type constraint.

A similar approach is possible for example constraints, which partially specify the behavior of a function using input-output pairs. For example, we may further specify hole $\bullet_0$, to try and synthesize a program that doubles each value in a list:[1]

$$\bullet_0 \vDash \{\, [0,1,2] \mapsto [0,2,4] \,\}$$

Now, when introducing map, we expect its argument $\bullet_1$ to have three example constraints, representing the doubling of a natural number:

$$\bullet_0 \vDash \{\, [0,1,2] \mapsto [0,2,4] \,\} \quad \xrightarrow{\mathsf{map}} \quad \mathsf{map}\ (\bullet_1 \vDash \left\{ \begin{array}{c} 0 \mapsto 0 \\ 1 \mapsto 2 \\ 2 \mapsto 4 \end{array} \right\})$$

Similar to type constraints, we want example constraints to be correctly propagated through each hole filling, such that example information is preserved. Unlike with type constraints, which are propagated through hole fillings using type checking/inference, it is not obvious how to propagate example constraints through arbitrary functions. Typically, synthesizers define propagation of example constraints for a hand-picked set of functions and language constructs. Feser et al. [3] define example propagation for a set of combinators, including map and foldr, for their synthesizer λ^2. Limited to this set of combinators, λ^2 excels at composition but lacks in generality. MYTH [4,10], and by extension SMYTH [8], takes a more general approach, in exchange for compositionality, defining example propagation for basic language constructs, including constructors and pattern matches.

Presenting SCRYBE. In this paper, we explore how the techniques of λ^2 and SMYTH can be combined to create a general-purpose, compositional example-driven synthesizer, which we will call SCRYBE. Figure 1 shows four different interactions with SCRYBE, where the function dupli is synthesized with different sets

[1] In this example, as well as in the rest of this paper, we leave type constraints implicit.

```
{-# USE ... #-}           assert dupli [] ⊨ []
dupli :: List a → List a  assert dupli [0] ⊨ [0,0]
dupli xs = ●              assert dupli [0,1] ⊨ [0,0,1,1]
```

```
{-# USE foldr #-}
● ↦ foldr (λx r. x : x : r) [] xs
```

```
{-# USE concat, map #-}
● ↦ concat (map (λx. [x, x]) xs)
```

```
{-# USE foldl, (++) #-}
● ↦ foldl (λr x. r ++ [x, x]) [] xs
```

```
{-# USE interleave #-}
● ↦ interleave xs xs
```

Fig. 1. (Top) A program sketch in SCRYBE for synthesizing the function dupli, which duplicates each value in a list. (Bottom) Different synthesis results returned by SCRYBE, for different sets of included functions.

of functions. SCRYBE is able to propagate examples through all of the provided functions using live bidirectional evaluation as introduced by Lubin et al. [8] for their synthesizer SMYTH, originally intended to support sketching [13,14]. By choosing the right set of functions (for example, the set of combinators used in λ^2), SCRYBE is able to cover different synthesis domains. Additionally, allowing the programmer to choose this set of functions opens up a new way for the programmer to express their intent to the synthesizer, without going out of their way to provide an exact specification.

Main Contributions. The contributions of this paper are as follows:

- We give an overview of example propagation and how it can be used to perform program synthesis (Sect. 2).
- We show how live bidirectional evaluation as introduced by Lubin et al. [8] allows arbitrary sets of functions to be used as refinements during program synthesis (Sect. 3).
- We present SCRYBE, an extension of SMYTH [8], and evaluate it against existing benchmarks from different synthesis domains (Sect. 4).

2 Example Propagation

Example constraints give a specification of a function in terms of input-output pairs. For example, the following constraint represents the function mult that multiplies two numbers.

$$\left\{\begin{array}{l} 0\ \ 1 \mapsto 0 \\ 1\ \ 1 \mapsto 1 \\ 2\ \ 3 \mapsto 6 \end{array}\right\}$$

The constraint consists of three input-output examples. Each arrow ($\mapsto$) maps the inputs on its left to the output on its right. A function can be checked against an example constraint by evaluating it on the inputs and matching the results against the corresponding outputs. During synthesis, we want to check that generated expressions adhere to this constraint. For example, to synthesize mult, we may generate a range of expressions of type $Nat \rightarrow Nat \rightarrow Nat$ and then check each against the example constraint. The expression $\lambda x\ y.$ double (plus x y) will be discarded, as it maps the inputs to 2, 4, and 10, respectively. It would be more efficient, however, to recognize that any expression of the form $\lambda x\ y.$ double e, for some expression e, can be discarded, since there is no natural number whose double is 1.

To discard incorrect expressions as early as possible, we incrementally construct a sketch, where each hole (denoted by $\bullet$) is annotated with an example constraint. Each time a hole is filled, its example constraint is propagated to the new holes and checked for contradictions. Let us start from a single hole $\bullet_0$. We refine the sketch by eta-expansion, binding the inputs to the variables x and y.

$$\bullet_0 \models \left\{\begin{matrix} 0 & 1 \mapsto 0 \\ 1 & 1 \mapsto 1 \\ 2 & 3 \mapsto 6 \end{matrix}\right\} \xrightarrow{\textbf{eta-expand}} \lambda x\ y.\ (\bullet_1 \models \left\{\begin{array}{cc|c} x & y & \\ 0 & 1 & 0 \\ 1 & 1 & 1 \\ 2 & 3 & 6 \end{array}\right\})$$

A new hole $\bullet_1$ is introduced, annotated with a constraint that captures the values of x and y. Example propagation through double should be able to recognize that the value 1 is not in the codomain of double, so that the hole filling $\boxed{\bullet_1 \mapsto \text{double}\ \bullet_2}$ can be discarded.

2.1 Program Synthesis Using Example Propagation

Program synthesizers based on example propagation iteratively build a program by filling holes. At each iteration, the synthesizer may choose to fill a hole using either a *refinement* or a *guess*. A *refinement* is an expression for which example propagation is defined. For example, eta-expansion is a refinement, as shown in the previous example. To propagate an example constraint through a lambda abstraction, we simply bind the inputs to the newly introduced variables. A *guess* is an expression for which example propagation is *not* defined. The new holes introduced by a guess will not have example constraints. Once you start guessing, you have to keep guessing! Only when all holes introduced by guessing are filled can the expression be checked against the example constraint. In a sense, guessing comes down to brute-force enumerative search. Refinements are preferred over guesses, since they preserve constraint information. It is, however, not feasible to define example propagation for every possible expression.

2.2 Handcrafted Example Propagation

One way to implement example propagation is to use handcrafted propagation rules on a per function basis. Consider, for example, map, which maps a function over a list. Refinement using map replaces a constraint on a list with constraints

$$
\begin{array}{lccc}
& \textsf{inc} & \textsf{compress} & \textsf{reverse} \\
\bullet_0 \models & \{\ [0,1,2] \mapsto [1,2,3]\ \} & \{\ [0,0] \mapsto [0]\ \} & \{\ [0,0,1] \mapsto [1,0,0]\ \} \\
\bullet_1 \models & \left\{ \begin{array}{c} 0 \mapsto 1 \\ 1 \mapsto 2 \\ 2 \mapsto 3 \end{array} \right\} & \underbrace{\left\{ \begin{array}{c} 0 \mapsto 0 \\ 0 \mapsto \end{array} \right\}}_{=\bot} & \underbrace{\left\{ \begin{array}{c} 0 \mapsto 1 \\ 0 \mapsto 0 \\ 1 \mapsto 0 \end{array} \right\}}_{=\bot}
\end{array}
$$

Fig. 2. Inputs and outputs for example propagation through the hole filling $\boxed{\bullet_0 \mapsto \textsf{map}\ \bullet_1}$, for example constraints taken from the functions inc, compress, and reverse. The latter two cannot be implemented using map, which is reflected in the contradictory constraints: for compress there is a length mismatch, and for reverse the same input is mapped to different outputs.

on its elements, while checking that the input and output lists have equal length and that no value in the input list is mapped to different values in the output list. Figure 2 illustrates how specific input-output examples (taken from constraints on common list functions) are propagated through map. Note that synthesis works on many input-output examples at the same time. The function inc, which increments each number in a list by one, can be implemented using map. As such, example propagation succeeds, resulting in a constraint that represents incrementing a number by one. The function compress, which removes consecutive duplicates from a list, cannot be implemented using map, since the input and output lists can have different lengths. As such, example propagation fails, as seen in Fig. 2. The function reverse, which reverses a list, has input and output lists of the same length. It can, however, not be implemented using map, as map cannot take the positions of elements in a list into account. This is reflected in the example in Fig. 2, where the resulting constraint is inconsistent, mapping 0 to two different values.

For their tool λ^2, Feser et al. [3] provide such handcrafted propagation in terms of deduction rules for a set of combinators, including map, foldr, and filter. This allows λ^2 to efficiently synthesize complex functions in terms of these combinators. For example, λ^2 is able to synthesize a function computing the Cartesian product in terms of foldr, believed to be the first functional pearl [2]. Feser et al. [3] show that synthesis using example propagation is feasible, but λ^2 is not general purpose. Many synthesis problems require other recursion schemes or are defined over different types. Example propagation can be added for other functions in a similar fashion by adding new deduction rules, but this is very laborious work.

2.3 Example Propagation for Algebraic Datatypes

As shown by Osera and Zdancewic [10] in their synthesizer Myth, example constraints can be propagated through constructors and pattern matches, as long as they operate on algebraic datatypes. To illustrate this, we will use Peano-style natural numbers (we use the literals 0, 1, 2, etc. as syntactic sugar):

$$\textbf{data}\ \mathit{Nat} = \textsc{Zero} \mid \textsc{Succ}\ \mathit{Nat}$$

Constructors. To propagate a constraint through a constructor, we have to check that all possible outputs agree with this constructor. For example, the constraint $\{\,\textsc{Zero}\,\}$ can be refined by the constructor Zero. No constraints need to be propagated, since Zero has no arguments. In the next example, the constraint on $\bullet_0$ has multiple possible outputs, depending on the value of x.

$$\bullet_0 \models \left\{\begin{array}{c|l} x & \\ \ldots & \textsc{Succ Zero} \\ \ldots & \textsc{Succ (Succ Zero)} \end{array}\right\} \xrightarrow{\textsc{Succ}} \textsc{Succ}\ (\bullet_1 \models \left\{\begin{array}{c|l} x & \\ \ldots & \textsc{Zero} \\ \ldots & \textsc{Succ Zero} \end{array}\right\})$$

Since every possible output is a successor, the constraint can be propagated through Succ by removing one Succ constructor from each output, i.e. decreasing each output by one. The resulting constraint on $\bullet_1$ cannot be refined by a constructor, since the outputs do not all agree.

Pattern Matching. The elimination of constructors (i.e. pattern matching) is a bit more involved. For now, we will only consider non-nested pattern matches, where the scrutinee has no holes. Consider the following example, wherein the sketch double $= \lambda n.\ \bullet_0$ is refined by propagating the constraint on $\bullet_0$ through a pattern match on the local variable n.

$$\bullet_0 \models \left\{\begin{array}{c|c} n & \\ 0 & 0 \\ 1 & 2 \\ 2 & 4 \end{array}\right\} \xrightarrow{\textbf{pattern match}} \begin{array}{l} \textbf{case}\ n\ \textbf{of} \\ \quad \textsc{Zero} \quad \rightarrow \bullet_1 \models \{\,0\,\} \\ \quad \textsc{Succ}\ m \rightarrow \bullet_2 \models \left\{\begin{array}{c|c} m & \\ 0 & 2 \\ 1 & 4 \end{array}\right\} \end{array}$$

Pattern matching on n creates two branches, one for each constructor of *Nat*, with holes on the right-hand side. The constraint on $\bullet_0$ is propagated to each branch by splitting up the constraint based on the value of n. For brevity, we leave n out of the new constraints. The newly introduced variable m is exactly one less than n, i.e. one Succ constructor is stripped away.

Tying the Knot. By combining example propagation for algebraic datatypes with structural recursion, Myth is able to perform general-purpose, propagation-based synthesis. To illustrate this, we show how Myth synthesizes the function double, starting from the previous sketch. Hole $\bullet_1$ is easily refined with Zero. Hole $\bullet_2$ can be refined with Succ twice, since every output is at least 2:

$$\bullet_2 \models \left\{\begin{array}{c|c} m & \\ 0 & 2 \\ 1 & 4 \end{array}\right\} \xrightarrow{\textsc{Succ}} \ldots \xrightarrow{\textsc{Succ}} \textsc{Succ}\ (\textsc{Succ}\ (\bullet_3 \models \left\{\begin{array}{c|c} m & \\ 0 & 0 \\ 1 & 2 \end{array}\right\}))$$

At this point, to tie the knot, MYTH should introduce the recursive call double m. Note, however, that double is not yet implemented, so we cannot directly test the correctness of this guess. We can, however, use the original constraint (on $\bullet_0$) as a partial implementation of double. The example constraint on $\bullet_3$ is a subset of this original constraint, with m substituted for n. This implies that double m is a valid refinement. This property of example constraints, i.e. that the specification for recursive calls is a subset of the original constraint, is known as *trace completeness* [4], and is a prerequisite for synthesizing recursive functions in MYTH.

2.4 Evaluation-Based Example Propagation

In their synthesizer SMYTH, Lubin et al. [8] extend MYTH with sketching, i.e. program synthesis starting from a sketch, a program containing holes. Lubin et al. define evaluation-based example propagation, in order to propagate global example constraints through the sketch, after which MYTH-style synthesis takes over using the local example constraints.

Take for example the constraint $\{\,[0,1,2] \mapsto [0,2,4]\,\}$, which represents doubling each number in a list. The programmer may provide the sketch map $\bullet$ as a starting point for the synthesis procedure. Unlike λ^2, SMYTH does not provide a handcrafted rule for map. Instead, SMYTH determines how examples are propagated through functions based on their implementation.

The crucial idea is that the sketch is first evaluated, essentially inlining all function calls[2] until only simple language constructs remain, each of which supports example propagation. Omar et al. [9] describe how to evaluate an expression containing holes using live evaluation. The sketch is applied to the provided input, after which map is inlined and evaluated as far as possible:

$$\mathsf{map}\ \bullet\ [0,1,2] \rightsquigarrow [\bullet\ 0,\ \bullet\ 1,\ \bullet\ 2]$$

At this point, the constraint can be propagated through the resulting expression. Lubin et al. [8] extend MYTH-style example propagation to work for the primitives returned by live evaluation. The constraint is propagated through the result of live evaluation.

$$[\bullet\ 0,\ \bullet\ 1,\ \bullet\ 2] \models \{\,[0,2,4]\,\} \quad \longrightarrow^* \quad \begin{array}{l} [\,(\bullet \models \{\,0 \mapsto 0\,\})\ 0 \\ ,(\bullet \models \{\,1 \mapsto 2\,\})\ 1 \\ ,(\bullet \models \{\,2 \mapsto 4\,\})\ 2\,] \end{array}$$

The constraints propagated to the different occurrences of $\bullet$ in the evaluated expression can then be collected and combined to compute a constraint for $\bullet$ in the input sketch.

$$\mathsf{map}\ (\bullet \models \left\{\begin{array}{c} 0 \mapsto 0 \\ 1 \mapsto 2 \\ 2 \mapsto 4 \end{array}\right\})$$

[2] Note that function calls within the branches of a stuck pattern match are not inlined.

$$\frac{\varphi_0 \neq \bot}{\mathcal{P} \leftarrow \{\bullet_0\}}\ \text{INIT}$$

$$\frac{p \in \mathcal{P} \qquad \bullet_i \in \mathit{holes}(p) \qquad f \in \mathcal{D}_i \qquad \bullet_i \models \varphi_i \xrightarrow{f} f(\bullet_1 \models \varphi_1, \ldots, \bullet_n \models \varphi_n)}{\mathcal{P} \leftarrow \mathcal{P} \cup \{[\bullet_i \mapsto f(\bullet_1, \ldots, \bullet_n)]p\}}\ \text{EXPAND}$$

$$\frac{p \in \mathcal{P} \qquad \mathit{holes}(p) = \emptyset}{p \text{ is a solution}}\ \text{FINAL}$$

Fig. 3. Program synthesis using example propagation as a set of guarded rules that can be applied non-deterministically.

$\bullet_0 \mapsto \mathsf{map}\ (\lambda x.\ \bullet_1)\ \bullet_2$
$\bullet_1 \mapsto \textsc{Succ}\ \bullet_3$
$\bullet_2 \mapsto xs$
$\bullet_3 \mapsto x$

Fig. 4. A set of hole fillings synthesizing an expression that increments each value in a list, starting from the sketch $\lambda xs.\ \bullet_0$.

This kind of example propagation based on evaluation is called live bidirectional evaluation. For a full description, see Lubin et al. [8]. SMYTH uses live bidirectional evaluation to extend MYTH with sketching. Note, however, that SMYTH does not use live bidirectional evaluation to introduce refinements during synthesis.

3 Program Synthesis Using Example Propagation

Inspired by Smith and Albarghouthi [12], we give a high-level overview of program synthesis using example propagation as a set of guarded rules that can be applied non-deterministically, shown in Fig. 3. We keep track of a set of candidate programs $\mathcal{P}$, which is initialized by the rule INIT and then expanded by the rule EXPAND until the rule FINAL applies, returning a solution.

The rule INIT initializes $\mathcal{P}$ with a single hole $\bullet_0$, given that the initial constraint φ_0 is not contradictory. Each invocation of the rule EXPAND non-deterministically picks a program p from $\mathcal{P}$ and fills one of its holes $\bullet_i$ using a refinement f from the domain $\mathcal{D}_i$, given that the constraint φ_i can be propagated through f. The new program is considered a valid candidate and added to $\mathcal{P}$. As an invariant, $\mathcal{P}$ only contains programs that do not conflict with the original constraint φ_0. A solution to the synthesis problem is therefore simply any program $p \in \mathcal{P}$ that contains no holes.

To implement a synthesizer according to these rules, we have to make the non-deterministic choices explicit: we have to decide in which order programs are expanded ($p \in \mathcal{P}$); which holes are selected for expansion ($\bullet_i \in \mathit{holes}(p)$); and how refinements are chosen ($f \in \mathcal{D}_i$). How each of these choices is made is described in Sects. 3.1 to 3.3 respectively.

3.1 Weighted Search

To decide which candidate expressions are selected for expansion, we define an order on expressions by assigning a weight to each of them. We implement $\mathcal{P}$

as a priority queue and select expressions for expansion in increasing order of their weight. The weight of an expression can be seen as a heuristic to guide the synthesis search. Intuitively, we want expressions that are closer to a possible solution to have a lower weight.

Firstly, we assign a weight to each *application* in an expression. This leads to a preference for smaller expression, which converge more quickly and are less prone to overfitting. Secondly, we assign a weight to each *lambda abstraction*, since each lambda abstraction introduces a variable, increasing the amount of possible holes fillings. Additionally, this disincentivizes an overuse of higher-order functions, as they are always eta-expanded. Lastly, we assign a weight to complex *hole constraints*, i.e. hole constraints that took very long to compute or that have a large number of disjunctions. We assume that smaller, simpler hole constraints are more easily resolved during synthesis and are thus closer to a solution.

3.2 Hole Order

Choosing in which order holes are filled during synthesis is a bit more involved. Consider, for example, the hole fillings in Fig. 4, synthesizing the expression $\lambda xs.\ \mathsf{map}\ (\lambda x.\ \textsc{Succ}\ x)\ xs$ starting from $\lambda xs.\ \bullet_0$. There are three different synthesis paths that lead to this result, depending on which holes are filled first. More specifically, $\bullet_2$ can be filled independently of $\bullet_1$ and $\bullet_3$, so it could be filled before, between, or after them. To avoid generating the same expression three times, we should fix the order in which holes are filled, so that there is a unique path to every possible expression.

Because our techniques rely heavily on evaluation, we let evaluation guide the hole order. After filling hole $\bullet_0$, we live evaluate.

$$\mathsf{map}\ (\lambda x.\ \bullet_1)\ \bullet_2 \rightsquigarrow \mathbf{case}\ \bullet_2\ \mathbf{of}\ \ldots$$

At this point, evaluation cannot continue, because we do not know which pattern $\bullet_2$ will be matched on. We say that $\bullet_2$ blocks the evaluation. By filling $\bullet_2$, the pattern match may resolve and generate new example constraints for $\bullet_1$. Conversely, filling $\bullet_1$ does not introduce any new constraints. Hence, we always fill blocking holes first. Blocking holes are easily computed by live evaluating the expression against the example constraints.

3.3 Generating Hole Fillings

Hole fillings depend on the local context and the type of a hole and may consist of constructors, pattern matches, variables, and function calls. To avoid synthesizing multiple equivalent expressions, we will only generate expressions in *β-normal, η-long* form. An expression is in *β-normal, η-long* form exactly if no η-expansions or β-reductions are possible. During synthesis, we guarantee *β-normal, η-long* form by greedily η-expanding newly introduced holes and always fully applying functions, variables, and constructors. Consider, for example, the function map. To use map as a refinement, it is applied to two holes, the first of which is η-expanded:

$$\mathsf{map}\ (\lambda x.\ \bullet_0)\ \bullet_1$$

We add some syntactic restrictions to the generated expressions: pattern matches are non-nested and exhaustive; and we do not allow constructors in the recursive argument of recursion schemes such as foldr. These are similar to the restrictions on pattern matching and structural recursion in MYTH [4,10] and SMYTH [8]. Additionally, we disallow expressions that are not in normal form, somewhat similar to equivalence reduction as described by Smith and Albarghouthi [12]. Currently, our tool provides a handcrafted set of expressions that are not in normal form, which are prohibited during synthesis. Ideally, these sets of disallowed expressions would be taken from an existing data set (such as HLint[3]), or approximated using evaluation-based techniques such as QUICKSPEC [1].

3.4 Pruning the Program Space

For a program $p \in \mathcal{P}$ and a hole $\bullet_i \in holes(p)$, we generate a set of possible hole fillings based on the hole domain $\mathcal{D}_i$. For each of these hole fillings, we try to apply the EXPAND rule. To do so, we must ensure that the constraint φ_i on $\bullet_i$ can be propagated through the hole filling. We use evaluation-based example propagation in the style of SMYTH to compute hole constraints for the newly introduced holes and check these for consistency. If example propagation fails, we do not extend $\mathcal{P}$, essentially pruning the program space.

Diverging Constraints. Unfortunately, example propagation is not feasible for all possible expressions. Consider, for instance, the function sum. If we try to propagate a constraint through sum $\bullet$, we first use live evaluation, resulting in the following partially evaluated result, with $\bullet$ in a scrutinized position:

$$\begin{array}{l} \textbf{case}\ \bullet\ \textbf{of} \\ \quad [\,]\quad\ \ \rightarrow 0 \\ \quad x : xs \rightarrow \mathsf{plus}\ x\ (\mathsf{sum}\ xs) \end{array}$$

Unlike MYTH, SMYTH (and by extension SCRYBE) allows examples to be propagated through pattern matches whose scrutinee may contain holes, considering each branch separately under the assumption that the scrutinee evaluates to the corresponding pattern. This introduces disjunctions in the example constraint. Propagating $\{\,\text{ZERO}\,\}$ through the previous expression results in a constraint that cannot be finitely captured in our constraint language:

$$(\bullet \vDash \{\,[\,]\,\}) \vee (\bullet \vDash \{\,[\text{ZERO}]\,\}) \vee (\bullet \vDash \{\,[\text{ZERO}, \text{ZERO}]\,\}) \vee \ldots$$

Without extending the constraint language it is impossible to compute such a constraint. Instead, we try to recognize that example propagation diverges, by

[3] https://github.com/ndmitchell/hlint.

setting a maximum to the amount of recursive calls allowed during example propagation. If the maximum recursion depth is reached, we cancel example propagation.

Since example propagation through sum $\bullet$ always diverges, we could decide to disallow it as a hole filling. This is, however, too restrictive, as example propagation becomes feasible again when the length of the argument to sum is no longer unrestricted. Take, for example, the following constraint, representing counting the number of TRUEs in a list, and a possible series of hole fillings:

$$\bullet_0 \vDash \left\{ \begin{array}{c|c} xs & \\ [\text{FALSE}] & 0 \\ [\text{FALSE}, \text{TRUE}] & 1 \\ [\text{TRUE}, \text{TRUE}] & 2 \end{array} \right\} \qquad \boxed{\begin{array}{l} \bullet_0 \mapsto \mathsf{sum}\ \bullet_1 \\ \bullet_1 \mapsto \mathsf{map}\ (\lambda x.\ \bullet_2)\ \bullet_3 \\ \bullet_3 \mapsto xs \end{array}}$$

Trying to propagate through $\boxed{\bullet_0 \mapsto \mathsf{sum}\ \bullet_1}$ diverges, since $\bullet_1$ could be a list of any length. At this point, we could decide to disregard this hole filling, but this would incorrectly prune away a valid solution. Instead, we allow synthesis to continue guessing hole fillings, until we get back on the right track: after guessing $\boxed{\bullet_1 \mapsto \mathsf{map}\ (\lambda x.\ \bullet_2)\ \bullet_3}$ and $\boxed{\bullet_3 \mapsto xs}$, the length of the argument to sum becomes restricted and example propagation no longer diverges:

$$\mathsf{sum}\ (\mathsf{map}\ (\lambda x.\ \bullet_2 \vDash \left\{ \begin{array}{c|c} x & \\ \text{FALSE} & 0 \\ \text{TRUE} & 1 \end{array} \right\})\ xs)$$

At this point, synthesis easily finishes by pattern matching on x. Note that, unlike λ^2, MYTH, and SMYTH, SCRYBE is able to interleave refinements and guesses.

Exponential Constraints. Even if example propagation does not diverge, it still might take too long to compute or generate a disproportionally large constraint, slowing down the synthesis procedure. Lubin et al. [8] compute the falsifiability of an example constraint by first transforming it to disjunctive normal form (DNF), which may lead to exponential growth of the constraint size. For example, consider the function or, defined as follows:

$$\begin{array}{l} \mathsf{or} = \lambda a\ b.\ \textbf{case}\ a\ \textbf{of} \\ \quad \text{FALSE} \rightarrow b \\ \quad \text{TRUE} \ \rightarrow \text{TRUE} \end{array}$$

Propagating the example constraint $\{\,\text{TRUE}\,\}$ through the expression or $\bullet_0\ \bullet_1$ puts the hole $\bullet_0$ in a scrutinized position, resulting in the following constraint:

$$(\bullet_0 \vDash \{\,\text{FALSE}\,\} \wedge \bullet_1 \vDash \{\,\text{TRUE}\,\}) \vee \bullet_0 \vDash \{\,\text{TRUE}\,\}$$

This constraint has size three (the number of hole occurrences). We can extend this example by mapping it over a list of length n as follows:

$$\mathsf{map}\ (\lambda x.\ \mathsf{or}\ \bullet_0\ \bullet_1)\ [0, 1, 2, \ldots] \vDash \{\,[\text{TRUE}, \text{TRUE}, \text{TRUE}, \ldots]\,\}$$

Propagation generates a conjunction of n constraints that are all exactly the same apart from their local context, which differs in the value of x. This constraint, unsurprisingly, has size $3n$. Computing the disjunctive normal form of this constraint, however, results in a constraint of size of $2^n \times \frac{3}{2}n$, which is exponential.

In some cases, generating such a large constraint may cause example propagation to reach the maximum recursion depth. In other cases, example propagation succeeds, but returns such a large constraint that subsequent refinements will take too long to compute. In both cases, we treat it the same as diverging example propagation.

4 Evaluation

We have implemented our synthesis algorithm in the tool Scrybe[4]. To evaluate Scrybe, we combine the benchmarks of Myth [10] and λ^2 [3]. We ran the Scrybe benchmarks using the Criterion[5] library on an HP Elitebook 850 G6 with an Intel® Core™ i7-8565U CPU (1.80 GHz) and 32 GB of RAM.

This evaluation is not intended to compare our technique directly with previous techniques in terms of efficiency, but rather to get insight into the effectiveness of example propagation as a pruning technique, as well show the wide range of synthesis problems that Scrybe can handle.

For ease of readability, the benchmark suite is split up into a set of functions operating on lists (Table 1) and a set of functions operating on binary trees (Table 2). We have excluded functions operating on just booleans or natural numbers, as these are all trivial and synthesize in a few milliseconds. For consistency, and to avoid naming conflicts, the names of some of the benchmarks are changed to better reflect the corresponding functions in the Haskell prelude. To avoid confusion, each benchmark function comes with a short description.

Each row describes a single synthesis problem in terms of a function that needs to be synthesized. The first two columns give the name and a short description of this function. The third and fourth columns show, in milliseconds, the average time Scrybe takes to correctly synthesize the function with example propagation (*EP*) and without example propagation (*NoEP*), respectively. Some functions may fail to synthesize ($\bot$) within 10 s and some cannot straightforwardly be represented in our language (-). The last three columns show, for Myth, Smyth, and λ^2, respectively, whether the function synthesizes (✓), fails to synthesize (✗), or is not included in their benchmark (-).

Each benchmark uses the same context as the original benchmark (if the benchmark occurs in both Myth and λ^2, the context from the Myth benchmark is chosen). Additionally, benchmarks from Myth use a catamorphism in place of structural recursion, except for list_set_insert and tree_insert, which use a paramorphism instead.

4 https://github.com/NiekM/scrybe.
5 https://github.com/haskell/criterion.

Table 1. Benchmark for functions acting on lists. Each row describes a single benchmark task and the time it takes for each function to synthesize with example propagation (*EP*) and without (*NoEP*) respectively. Some tasks cannot be synthesized within 10 s (⊥) and others are omitted, since they cannot straightforwardly be translated to our language (-).

Name	Description	Runtime		Myth	Smyth	λ^2
		EP	*NoEP*			
list_add	Increment each value in a list by n	4.70	5.65	-	-	✓
list_append	Append two lists	13.35	41.36	✓	✓	✓
list_cartesian	The cartesian product	⊥	⊥	-	-	✓
list_compress	Remove consecutive duplicates from a list	⊥	⊥	✓	✗	-
list_copy_first	Replace each element in a list with the first	30.71	20.55	-	-	✓
list_copy_last	Replace each element in a list with the last	14.89	28.00	-	-	✓
list_delete_max	Remove the largest numbers from a list	16.05	31.95	-	-	✓
list_delete_maxs*	Remove the largest numbers from a list of lists	1875.43	⊥	-	-	✓
list_drop‡	All but the first n elements of a list	554.96	⊥	✓	✓	-
list_dupli	Duplicate each element in a list	6.43	7.60	✓	✓	✓
list_evens	Remove any odd numbers from a list	1.84	2.62	-	-	✓
list_even_parity	Whether a list has an odd number of Trues	42.72	201.73	✓	✗	-
list_filter	The elements in a list that satisfy p	80.25	162.70	✓	✓	-
list_flatten	Flatten a list of lists	7.54	8.12	✓	✓	✓
list_fold	A catamorphism over a list	9.48	6.43	✓	✓	-
list_head†	The first element of a list	1.40	2.20	✓	✓	-
list_inc	Increment each value in a list by one	5.38	24.36	✓	✓	-
list_incs	Increment each value in a list of lists by one	12.79	19.73	-	-	✓
list_index‡	Index a list starting at zero	79.01	2956.53	✓	✓	-
list_init†	All but the last element of a list	115.59	453.28	-	-	✓
list_last†	The last element of a list	9.63	15.14	✓	✓	✓
list_length	The number of elements in a list	1.33	2.16	✓	✓	✓
list_map	Map a function over a list	2.82	4.75	✓	✓	✓
list_maximum	The largest number in a list	120.38	231.44	-	-	✓
list_member	Whether a number occurs in a list	212.45	1145.07	-	-	✓
list_nub	Remove duplicates from a list	450.56	9245.26	-	-	✓
list_reverse*	Reverse a list	131.04	574.82	✓	✓	-
list_set_insert	Insert an element in a set	⊥	⊥	✓	✓	-
list_shiftl	Shift all elements in a list to the left	723.97	525.29	-	-	✓
list_shiftr	Shift all elements in a list to the right	369.27	620.67	-	-	✓
list_snoc	Add an element to the end of a list	6.77	55.76	✓	✓	✓
list_sum	The sum of all numbers in a list	4.36	17.05	✓	✓	✓
list_sums	The sum of each nested list in a list of lists	69.71	677.12	-	-	✓
list_swap*	Swap the elements in a list pairwise	-	-	✓	✓	-
list_tail†	All but the first element of a list	1.65	5.43	✓	✓	-
list_take‡	The first n elements of a list	462.50	⊥	✓	✓	-
list_to_set	Sort a list, removing duplicates	37.06	41.18	✓	✓	-

The benchmarks list_head, list_tail, list_init, and list_last are all partial functions (marked †). We do not support partial functions, and therefore these functions are replaced by their total counterparts, by wrapping their return type in *Maybe*. For example, list_last is defined as follows, where the outlined hole filling is the result returned by Scrybe (input-output constraints are omitted for brevity):

Table 2. Benchmark for functions acting on binary trees. Each row describes a single benchmark task and the time it takes for each function to synthesize with example propagation (*EP*) and without (*NoEP*) respectively. Some tasks cannot be synthesized within 10 s (⊥) and others are omitted, since they cannot straightforwardly be translated to our language (-).

Name	Description	Runtime		Myth	Smyth	λ^2
		EP	*NoEP*			
tree_cons	Prepend an element to each list in a tree of lists	3.38	5.08	-	-	✓
tree_flatten	Flatten a tree of lists into a list	68.34	72.29	-	-	✓
tree_height	The height of a tree	7.01	34.89	-	-	✓
tree_inc	Increment each element in a tree by one	1.40	2.21	-	-	✓
tree_inorder	Inorder traversal of a tree	15.77	18.28	✓	✓	✓
tree_insert	Insert an element in a binary tree	⊥	⊥	✓	✗	-
tree_leaves	The number of leaves in a tree	14.77	45.67	✓	✓	✓
tree_level‡	The number of nodes at depth n	⊥	⊥	✓	✗	-
tree_map	Map a function over a tree	3.38	11.77	✓	✓	-
tree_maximum	The largest number in a tree	25.02	114.68	-	-	✓
tree_member	Whether a number occurs in a tree	907.81	⊥	-	-	✓
tree_postorder	Postorder traversal of a tree	19.50	44.79	✓	✗	-
tree_preorder	Preorder traversal of a tree	9.09	21.10	✓	✓	-
tree_search	Whether a number occurs in a tree of lists	1218.10	5253.07	-	-	✓
tree_select	All nodes in a tree that satisfy p	652.12	1268.76	-	-	✓
tree_size	The number of nodes in a tree	31.48	252.63	✓	✓	✓
tree_snoc	Append an element to each list in a tree of lists	57.13	2156.82	-	-	✓
tree_sum	The sum of all nodes in a tree	18.21	89.11	-	-	✓
tree_sum_lists	The sum of each list in a tree of lists	19.32	285.03	-	-	✓
tree_sum_trees	The sum of each tree in a list of trees	1000.26	⊥	-	-	✓

{-# USE foldr #-}
list_last :: $\mathit{List}\ a \to \mathit{Maybe}\ a$
list_last $xs = \bullet$

$$\bullet \mapsto \text{foldr}\ (\lambda x\ r.\ \textbf{case}\ r\ \textbf{of}\ \begin{array}{l}\textsc{Nothing} \to \textsc{Just}\ x \\ \textsc{Just}\ y \quad\ \to r\end{array})\ \textsc{Nothing}\ xs$$

The benchmarks list_drop, list_index, list_take, and tree_level (marked ‡) all recurse over two datatypes at the same time. As such, they cannot be implemented using foldr as it is used in Sect. 3.3. Instead, we provide a specialized version of foldr that takes an extra argument. For example, for list_take:

{-# USE foldr :: $(a \to (c \to b) \to (c \to b)) \to (c \to b) \to \mathit{List}\ a \to c \to b$ #-}
list_take :: $\mathit{Nat} \to \mathit{List}\ a \to \mathit{List}\ a$
list_take $n\ xs = \bullet$

$$\bullet \mapsto \text{foldr}\ (\lambda x\ r\ m.\ \textbf{case}\ m\ \textbf{of}\ \begin{array}{l}\textsc{Zero} \quad\ \to [\,] \\ \textsc{Succ}\ o \to (x : r\ o)\end{array})\ (\lambda\_.\ [\,])\ xs\ n$$

A few functions (marked *) could not straightforwardly be translated to our approach:

- Function list_delete_maxs replaces the function list_delete_mins from λ^2. The original list_delete_mins requires a total function in scope that returns the minimum number in a list. This is not possible for natural numbers, as there is no obvious number to return for the empty list.
- Function list_swap uses nested pattern matching on the input list, which is not possible to mimic using a fold.
- Function list_reverse combines a set of benchmarks from MYTH that synthesize reverse using different techniques, which are not easily translated to our language.

4.1 Results

SCRYBE is able to synthesize most of the combined benchmarks of MYTH and λ^2, with a median runtime of 19 ms. Furthermore, synthesis with example propagation is on average 5.1 times as fast as without example propagation, disregarding the benchmarks where synthesis without example propagation failed. λ^2 noticed a similar improvement (6 times as fast) for example propagation based on automated deduction, which indicates that example propagation using live-bidirectional evaluation is similar in effectiveness, while being more general.

Some functions benefit especially from example propagation, in particular those that can be decomposed into smaller problems. Take, for example, the function tree_snoc, which can be decomposed into list_snoc and tree_map. By preserving example constraints between these subproblems, SCRYBE greatly reduces the search space.

$$\{-\#\ \text{USE mapTree}, \text{foldr}, \ldots\ \#-\}$$
$$\text{tree\_snoc} :: a \rightarrow \mathit{Tree}\ (\mathit{List}\ a) \rightarrow \mathit{Tree}\ (\mathit{List}\ a)$$
$$\text{tree\_snoc}\ x\ t = \bullet$$
$$\boxed{\bullet \mapsto \text{mapTree}\ (\lambda xs.\ \text{foldr}\ (\lambda y\ r.\ y : r)\ [x]\ xs)\ t}$$

On the other hand, for some functions, such as list_shiftl, synthesis is noticeably faster without example propagation, showing that the overhead of example propagation sometimes outweighs the benefits. This indicates that it might be helpful to use some heuristics to decide when example propagation is beneficial. A few functions that fail to synthesize, such as list_compress, do synthesize when a simple sketch is provided:

$$\{-\#\ \text{USE foldr}, (\equiv), \ldots\ \#-\}$$
$$\text{compress} :: \mathit{List\ Nat} \rightarrow \mathit{List\ Nat}$$
$$\text{compress}\ xs = \text{foldr}\ (\lambda x\ r.\ \bullet_0)\ \bullet_1$$

$$\begin{array}{|l|}\hline \bullet_0 \mapsto x : \textbf{case}\ r\ \textbf{of} \\ \quad [\,] \quad\quad \rightarrow r \\ \quad y : ys \rightarrow \textbf{if}\ x \equiv y\ \textbf{then}\ ys\ \textbf{else}\ r \\ \bullet_1 \mapsto [\,] \\ \hline \end{array}$$

Since our evaluation is not aimed at sketching, we still denote list_compress as failing ($\bot$) in the benchmark.

5 Conclusion

We presented an approach to program synthesis using example propagation that specializes in compositionality, by allowing arbitrary functions to be used as refinement steps. One of the key ideas is holding on to constraint information as long as possible, rather than resorting to brute-force, enumerative search. Our experiments show that we are able to synthesize a wide range of synthesis problems from different synthesis domains.

There are many avenues for future research. One direction we wish to explore is to replace the currently ad hoc constraint solver with a more general purpose SMT solver. Our hope is that this paves the way for the addition of primitive data types such as integers and floating point numbers.

Acknowledgements. We would like to thank Alex Gerdes, Koen Claessen, and the anonymous reviewers of HATRA 2022 and PADL 2023 for their supportive comments and constructive feedback.

References

1. Claessen, K., Smallbone, N., Hughes, J.: QuickSpec: guessing formal specifications using testing. In: Fraser, G., Gargantini, A. (eds.) TAP 2010. LNCS, vol. 6143, pp. 6–21. Springer, Heidelberg (2010). https://doi.org/10.1007/978-3-642-13977-2_3
2. Danvy, O., Spivey, M.: On Barron and Strachey's cartesian product function. In: Proceedings of the 12th ACM SIGPLAN International Conference on Functional Programming, ICFP 2007, pp. 41–46. Association for Computing Machinery, New York (2007). https://doi.org/10.1145/1291151.1291161
3. Feser, J.K., Chaudhuri, S., Dillig, I.: Synthesizing data structure transformations from input-output examples. ACM SIGPLAN Not. **50**(6), 229–239 (2015). https://doi.org/10.1145/2813885.2737977
4. Frankle, J., Osera, P.M., Walker, D., Zdancewic, S.: Example-directed synthesis: a type-theoretic interpretation. SIGPLAN Not. **51**(1), 802–815 (2016). https://doi.org/10.1145/2914770.2837629
5. Guo, Z., et al.: Program synthesis by type-guided abstraction refinement. Proc. ACM Program. Lang. **4**(POPL) (2019). https://doi.org/10.1145/3371080
6. Katayama, S.: Efficient exhaustive generation of functional programs using Monte-Carlo search with iterative deepening. In: Ho, T.-B., Zhou, Z.-H. (eds.) PRICAI 2008. LNCS (LNAI), vol. 5351, pp. 199–210. Springer, Heidelberg (2008). https://doi.org/10.1007/978-3-540-89197-0_21
7. Koppel, J., Guo, Z., de Vries, E., Solar-Lezama, A., Polikarpova, N.: Searching entangled program spaces. Proc. ACM Program. Lang. **6**(ICFP) (2022). https://doi.org/10.1145/3547622
8. Lubin, J., Collins, N., Omar, C., Chugh, R.: Program sketching with live bidirectional evaluation. Proc. ACM Program. Lang. **4**(ICFP) (2020). https://doi.org/10.1145/3408991
9. Omar, C., Voysey, I., Chugh, R., Hammer, M.A.: Live functional programming with typed holes. Proc. ACM Program. Lang. **3**(POPL) (2019). https://doi.org/10.1145/3290327

10. Osera, P.M., Zdancewic, S.: Type-and-example-directed program synthesis. ACM SIGPLAN Not. **50**(6), 619–630 (2015). https://doi.org/10.1145/2813885.2738007
11. Peleg, H., Gabay, R., Itzhaky, S., Yahav, E.: Programming with a read-eval-synth loop. Proc. ACM Program. Lang. **4**(OOPSLA) (2020). https://doi.org/10.1145/3428227
12. Smith, C., Albarghouthi, A.: Program synthesis with equivalence reduction. In: Enea, C., Piskac, R. (eds.) VMCAI 2019. LNCS, vol. 11388, pp. 24–47. Springer, Cham (2019). https://doi.org/10.1007/978-3-030-11245-5_2
13. Solar-Lezama, A.: Program synthesis by sketching. Ph.D. thesis, Berkeley (2008)
14. Solar-Lezama, A.: The sketching approach to program synthesis. In: Hu, Z. (ed.) APLAS 2009. LNCS, vol. 5904, pp. 4–13. Springer, Heidelberg (2009). https://doi.org/10.1007/978-3-642-10672-9_3

Embedding Functional Logic Programming in Haskell via a Compiler Plugin

Kai-Oliver Prott[1(✉)], Finn Teegen[1], and Jan Christiansen[2]

[1] Kiel University, Kiel, Germany
{kpr,fte}@informatik.uni-kiel.de
[2] Flensburg University of Applied Sciences, Flensburg, Germany
jan.christiansen@hs-flensburg.de

Abstract. We present a technique to embed a functional logic language in Haskell using a GHC plugin. Our approach is based on a monadic lifting that models the functional logic semantics explicitly. Using a GHC plugin, we get many language extensions that GHC provides for free in the embedded language. As a result, we obtain a seamless embedding of a functional logic language, without having to implement a full compiler. We briefly show that our approach can be used to embed other domain-specific languages as well. Furthermore, we can use such a plugin to build a full blown compiler for our language.

Keywords: Functional programming · Logic programming · DSL · Haskell · GHC plugin · Monadic transformation

1 Introduction

Writing a compiler for a programming language is an elaborate task. For example, we have to write a parser, name resolution, and a type checker. In contrast, embedding a domain-specific language (DSL) within an already existing host language is much easier. If the host language is a functional programming language, such an embedding often comes with the downside of monadic overhead. To get the best of both worlds, we use a compiler plugin to embed a domain-specific language that does not require the user to write their code in a monadic syntax. Concretely, we implement a plugin for the Glasgow Haskell Compiler (GHC) that allows for a seamless embedding of functional logic programming in Haskell.[1] An additional benefit of our approach is that our embedded language supports a lot of Haskell's language extensions without additional effort.

Our lazy functional logic language is loosely based on the programming language Curry [2,19]. However, instead of Curry's syntax, we have to restrict ourselves to Haskell's syntax. This is because at the moment[2] a GHC plugin cannot

[1] Available at https://github.com/cau-placc/ghc-language-plugin.
[2] We plan to change this; see GHC issue 22401 [30].

M. Hanus and D. Inclezan (Eds.): PADL 2023, LNCS 13880, pp. 37–55, 2023.
https://doi.org/10.1007/978-3-031-24841-2_3

alter the static syntax accepted by GHC. Moreover, overlapping patterns in our language use Haskell's semantics of choosing the first rule instead of introducing non-determinism as Curry does. In return we save a significant amount of work with our approach compared to implementing a full compiler for Curry. One of the compilers for the Curry programming language, the KiCS2 [7], translates Curry into Haskell. While the KiCS2 has around 120,000 lines of Haskell code, by using a GHC plugin we get the most basic functionality with approximately 7,500 lines of code. We have to be careful with these figures because we do not provide the same feature set. For example, KiCS2 provides free variables and unification whereas our plugin instead provides multi-parameter type classes. However, by adapting the underlying implementation, we could provide the remaining features of KiCS2 as well. Besides reducing the amount of code, when using a plugin we do not have to implement features that are already implemented in GHC from scratch. As examples, it took two master's theses to add multi-parameter type (constructor) classes to the KiCS2 [25,34] while we get them for free.

The following example demonstrates the basic features of our approach using the Curry-like language. It enumerates permutations, the "hello world" example of the Curry programming language community. Here and in the following we label code blocks as follows: `Source` for code that is written in the embedded language, `Target` for corresponding Haskell code, which is generated by our plugin, and `Plugin` for code that is provided by us as part of our plugin.

```
{-# OPTIONS_GHC -fplugin Plugin.CurryPlugin #-}                    Source
module Example where

insert :: a → [a] → [a]
insert e []       = [e]
insert e (y : ys) = (e : y : ys) ? (y : insert e ys)

permutations :: [a] → [a]
permutations []       = []
permutations (x : xs) = insert x (permutations xs)
```

The first line activates our plugin. The function *insert* uses the (?) operator for introducing non-determinism to insert a given element at an arbitrary position in a list. That is, *insert* either adds *e* to the front of the list or puts *y* in front of the result of the recursive call. Building upon this, *permutations* non-deterministically computes a permutation of its input list by inserting the head of the list anywhere into a permutation of the remaining list.

Having the ability to write functional logic code in direct-style, i.e., without explicit monadic abstractions, is a great advantage. The code in direct-style is more concise as well as more readable, and makes it easy to compose non-deterministic and deterministic functions. To enable writing code in direct-style, our plugin transforms the code from direct-style to an explicit monadic representation behind the scenes. This transformation is type-safe and therefore allows for a seamless embedding of a functional logic language within Haskell.

To use the non-deterministic *permutations* operation in a Haskell program, we need to capture the non-determinism and convert the function to work on ordinary lists.

ghci> eval1 permutations $[1, 2, 3]$
$[[1, 2, 3], [1, 3, 2], [2, 1, 3], [2, 3, 1], [3, 1, 2], [3, 2, 1]]$

Here, *eval1* is essentially a wrapper that converts the 1-ary function *permutations* into a Haskell function that returns a list of all results and without non-determinism nested in data structures.[3] The result is a list of all permutations of the given input list.

We present the following contributions.

- We show how a GHC plugin can be used to embed a functional logic language in Haskell. By embedding the language via a plugin (Sect. 2, Sect. 3), we obtain the ability to write functional logic definitions, and can use the code written in the embedded language in Haskell.
- Besides saving the implementation of a parser and a type checker, we get additional language features like type classes, do-notation, bang patterns and functional dependencies for the functional logic language for free (Sect. 4).
- Although we demonstrate the embedding of a functional logic language in this paper, we argue that our approach can be generalized to languages other domain-specific languages as well (Sect. 5).

2 From Direct-Style to Explicit Monadic Style

Recall from Sect. 1 the implementation of a *permutations* functions in direct-style. Since Haskell does not natively support non-determinism, we need a way to represent our embedded programs in Haskell. To do this, we transform direct-style code to its explicit monadic counterpart. In this section, we will show the transformation rules from direct-style to monadic style.

2.1 A Lazy Functional-Logic Monad

Before we go into detail with the transformation, we need to introduce our functional logic language and its monadic implementation.

As seen before, our language provides non-deterministic branching. Since Haskell uses call-by-need evaluation, we want our language to be non-strict as well. In the context of functional logic programming, we also have to decide between using call-time choice or run-time choice [22]. Consider the function *double* $x = x + x$ applied to $1 \mathbin{?} 2$. With run-time choice, we would get the results 2, 3, and 4. The choice for the non-deterministic value in the argument of *double* is made independently for each occurrence of x. However, obtaining the odd number 3 as a result of *double* is unintuitive. With call-time choice the choice

[3] Using Template Haskell, we also provide an arity-independent wrapper function where one can specify a search strategy as well.

for x is shared between all occurrences. Thus, call-time choice only produces the result 2 and 4. For our language we settled for call-time choice since it is more intuitive in many cases [2].

To model a functional logic language with call-by-need semantics and call-time choice, we use a technique presented by Fischer et al. [16]. In their paper, Fischer et al. use an operator *share* to turn a monadic computation into a computation that is evaluated at most once and is, thus, lazy. The *share* operator is part of the type class *MonadShare* that is defined as follows.[4]

```
class Monad m ⇒ MonadShare m where                                    Plugin
  share :: Shareable m a ⇒ m a → m (m a)
class Shareable m a where
  shareArgs :: a → m a
```

Intuitively, *share* registers its argument as a thunk in some kind of state and returns a new computation as a replacement for the shared expression. Whenever the returned computation is executed, it looks up if a value has been computed for the thunk. If this is the case, the said value is returned. Otherwise, the thunk is executed and the result value is saved for the future.

The additional type class *Shareable* needs to be defined for all monadically transformed data types. The class method *shareArgs* applies *share* recursively to the components of the transformed data types. Because data constructors can contain non-deterministic operations in their arguments, constructors take computations as arguments as well. This approach is well-known from modeling the non-strictness of Haskell in an actually pure programming language like Agda [1]. The usefulness of interleaving data types with computational effects has also been observed in more general settings [3].

Our lazy non-determinism monad *ND* is implemented using techniques from Fischer et al. [16]. The implementation provides the following two operations to introduce non-deterministic choices and failures by using the *Alternative* instance of our monad.

```
(?) :: a → a → a                                                      Source
failed :: a
```

We omit the full implementation from the paper, because one could use any monadic implementation for a lazy functional logic language. In the future, we want to use the implementation from [20] for our language. However, the focus of this work is the embedding of the language, not the encoding of it.

2.2 The Transformation

Now we will show our transformation rules to compile direct-style to monadic style. In the context of a GHC plugin, we have used a similar transformation of Haskell code [35] to generate inverses of Haskell functions. A notable difference

[4] The version presented in [16] has a more general type.

is that we include the modification with *share* to model call-by-need instead of just call-by-name.

We will use the *ND* type constructor in our transformation, although it works with any monadic type constructor that provides an implementation of *share*. To increase readability of monadically transformed function types, we will use the following type.

$$\textbf{newtype}\ a \rightarrow_{ND} b = \mathit{Func}\ (\mathit{ND}\ a \rightarrow \mathit{ND}\ b)$$ Plugin

2.3 Types

Our transformation on types replaces all type constructors with their non-deterministic counterparts. By replacing the type constructor $\rightarrow$ with $\rightarrow_{ND}$, the result and arguments of each functions are effectively wrapped in *ND* as well. Any quantifiers and constraints of a transformed type remain at the beginning of the type signature to avoid impredicativity. Additionally, we also wrap the outer type of a function in *ND* to allow an operation to introduce non-determinism before applying any arguments. In order to keep the original definitions of types available for interfacing the plugin with Haskell (Sect. 3.5), it is necessary to rename (type) constructors. This way we avoid name clashes by altering names of identifiers. Figure 1 presents the transformation of types, including the renaming of type constructors. An example for our transformation on types is given below.[5]

$$[\![\forall \alpha_1 \ \ldots \ \alpha_n . \varphi \Rightarrow \tau]\!]^t := \forall \alpha_1 \ \ldots \ \alpha_n . [\![\varphi]\!]^t \Rightarrow \mathit{ND}\ [\![\tau]\!]^i \qquad \text{(Polymorphic type)}$$

$$[\![\langle \kappa_1\ \tau_1, \ldots, \kappa_n\ \tau_n \rangle]\!]^t := \langle \text{rename}(\kappa_1)\ [\![\tau_1]\!]^i, \ldots, \text{rename}(\kappa_n)\ [\![\tau_n]\!]^i \rangle \qquad \text{(Context)}$$

$$[\![\tau_1 \rightarrow \tau_2]\!]^i := [\![\tau_1]\!]^i \rightarrow_{ND} [\![\tau_2]\!]^i \qquad \text{(Function type)}$$

$$[\![\tau_1\ \tau_2]\!]^i := [\![\tau_1]\!]^i\ [\![\tau_2]\!]^i \qquad \text{(Type application)}$$

$$[\![\chi]\!]^i := \text{rename}(\chi) \qquad \text{(Type constructor)}$$

$$[\![\alpha]\!]^i := \alpha \qquad \text{(Type variable)}$$

Fig. 1. Type lifting $[\![\cdot]\!]^t$

$\mathit{double} :: \mathit{Int} \rightarrow \mathit{Int}$ Source $\qquad$ $\mathit{double} :: \mathit{ND}\ (\mathit{Int} \rightarrow_{ND} \mathit{Int})$ Target

2.4 Data Type Declarations

To model non-strictness of data constructors, we need to modify most data type definitions and lift every constructor. As the (partial or full) application of a

[5] The transformed version of *Int* is called *Int* as well.

constructor can never introduce any non-determinism by itself, we neither have to transform the result type of the constructor nor wrap the function arrow. This allows us to only transform the parameters of constructors, because they are the only potential sources of non-determinism in a data type (Fig. 2). The following code shows how a language plugin transforms the list data type. To improve readability we use regular algebraic data type syntax for the list type and its constructors.

$$\begin{aligned}
[\![\ \mathbf{data}\ D\ \alpha_1\ \dots\ \alpha_n = C_1 \mid \dots \mid C_n]\!]^d &:= \mathbf{data}\ \mathrm{rename}(D)\ \alpha_1\ \dots\ \alpha_n \\
&\quad = [\![C_1]\!]^c \mid \dots \mid [\![C_n]\!]^c && \text{(Data type)} \\
[\![C\ \tau_1\ \dots\ \tau_n]\!]^c &:= \mathrm{rename}(C)\ [\![\tau_1]\!]^t\ \dots\ [\![\tau_n]\!]^t && \text{(Constructor)}
\end{aligned}$$

Fig. 2. Data type lifting $[\![\cdot]\!]^d$

$$\mathbf{data}\ \mathit{List}\ a = \mathit{Nil} \mid \mathit{Cons}\ a\ (\mathit{List}\ a) \qquad \texttt{Source}$$

$$\mathbf{data}\ \mathit{List}_{ND}\ a = \mathit{Nil}_{ND} \mid \mathit{Cons}_{ND}\ (\mathit{ND}\ a)\ (\mathit{ND}\ (\mathit{List}_{ND}\ a)) \qquad \texttt{Target}$$

2.5 Functions

The type of a transformed function serves as a guide for the transformation of functions and expressions. We derive three rules for our transformation of expressions from the one of types:

1. Each function arrow is wrapped in *ND*. Therefore, function definitions are replaced by constants that use multiple unary lambda expressions to introduce the arguments of the original function. All lambda expressions are wrapped in a *return* because function arrows are wrapped in a *ND* type.
2. We have to extract a value from the monad using ($\gg\!=$) before we can pattern match on it. As an implication, we cannot pattern match directly on the argument of a lambda or use nested pattern matching, as arguments of a lambda and nested values are potentially non-deterministic and have to be extracted using ($\gg\!=$) again.
3. Before applying a function to an argument, we first have to extract the function from the monad using ($\gg\!=$). As each function arrow is wrapped separately, we extract each parameter of the original function.

Implementing this kind of transformation in one pass over a complex Haskell program is challenging. Thus, we simplify each program first. As the most difficulties arise from pattern matching, we perform pattern match compilation to get rid of complex and nested patterns. In contrast to the transformation done by GHC, we want a Haskell AST instead of Core code as output. Thus, we cannot re-use GHC's existing implementation and use a custom transformation that is similar to the one presented in [39]. Note that we still preserve any non-exhaustiveness warnings and similar for pattern matching by applying GHC's pattern match checker before our transformation. For the remainder of this subsection we assume that our program is already desugared, that is, all pattern matching is performed with non-nested patterns in case expressions.

$$
\begin{aligned}
[\![v]\!] &:= v && \text{(Variable)}\\
[\![\lambda x \to e]\!] &:= \mathit{return}\ (\mathit{Func}\ (\lambda y \to \mathrm{alias}(y, x, [\![e]\!]))) && \text{(Abstraction)}\\
[\![e_1\ e_2]\!] &:= [\![e_1]\!] \ggg\!= (\lambda(\mathit{Func}\, f) \to f\ [\![e_2]\!]) && \text{(Application)}\\
[\![C]\!] &:= \mathit{return}\ (\mathit{Func}\ (\lambda y_1 \to \ldots\\
&\qquad \mathit{return}\ (\mathit{Func}\ (\lambda y_n \to \ldots\\
&\qquad \mathit{return}\ (\mathrm{rename}(C)\ y_1\ \ldots\ y_n))))) && \text{(Constructor)}\\
[\![\,\mathbf{case}\ e\ \mathbf{of}\ \{\, br_1; \ldots; br_n \}]\!] &:= [\![e]\!] \ggg\!= (\lambda y \to \mathbf{case}\ y\ \mathbf{of}\ \{[\![br_1]\!]^b; \ldots; [\![br_n]\!]^b\})\\
& && \text{(Case Expression)}\\
[\![C\ x_1\ \ldots\ x_n \to e]\!]^b &:= \mathrm{rename}(C)\ y_1\ \ldots\ y_n \to\\
&\qquad \mathrm{alias}(y_1, x_1, \ldots\ \mathrm{alias}(y_n, x_n, [\![e]\!])) && \text{(Case Branch)}
\end{aligned}
$$

$$
\mathrm{alias}(v_{new}, v_{old}, e) := \begin{cases} \mathit{share}\ v_{new} \ggg\!= \lambda v_{old} \to e & \text{if } v_{old} \text{ occurs at least twice in } e\\ \mathbf{let}\ v_{old} = v_{new}\ \mathbf{in}\ e & \text{otherwise} \end{cases} \qquad \text{(Aliasing)}
$$

Fig. 3. Expression lifting $[\![\cdot]\!]$ (y and y_1 to y_n are fresh variables)

Figure 3 presents the rules of our transformation. Note that any bindings introduced by a lambda or case expression are shared using the alias rule. We have omitted rules for let declarations, since recursive bindings are a bit tricky to get right. They are supported by our implementation as long as the monadic type has a sensible *MonadFix* instance. This is the case for our simple implementation.

To give an example of our whole lifting, the *permutations* example from the introduction with minor simplifications looks as follows.

```
insertND :: ND (a →ND ListND a →ND ListND a)                                Target
insertND = return (Func (λarg1 → share arg1 ≫= λe →
             return (Func (λarg2 → share arg2 ≫= λxs → xs ≫= λcase
  NilND         → return (ConsND e (return NilND))
  ConsND y ys → return (ConsND e (return (ConsND y ys))) <|>
                return (ConsND y (insertND ≫= λ(Func f1) → f1 e
                                           ≫= λ(Func f2) → f2 ys))))))

permutationsND :: ND (ListND a →ND ListND a)
permutationsND = return (Func (λarg → share arg ≫= λarg' → arg' ≫= λcase
  NilND         → return NilND
  ConsND x xs → insertND ≫= λ(Func f1) → f1 x ≫= λ(Func f2) →
                  f2 (permutationsND ≫= λ(Func f3) → f3 xs)))
```

The transformation made the code significantly more complex, where most of the complexity arises from the modeling of laziness. This also emphasizes that writing code in direct-style is more readable for lazy functional logic programming.

3 Plugin Core

In this section, we motivate the use of a plugin and discuss its core. In particular we present how the transformation to monadic code is achieved.

3.1 Why a GHC Plugin?

The GHC allows extending its functionality using a compiler plugin API. Using this API, we can change the behavior of the type check and analyze or transform code at various intermediate stages.

There are some alternatives to using a plugin, for example, metaprogramming via Template Haskell [32]. One crucial point is that our transformation (e.g. in Sect. 3.4) requires type information that is not present when using Template Haskell. Even with typed Template Haskell, we do not get all type annotations that we need for the transformation. Additionally, we aim to provide good quality error messages produced by GHC itself (Sect. 3.3), for which we have no idea how to properly achieve this with a transformation using (typed) Template Haskell. We could do the transformation on the level of GHC's intermediate language Core, which would eliminate the need for our own pattern match transformation and reduce the amount of syntax constructs we have to take into account. But a short evaluation showed that some aspects of the transformation are harder to achieve on core level, for example, generating additional instances, modifying data types, and adding additional type constraints to functions.

3.2 Plugin Phases

The plugin consists of three sub-plugins, each having a different purpose and using a different extension point of GHC's plugin API as shown in Fig. 4. The "Import Check" and the "Import Constraint Solver" are both concerned with the correct handling of imports. The former checks that each imported module has been compiled with the same language plugin, while the latter resolves all type mismatches that arise from the transformation (Sect. 3.3). At last, the main work is done in the "Transformation" phase, which is again divided into four sub-phases:

1. Transforming data types to model non-strictness of data constructors. Type classes are also transformed in this sub-phase (not shown in this paper).
2. Compiling and simplifying pattern matching to make the implementation of sub-phase 4 simpler.

3. Deriving instances of internal type classes for all data types, because our transformation requires each data type to have certain instances (like *Shareable* to model sharing of non-deterministic computations).
4. Transforming function definitions to achieve our desired semantics. Here we use the rules shown in Sect. 2.2.

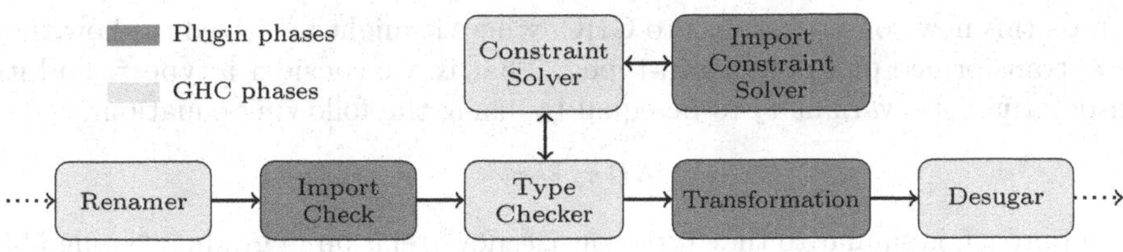

Fig. 4. Extension points used by our plugin

3.3 Handling of Imported Definitions

We consider the following constant where *double* is imported from a different module.

Source

$$example :: Int$$
$$example = double\ (23 :: Int)$$

If the definition was not imported from a module that has already been compiled by the plugin, it will not be compatible with the current module. Therefore we check that each module only imports plugin-compiled modules and abort compilation with an error message otherwise. A programmer can explicitly mark a non-plugin module as safe to be imported, but its the programmers obligation to ensure that all functions in that module have plugin-compatible types. Let us assume the import of *double* passes the "Import Check". Before our transformation takes place, GHC type checks the whole module. The definition of *example* will cause a type error, because the non-determinism is explicit in the transformed variant $double :: ND\ (Int \rightarrow_{ND} Int)$ that has already been compiled by the plugin. In fact, any imported function (and data type) has a lifted type and, thus, using such a function will be rejected by GHC's type check. However, the transformed version of *example* will later be type correct and we thus want to accept the definition.

To overcome this issue we use a constraint solver plugin. When GHC type checks *example*, it realizes that *double* should have type $Int \rightarrow Int$ from looking at the argument $23 :: Int$ and the result type of *example*. However, *double* was imported and its type is known to be $ND\ (Int \rightarrow_{ND} Int)$. The type checker will now create an equality constraint[6]

$$ND\ (Int \rightarrow_{ND} Int) \sim (Int \rightarrow Int)$$

[6] The operator ($\sim$) denotes homogeneous (both kinds are the same) equality of types.

and sends it to GHC's constraint solver to (dis-)prove. Naturally, GHC cannot make any progress proving this equality and instead forwards it to our constraint solver plugin. There we un-transform the type, that is, we remove the outer *ND* and un-transform the transformed function arrow ($\rightarrow_{ND}$) to obtain our new left-hand side for the equality constraint:

$$(Int \rightarrow Int) \sim (Int \rightarrow Int).$$

We pass this new constraint back to GHC, where it might be solved to show that the untransformed program is well-typed. That is, we consider a type τ_1 and its non-deterministic variant τ_2 to be equal by using the following equation.

$$ND\ \tau_2 \sim \tau_1$$

This approach is similar to the "type consistency" relation in gradual typing [33]. For details about GHC's constraint solver, we refer the interested reader to [36].

Besides making transformed and untransformed types compatible the constraint solver plugin also needs to automatically solve any constraints that mention the *Shareable* type class. GHC cannot always discharge these constraints before our main transformation has taken place.

Type Errors. Our plugin ensures that ill-typed programs are not accepted and returns a meaningful type error. Consider the following ill-typed program.

```
typeError :: Int                                   Source
typeError = double True
```

Using our plugin, GHC generates the following type error as if *double* actually had the untransformed type.

- Couldn't match expected type *Int* with actual type *Bool*
- In the first argument of *double*, namely *True*
  In the expression: *double True*
  In an equation for *typeError*: *typeError* = *double True*

In fact, any error messages produced by GHC for modules with our plugin activated are (mostly) the same as if the user would have written the identical code in "vanilla" Haskell. There are only a few deliberate exceptions where we augment certain errors or warnings with more information. We achieve these error messages by transforming the code after the type check and by deliberately untransforming the types of imported functions.

3.4 Sharing Polymorphic Expressions

We have already introduced the *share* operator at the beginning of the section. For call-by-need evaluation, this operator is implemented as presented in [16] and requires a type class constraint *Shareable* $(m :: \mathit{Type} \rightarrow \mathit{Type})$ $(a :: \mathit{Type})$. The constraint enforces that values of type a can be shared in the monad m. Instances of this type class are automatically derived by the plugin for every user-defined data type via generic deriving [26].

Ensuring that all types in a type signature are *Shareable* is easy to achieve if the type signature is monomorphic or only contains type variables with the simple kind *Type*. For variables with other kinds, e.g., $f :: \mathit{Type} \rightarrow \mathit{Type}$, we cannot add a constraint for that variable as *Shareable* $m\ f$ is ill-kinded. So, what can one do if Haskell's type and class system is not powerful enough? Just enable even more language extensions![7] In our case, we can solve the problem by using the extension *QuantifiedConstraints* [6]. What we really want to demand for f in the example above is that, for every type x that is *Shareable*, the type $f\ x$ should be *Shareable* as well. We can express exactly this requirement by using a $\forall$ quantifier in our constraint. This is demonstrated by the following example.

Source

$$\mathit{void} :: \mathit{Functor}\ f \Rightarrow f\ a \rightarrow f\ ()$$

Target

$$\begin{aligned}\mathit{void} :: (&\mathit{Functor}\ f, \mathit{Shareable}\ \mathit{ND}\ a\\ &, \forall x.\mathit{Shareable}\ \mathit{ND}\ x \Rightarrow \mathit{Shareable}\ \mathit{ND}\ (f\ x))\\ &\Rightarrow \mathit{ND}\ (f\ a \rightarrow_{\mathit{ND}} f\ ())\end{aligned}$$

We can extend this scheme to even more complex kinds, as long as it only consists of the kind of ordinary Haskell values (*Type*) and the arrow kind constructor ($\rightarrow$). Language extensions that allow the usage of other kinds are not supported at the moment. To add *Shareable* constraints in types, we replace the first rule in Fig. 1 by the rules in Fig. 5.

$$[\![\forall\alpha_1\ \ldots\ \alpha_n.\varphi \Rightarrow \tau]\!]^t := \forall\alpha_1\ \ldots\ \alpha_n.([\![\varphi]\!]^t \cup \langle[\![\alpha_1]\!]^s, \ldots, [\![\alpha_n]\!]^s\rangle) \Rightarrow [\![\tau]\!]^t \qquad \text{(Polymorphic type)}$$

$$[\![v :: (\kappa_1 \rightarrow \ldots \rightarrow \kappa_n)]\!]^s := \forall(\beta_1 :: \kappa_1)\ \ldots\ (\beta_{n-1} :: \kappa_{n-1}).\langle[\![\beta_1]\!]^s, \ldots, [\![\beta_{n-1}]\!]^s\rangle \Rightarrow \mathit{Shareable}\ \mathit{ND}\ (v\ \beta_1\ \ldots\ \beta_{n-1}) \qquad \text{(Type variable of kind } \kappa_1 \rightarrow \ldots \rightarrow \kappa_n)$$

$$[\![v :: \mathit{Type}]\!]^s := \mathit{Shareable}\ \mathit{ND}\ v \qquad \text{(Type variable of kind } \mathit{Type})$$

Fig. 5. Type lifting $[\![\cdot]\!]^t$ with *Shareable* constraints ($\cup$ joins two contexts, $\beta_1\ \ldots\ \beta_{n-1}$ are fresh variables)

3.5 Interfacing with Haskell

At the end of the day, we want to call the functions written in our embedded language from within Haskell and extract the non-deterministic results in some kind of data structure. Because data types are transformed to support non-strict data constructors, we do not only have to encapsulate non-determinism at the

[7] Use this advice at your own risk.

top-level of a result type, but effects that are nested inside of data structures as well. To trigger all non-determinism inside a data structure, we convert data types to some kind of normal form, where only top-level non-determinism is possible. We also translate data types defined in our embedded language into their original counterparts. This way, we can embed our new language within Haskell as a DSL.

Consider the following example with a list of n non-deterministic coins that are either *False* or *True*.

```
coin :: Bool                                                    Source
coin = False ? True
manyCoins :: Int → [Bool]
manyCoins n = [coin | _ ← [1 .. n]]
```

The transformed type will be $ND\ (Int \rightarrow_{ND} List_{ND}\ Bool_{ND})$. To use the result of *manyCoins* in plain Haskell, it is convenient to convert the list with non-deterministic components into an ordinary Haskell list. Therefore, we move any non-determinism occurring in the elements of the list to the outermost level.

The conversion to normal form is automatically derived for all data types by generating instances of the type class *NormalForm*.[8]

```
class NormalForm a b where                                      Plugin
  nf :: a → ND b
  embed :: b → a
```

Here, the type variable a will be a transformed type, while b will be the corresponding original definition. The function *nf* converts a value with nested non-determinism into its normal form. To convert a data type to normal form, we convert the arguments to normal form and chain the values together with (≫=). The result of *nf* contains non-determinism only at the outermost level and the data type is transformed back to its original definition.

We also support conversion in the other direction. The function *embed* converts an untransformed data type to its transformed representation by recursively embedding all constructor arguments. Note that this type class is not required for the monadic transformation itself, but for the interface between Haskell and our embedded language.

For the list data type and its non-deterministic variant $List_{ND}$ from Sect. 2.4, we define the *NormalForm* instance as follows.

```
instance NormalForm a b ⇒ NormalForm (List_ND a) (List b) where      Target
  nf Nil_ND           = return Nil
  nf (Cons_ND x xs)   = liftM2 Cons (x ≫= nf) (xs ≫= nf)
  embed Nil           = Nil_ND
  embed (Cons x xs)   = Cons_ND (return (embed x)) (return (embed xs))
```

[8] It is similar to the type class *Convertible* from [16].

3.6 Compatibility

The plugin makes heavy use of GHC's internal APIs and is thus only compatible with a specific GHC version. Development started with GHC 8.10, but some of the experiments we conducted with higher-rank types required us to enable the "Quick Look" approach for impredicative types [31] introduced in GHC 9.2. The difficulty of upgrading the plugin to support a new GHC version depends on the complexity of GHC's internal changes, but even the upgrade from 8.10 to 9.2 did not take long although there were significant changes to GHC under the hood.

Also note that new GHC language features are not supported automatically, even if we upgrade the plugin to the new GHC version. For example, consider Linear Haskell as introduced in GHC 9.0. With this version, GHC's internal representation for function types changed to accommodate linearity information. When we upgraded the plugin past this major version, we had to adapt to the new representation. However, without careful considerations during the transformation, the linearity information is lost or not accurately reflected in the final code. Thus, without explicitly treating the linearity information correctly (what that means exactly is still left to determine), the plugin does not support Linear Haskell properly. But when linear types are not used, the transformation succeeds for the new GHC version.

4 Inherited Features

In this section we discuss GHC features that our plugin supports for the corresponding embedded language as well. Some of these features are supported out-of-the-box, that is, just by using a GHC plugin to implement a language. A few type-level and type class extensions are among these features, for instance. Other features required us to make small adaptions to the plugin, for example, to add additional cases to the transformation.

To demonstrate the benefits of our approach, we revisit the example from Sect. 1. We consider a variant that uses a multi-parameter type class with a functional dependency [24] to abstract from list-like data structures and view patterns [14, 37] to define a generalized version of *permutations*. We cannot write this program using any of the available Curry compilers as none of them support the required language features. That is, using a plugin we can not only implement a non-deterministic language in 7,500 lines of code, we also get features for free that no other full-blown Curry compiler with more than 120,000 lines of code supports.

```
class ListLike l e | l → e where                                    Source
  nil :: l
  cons :: e → l → l
  uncons :: l → Maybe (e, l)
permutations :: ListLike l e ⇒ l → l
permutations (uncons → Nothing)      = nil
permutations (uncons → Just (x, xs)) = ins x (permutations xs)
```

$$\begin{aligned} \textbf{where } & ins\ e\ (uncons \to Nothing) &&= cons\ e\ nil \\ & ins\ e\ (uncons \to Just\ (x, xs)) &&= cons\ e\ (cons\ x\ xs)\ ?\ cons\ x\ (ins\ e\ xs) \end{aligned}$$

Ad-hoc polymorphism via type classes is one of Haskell's most useful features for writing concise programs. Properly supporting type classes as well as extensions surrounding them like multi-parameter type classes, functional dependencies and flexible contexts is straightforward. We can just rely on GHC for solving type class constraints and for implementing type classes via dictionary transformation [23,38]. Apart from minor technical differences, we can handle functions from a type class instance like any other function during our monadic transformation. As type classes are a complex feature, we were kind of surprised that a lot of Haskell's type class extensions are easy to support.

Most language extensions required little to no additional lines of code to be supported by our plugin. Some of them are among the most commonly used extensions on Hackage [9,10]. From the 30 most used extensions, 19 are currently supported by our plugin. Figure 6 lists a selection of these extensions and their respective rank in the list of most commonly used GHC extensions. The extensions in the column labeled "Supported out-of-the-box" are orthogonal to the plugin transformation and are supported without any additional effort. The column labeled "Small adaptions" in Fig. 6 lists extensions that are supported after minor changes. Most of these extensions add syntactic sugar that is simple enough to transform.

Supported out-of-the-box		Small adaptions
2. `FlexibleInstances`	5. `MultiParamTypeClasses`	1. `OverloadedStrings`
3. `FlexibleContexts`	13. `RecordWildCards`	15. `LambdaCase`
4. `ScopedTypeVariables`	14. `TypeOperators`	20. `TupleSections`

Fig. 6. Some extensions with their rank in most commonly used GHC extensions

Unsupported Features and Limits. Although our plugin can handle many language extensions, some of GHC's more feature-rich and expressive extensions remain unsupported. We can group the problems for each extension into three categories:

1. The extension enriches GHC's type system. Since our transformation works on the typed syntax tree, our transformation would have to consider more complex type expressions, which is challenging on a technical level to get right. This is the case for, e.g., type families and higher-rank types. For higher-rank types, one of the problems is that they lead to impredicativity in combination with our transformation.
2. The extension allows the definition of more flexible data types. Our plugin has to derive instances of some type classes (e.g., Sect. 3.5) automatically, which gets more challenging for complex data types as well. Examples for such an extension are existential data types.

3. The extension introduces a lot of syntactical sugar that requires a great deal of work to support. One prominent example for this category is the extension *RebindableSyntax*, which just introduces a lot of corner cases to consider during our monadic transformation. However, there is no reason for extensions in this category to be hard or impossible to support.

5 Embedding Other Languages

To show the applicability of our approach for different languages, we have embedded additional languages using a plugin: a strict functional language, a probabilistic language, and a language with dynamic information flow control. For brevity, we do not include these in the paper but provide them on GitHub.[9] All these additional languages must use Haskell's syntax too. However, we are working on lifting that restriction by extending the GHC plugin API with the required capabilities. Note that all plugins are built upon a generalized variant of the plugin we presented in this paper. Implementing the monads that model the effects of these languages is as simple or as complex as implementing a monad in Haskell – with the same type of errors and warnings, for example. Apart from the concrete effect implementation, almost all of our plugin code is shared between the three examples. The implementations of these languages are approximately 400 lines of code each.

Note that the additional languages do not use call-by-need as their evaluation strategy. This is not a problem, however, as we generalized our monadic transformation to model call-by-value and call-by-name semantics as well. These languages also need different implementations for the *share* operator. Details on implementing *share* for different evaluation strategies can be found in [27].

A major goal for future research is to embed even more languages. We plan to enlarge the class of languages that can be embedded using our plugin. Currently, we can use all languages whose semantics can be modeled using a standard monad that are not parametrized over an additional parameter. For example, at the moment we cannot use our plugin to model a language with algebraic effects on the basis of a free monad. We also plan to investigate the generalization to more different forms of monads like graded or indexed monads.

6 Related Work

There are three different groups of related work.

6.1 GHC Plugins

The first group of related work uses GHC plugins. The GHC plugin infrastructure has been used to extend Haskell with different kinds of features. Using constraint solver plugins, Gundry [18] adds support for units of measure and Diatchki

[9] See README file in the linked repository from Footnote 3.

[12] adds type-level functions on natural numbers. Di Napoli et al. [11] use a combination of plugins to extend Haskell with liquid types and Breitner [8] uses a combination of Template Haskell and a Core plugin to provide inspection testing. We also use a combination of plugins, we use a constraint solver plugin to recover from type errors and a source plugin [29] that uses the renamer and the type checker extension point for the transformation. However, none of the other plugins transform the compiled code to such an extent as our plugin since they do not aim for a thorough transformation. The only plugin that is similar is one that compiles GHC Core code to abstractions from category theory [13]. The approach is suitable for the definition of DSLs, but only supports monomorphic functions and values. In contrast, our plugin aims to model a full functional programming language with implicit effects.

6.2 Monadic Intermediate Languages

Our transformation basically models the denotational semantics of a language explicitly. Peyton Jones et al. [28] have applied a similar approach to model Haskell and ML in a single monadic intermediate language. However, although the underlying idea bears some resemblance, the contributions of our work are different. While existing work focuses on common intermediate languages, we use Haskell and GHC for more than just the compiler backend.

Transforming an impure functional language into monadic code has been applied to model Haskell in various proof assistants. For example, a Core plugin for GHC is used by [1] to generate Agda code that captures Haskell's semantics via an explicit monadic effect.

6.3 Embedding Languages

Last but not least in the Scheme and Racket family of programming languages users can define new languages by using advanced macro systems [17]. For example, there exists a lazy variant of PLT Scheme for teaching [4] and an implementation of miniKanren, a logic programming language [21]. Felleisen et al. [15] even aim for language-oriented programming that facilitates the creation of DSLs, the development in these languages and the integration of multiple of these languages. We do not aim for language-oriented programming but to simplify writing and embedding research languages within a different ecosystem, namely the GHC with its statically typed, purely functional programming language. We target all those research prototype languages that could as well be implemented as a monadic variant of Haskell without the syntactic monadic overhead.

7 Conclusion

In this paper, we have shown how to embed a functional logic language into Haskell. Using a compiler plugin allows us to use direct-style syntax for the embedded language while retaining good quality error messages. In the future, we want to extend the approach as mentioned in Sect. 5 and add support for more of GHC's language extensions. For example, we plan to investigate the support of more current features like linear types [5] as well as extensions like higher-rank types.

Acknowledgements. We thank our anonymous reviewers and Michael Hanus for their helpful comments and fruitful discussions.

References

1. Abel, A., Benke, M., Bove, A., Hughes, J., Norell, U.: Verifying Haskell programs using constructive type theory. In: Proceedings of the 2005 ACM SIGPLAN Workshop on Haskell, pp. 62–73. ACM Press, New York (2005). https://doi.org/10.1145/1088348.1088355
2. Antoy, S., Hanus, M.: Functional logic programming. Commun. ACM **53**(4), 74 (2010). https://doi.org/10.1145/1721654.1721675
3. Atkey, R., Johann, P.: Interleaving data and effects. J. Funct. Program. **25** (2015). https://doi.org/10.1017/S0956796815000209
4. Barzilay, E., Clements, J.: Laziness without all the hard work: combining lazy and strict languages for teaching. In: Proceedings of the 2005 Workshop on Functional and Declarative Programming in Education, FDPE 2005, Tallinn, Estonia, p. 9. ACM Press (2005). https://doi.org/10.1145/1085114.1085118
5. Bernardy, J.P., Boespflug, M., Newton, R.R., Jones, S.P., Spiwack, A.: Linear Haskell: practical linearity in a higher-order polymorphic language. Proc. ACM Program. Lang. **2**(POPL), 1–29 (2018). https://doi.org/10.1145/3158093
6. Bottu, G.J., Karachalias, G., Schrijvers, T., Oliveira, B.C.D.S., Wadler, P.: Quantified class constraints. In: Proceedings of the 10th ACM SIGPLAN International Symposium on Haskell, Haskell 2017, pp. 148–161. Association for Computing Machinery, New York (2017). https://doi.org/10.1145/3122955.3122967
7. Braßel, B., Hanus, M., Peemöller, B., Reck, F.: KiCS2: a new compiler from curry to Haskell. In: Kuchen, H. (ed.) WFLP 2011. LNCS, vol. 6816, pp. 1–18. Springer, Heidelberg (2011). https://doi.org/10.1007/978-3-642-22531-4_1
8. Breitner, J.: A promise checked is a promise kept: inspection testing. In: Proceedings of the 11th ACM SIGPLAN International Symposium on Haskell, Haskell 2018, pp. 14–25. Association for Computing Machinery, New York (2018). https://doi.org/10.1145/3242744.3242748
9. Breitner, J.: GHC Extensions stats (2020). https://gist.github.com/nomeata/3d1a75f8ab8980f944fc8c845d6fb9a9
10. Breitner, J.: [GHC-steering-committee] Prelimary GHC extensions stats (2020). https://mail.haskell.org/pipermail/ghc-steering-committee/2020-November/001876.html
11. Di Napoli, A., Jhala, R., Löh, A., Vazou, N.: Liquid Haskell as a GHC Plugin - Haskell implementors workshop 2020 (2020). https://icfp20.sigplan.org/details/hiw-2020-papers/1/Liquid-Haskell-as-a-GHC-Plugin
12. Diatchki, I.S.: Improving Haskell types with SMT. In: Proceedings of the 2015 ACM SIGPLAN Symposium on Haskell, Haskell 2015, pp. 1–10. Association for Computing Machinery, New York (2015). https://doi.org/10.1145/2804302.2804307

13. Elliott, C.: Compiling to categories. Proc. ACM Program. Lang. **1**(ICFP) (2017). https://doi.org/10.1145/3110271
14. Erwig, M., Peyton Jones, S.: Pattern guards and transformational patterns. Electron. Notes Theor. Comput. Sci. **41**(1), 3 (2001). https://doi.org/10.1016/S1571-0661(05)80540-7
15. Felleisen, M., et al.: A programmable programming language. Commun. ACM **61**(3), 62–71 (2018). https://doi.org/10.1145/3127323
16. Fischer, S., Kiselyov, O., Shan, C.C.: Purely functional lazy nondeterministic programming. J. Funct. Program. **21**(4–5), 413–465 (2011). https://doi.org/10.1017/S0956796811000189
17. Flatt, M.: Composable and compilable macros: you want it when? In: Proceedings of the Seventh ACM SIGPLAN International Conference on Functional Programming, ICFP 2002, pp. 72–83. Association for Computing Machinery, New York (2002). https://doi.org/10.1145/581478.581486
18. Gundry, A.: A typechecker plugin for units of measure: domain-specific constraint solving in GHC Haskell. In: Proceedings of the 2015 ACM SIGPLAN Symposium on Haskell, Haskell 2015, pp. 11–22. Association for Computing Machinery, New York (2015). https://doi.org/10.1145/2804302.2804305
19. Hanus, M., Kuchen, H., Moreno-Navarro, J.: Curry: a truly functional logic language. In: Proceedings of the ILPS 1995 Workshop on Visions for the Future of Logic Programming, pp. 95–107 (1995)
20. Hanus, M., Prott, K.O., Teegen, F.: A monadic implementation of functional logic programs. In: Proceedings of the 24th International Symposium on Principles and Practice of Declarative Programming, PPDP 2022. Association for Computing Machinery, New York (2022). https://doi.org/10.1145/3551357.3551370
21. Hemann, J., Friedman, D.P., Byrd, W.E., Might, M.: A small embedding of logic programming with a simple complete search. In: Proceedings of the 12th Symposium on Dynamic Languages, DLS 2016, pp. 96–107. Association for Computing Machinery, New York (2016). https://doi.org/10.1145/2989225.2989230
22. Hussmann, H.: Nondeterministic algebraic specifications and nonconfluent term rewriting. J. Log. Program. **12**, 237–255 (1992). https://doi.org/10.1016/0743-1066(92)90026-Y
23. Jones, M.P.: A system of constructor classes: overloading and implicit higher-order polymorphism. In: Proceedings of the Conference on Functional Programming Languages and Computer Architecture - FPCA 1993, Copenhagen, Denmark, pp. 52–61. ACM Press (1993). https://doi.org/10.1145/165180.165190
24. Jones, M.P.: Type classes with functional dependencies. In: Smolka, G. (ed.) ESOP 2000. LNCS, vol. 1782, pp. 230–244. Springer, Heidelberg (2000). https://doi.org/10.1007/3-540-46425-5_15
25. Krüger, L.E.: Extending Curry with multi parameter type classes (in German). Master's thesis, Christian-Albrechts-Universität zu Kiel, Kiel, Germany (2021). https://www.informatik.uni-kiel.de/mh/lehre/abschlussarbeiten/msc/Krueger_Leif_Erik.pdf
26. Magalhães, J.P., Dijkstra, A., Jeuring, J., Löh, A.: A generic deriving mechanism for Haskell. In: Proceedings of the Third ACM Haskell Symposium on Haskell, Haskell 2010, pp. 37–48. Association for Computing Machinery, New York (2010). https://doi.org/10.1145/1863523.1863529
27. Petricek, T.: Evaluation strategies for monadic computations. In: Proceedings of Mathematically Structured Functional Programming, vol. 76, pp. 68–89 (2012). https://doi.org/10.4204/EPTCS.76.7

28. Peyton Jones, S., Shields, M., Launchbury, J., Tolmach, A.: Bridging the gulf: a common intermediate language for ML and Haskell. In: Proceedings of the 25th ACM SIGPLAN-SIGACT Symposium on Principles of Programming Languages, POPL 1998, pp. 49–61. Association for Computing Machinery, New York (1998). https://doi.org/10.1145/268946.268951
29. Pickering, M., Wu, N., Németh, B.: Working with source plugins. In: Proceedings of the 12th ACM SIGPLAN International Symposium on Haskell, Haskell 2019, Berlin, Germany, pp. 85–97. ACM Press (2019). https://doi.org/10.1145/3331545.3342599
30. Prott, K.O.: Plugin-swappable parser (2022). https://gitlab.haskell.org/ghc/ghc/-/issues/22401
31. Serrano, A., Hage, J., Peyton Jones, S., Vytiniotis, D.: A quick look at impredicativity. Proc. ACM Program. Lang. **4**(ICFP), 1–29 (2020). https://doi.org/10.1145/3408971
32. Sheard, T., Jones, S.P.: Template meta-programming for Haskell. In: Proceedings of the 2002 ACM SIGPLAN Workshop on Haskell, Haskell 2002, pp. 1–16. Association for Computing Machinery, New York (2002). https://doi.org/10.1145/581690.581691
33. Siek, J., Taha, W.: Gradual typing for functional languages. In: Proceedings of the 2006 Scheme and Functional Programming Workshop, pp. 81–92 (2006)
34. Teegen, F.: Extending Curry with type classes and type constructor classes (in German). Master's thesis, Christian-Albrechts-Universität zu Kiel, Kiel, Germany (2016). https://www.informatik.uni-kiel.de/mh/lehre/abschlussarbeiten/msc/Teegen.pdf
35. Teegen, F., Prott, K.O., Bunkenburg, N.: Haskell^{-1}: automatic function inversion in Haskell. In: Proceedings of the 14th ACM SIGPLAN International Symposium on Haskell, Haskell 2021, pp. 41–55. Association for Computing Machinery, New York (2021). https://doi.org/10.1145/3471874.3472982
36. Vytiniotis, D., Jones, S.P., Schrijvers, T., Sulzmann, M.: OutsideIn(X) Modular type inference with local assumptions. J. Funct. Program. **21**(4–5), 333–412 (2011). https://doi.org/10.1017/S0956796811000098
37. Wadler, P.: Views: a way for pattern matching to cohabit with data abstraction. In: Proceedings of the 14th ACM SIGACT-SIGPLAN Symposium on Principles of Programming Languages, POPL 1987, Munich, West Germany, pp. 307–313. ACM Press (1987). https://doi.org/10.1145/41625.41653
38. Wadler, P., Blott, S.: How to make ad-hoc polymorphism less ad hoc. In: Proceedings of the 16th ACM SIGPLAN-SIGACT Symposium on Principles of Programming Languages, POPL 1989, Austin, Texas, USA, pp. 60–76. ACM Press (1989). https://doi.org/10.1145/75277.75283
39. Wadler, P.: Efficient compilation of pattern-matching. In: Jones, S.L.P. (ed.) The Implementation of Functional Programming Languages, pp. 78–103. Prentice-Hall (1987). https://doi.org/10.1016/0141-9331(87)90510-2

Execution Time Program Verification with Tight Bounds

Ana Carolina Silva[1,3(✉)], Manuel Barbosa[1,2], and Mário Florido[1,3]

[1] Faculdade de Ciências, Universidade do Porto, Porto, Portugal
carolsilva.acfs@gmail.com
[2] INESC TEC, Porto, Portugal
[3] LIACC, Universidade do Porto, Porto, Portugal

Abstract. This paper presents a proof system for reasoning about execution time bounds for a core imperative programming language. Proof systems are defined for three different scenarios: approximations of the worst-case execution time, exact time reasoning, and less pessimistic execution time estimation using amortized analysis. We define a Hoare logic for the three cases and prove its soundness with respect to an annotated cost-aware operational semantics. Finally, we define a verification conditions (VC) generator that generates the goals needed to prove program correctness, cost, and termination. Those goals are then sent to the Easycrypt toolset for validation. The practicality of the proof system is demonstrated with an implementation in OCaml of the different modules needed to apply it to example programs. Our case studies are motivated by real-time and cryptographic software.

Keywords: Program verification · Execution time analysis · Amortized analysis · Hoare logic

1 Introduction

Program verification of properties determining resource consumption is a research topic which is fast growing in relevance. *Resource* is used in a broad sense: it can be time used to execute the program, memory used during program execution or even energy consumption. Resource consumption has a huge impact in areas such as real-time systems, critical systems relying on limited power sources, and the analysis of timing side-channels in cryptographic software.

A proof system for total correctness can be used to prove that a program execution terminates but it does not give any information about the resources it requires. In this paper we study extended proof systems for proving assertions about program behavior that may refer to the required resources and, in particular, to the execution time. Proof systems to prove bounds on the execution time of program execution have been defined before [6,19]. Inspired by the work developed by Barbosa *et al.* [6] we define an inference system to prove assertions

M. Hanus and D. Inclezan (Eds.): PADL 2023, LNCS 13880, pp. 56–72, 2023.
https://doi.org/10.1007/978-3-031-24841-2_4

of the form $\{\varphi\}C\{\psi|t\}$, meaning that if the execution of statement C is started in a state that validates the precondition φ then it terminates in a state that validates postcondition ψ and the required execution time is at most of magnitude t.

The main contribution of this paper is that our proof system is able to verify resource assumptions for three different scenarios: i. upper bounds on the required execution time, ii. amortized costs denoting less pessimistic bounds on the execution time and iii. exact costs for a fragment of the initial language with bounded recursion and a constrained form of conditional statements. The two last scenarios are a novel contribution of our system and we treat them in a unified way to enable their integrated use.

Assertions on program behavior that establish upper bounds on execution time may be useful for general programming, where one wants to prove safety conditions concerning the worst case program complexity. As in prior approaches, the tightness of the bound is not captured by the logic, and there is often a trade-off between the tightness of the proved bound and the required proof effort. Proofs that leverage amortized costs may be used in situations where trivially composing worst-case run-time bounds results in overly pessimistic analyses. This is particularly useful for algorithms where some components imply a significant cost in resources, whereas other components are not as costly. With amortized costs we may prove assertions about the aggregate use of costly and less costly operations, over the whole algorithm execution. Finally, the third class of assertions denoting exact costs are useful in scenarios where approximation of execution time is not enough to guarantee safety, as it happens for critical systems and real-time programming. Moreover, proving that the exact execution time of a program is an expression that does not depend on secret data provides a direct way to prove the absence of timing leakage, which is relevant in cryptographic implementations. To guarantee the ability to prove exact costs we necessarily have to restrict the programming language, thus, in this third scenario, programs have bounded recursion and conditional statements have to be used in situations where both branches have the same cost.

To demonstrate the practicality of our proof system we have implemented in OCaml the different modules needed to apply it to example programs. We then present several application examples motivated by real-time and cryptographic software.

The paper is organized as follows. Section 2 presents a simple imperative language with assignments, conditional statements, a *while* statement, and sequences of statements. It also presents a cost-aware operational semantics for this language. Section 3 contains the formalization of the base proof system in the form a Hoare logic considering triples of the form $\{\varphi\}C\{\psi|t\}$. This can be read as, executing C from a state σ that validates the precondition φ leads to a state that validates postcondition ψ and this execution costs at most t to complete. In Sect. 4 we present the amortized costs extension, and the exact cost extension is presented in Sects. 5. In Sect. 6 we present the soundness theorem for our logic, in the sense that conclusions of proof derivations (i.e. theorems) are valid partial correctness assertions. Section 7 briefly describes the implementation of a prototype for mechanizing our logic. Finally, we conclude and point some future work.

Related Work: One of the first and still one of the most relevant works on this topic is the one presented by Nielsen [19,20]. The author defines an axiomatic semantics for a simple imperative programming language and extends Hoare's logic so that the proof system would be capable of proving the magnitude of worst-case running time. Even though this work operates on a similar imperative language to the one we defined, it lacks the precision our logic provides since it only allows proving order-of-magnitude.

In 2014, Carbonneaux *et al.* [8] presented a system that verifies stack-space bounds of compiled machine code at the C level. This work is an extension of the CompCert C Compiler [17]. Carbonneaux *et al.* [9] continued their previous work on deriving worst-case resource bounds for C programs, but they now implemented a system that uses amortized analysis and abstract interpretations. These works focus on the compilation from C to assembly using quantitative logic, which does not serve the same purpose we are trying to achieve. With our work, we can prove tight bounds on imperative programs using an assisted proof system, where the user can help make the bounds as precise as possible. Additionally, our approach to defining the costs of default constructors makes our work easy to adjust to different resource analyses.

In 2018, Kaminski *et al.* defined a conservative extension of Nielsen's logic for deterministic programs [19,20] by developing a Weakest Precondition calculus for expected runtime on probabilistic programs [16]. Again this work is largely automated, which differs from our user-assisted approach. Since it reasons about probabilistic programs, it faces other challenges than the ones we are interested in this work. Haslbeck and Nipkow [12] analyze the works of Nielson [20], Carbonneaux *et al.* [8,9] and Atkey [4] and prove the soundness of their systems. While our system uses a Hoare logic to prove upper bounds on program cost, many existing systems are type-based [4,5,13–15,21,22,24,26]. These works highly differ from ours in the underlying proof method.

Some other works on static estimation of resource bounds use other methodologies other than axiomatic semantics or type theory: Albert *et al.* [1] focus on static analysis of Java bytecode, which infers cost and termination information. It takes a cost model for a user-defined resource as input and obtains an upper bound on the execution time of this resource. Gulwani *et al.* [11] compute symbolic bounds on the number of statements a procedure executes in terms of its input. Finally, Brockschmidt *et al.* [7] use Polynomial Rank Functions (PRF) to compute time bounds. This work considers small parts of the program in each step and incrementally improve the bound approximation.

The work by Barbosa *et al.* [6] extends the logic of EasyCrypt, allowing to prove properties on the cost of a program. Our work operates on a subset of EasyCrypt's language, but we extended the logic to use the potential method of amortized analysis and the ability to prove exact bounds, improving the precision of the achievable proof goals. We envison an integration of our extensions into EasyCrypt in the future.

2 Cost Instrumented Operational Semantics

We assume the syntactic structure of numbers $\langle int \rangle$ and identifiers $\langle ident \rangle$ is given. We also assume the usual formation rules for arithmetic expressions $\langle aexp \rangle$ and boolean expressions $\langle bexp \rangle$. We start by defining a core imperative language with the arithmetic and Boolean expressions and the statements in Fig. 1.

$$
\begin{array}{rl}
\langle statement \rangle ::= & \textbf{skip} \\
\mid & \langle ident \rangle = \langle aexp \rangle \\
\mid & \langle ident \rangle \ [\ \langle aexp \rangle\ ] = \langle aexp \rangle \\
\mid & \textbf{if}\ \langle bexp \rangle\ \textbf{then}\ \langle statement \rangle\ \textbf{else}\ \langle statement \rangle\ \textbf{done} \\
\mid & \textbf{while}\ \langle bexp \rangle\ \textbf{do}\ \langle statement \rangle\ \textbf{done} \\
\mid & \langle statement \rangle \ ;\ \langle statement \rangle
\end{array}
$$

Fig. 1. Syntax of core imperative language.

Our language will support annotations with preconditions and post-conditions in order to prove correctness, termination and time restrictions on our program. This involves extending the boolean expressions to a rich language of assertions about program states and execution time. Basically our assertions are boolean expressions extended with quantifiers over integer variables and implication. In our examples we will sometimes further extend the assertion language in various ways, but we will mostly omit the formal details because the meaning is clear from the context.

An example of a simple program in our language is shown in Fig. 2, where we note the time annotation in the right-hand side of the post-condition.

In order to be able to prove assertions on the execution time of programs in this language, we need to define a cost aware semantics. We define an operational semantics as $\langle S, \sigma \rangle \Downarrow^t \sigma'$, meaning that after executing statement S from state σ the final state was σ' and the execution time was t. Our semantics computes the exact cost of executing a program and is defined in Fig. 3. The rules are mostly self-explanatory; we note only that the *while* rule is computing one loop body at a time. The semantics for the cost of evaluating an expression are shown in Fig. 4. Here, C_{CST} corresponds to the cost of evaluating a constant, C_{VAR} the cost of evaluating a variable and so on. These rules are simultaneously used by our operational semantics, when executing our program, and by our axiomatic semantics, when proving time restrictions statically using our VC generator.

3 Cost-Aware Program Logic

We define a logic with triples in the form $\{\varphi\}C\{\psi|t\}$. This can be read as, executing C from a state σ that validates the precondition φ leads to a state that validates postcondition ψ and this execution costs at most t to complete. We will call these triples *total correctness assertions*.

```
{ r == x and q == 0 and y > 0 and x >= 0}
while r >= y do
  r = r - y;
  q = q + 1
end
{ x == r + y * q and r < y | 20x + 5 }
```

Fig. 2. Division algorithm implemented in our annotated imperative language.

$$[Skip] \quad \langle \text{skip}, s\rangle \rightarrow^{C_{SKIP}} s$$

$$[Assign] \quad \langle x = a, s\rangle \rightarrow^{\mathcal{TA}[\![a]\!]+C_{ASSIGN\_V}} s[\ \mathcal{A}[\![a]\!]s \ / \ x\ ]$$

$$[Array] \quad \langle x[i] = a, s\rangle \rightarrow^{\mathcal{TA}[\![a]\!]+\mathcal{TA}[\![i]\!]+C_{ASSIGN\_A}} s[\ \mathcal{A}[\![a]\!]s \ / \ x[\mathcal{A}[\![i]\!]s]\ ]$$

$$[seq] \quad \frac{\langle S_1, s\rangle \rightarrow^{t_1} s' \qquad \langle S_2, s'\rangle \rightarrow^{t_2} s''}{\langle S_1; S_2, s\rangle \rightarrow^{t_1+t_2} s''}$$

$$[if^{tt}] \quad \frac{\langle S_1, s\rangle \rightarrow^{t} s'}{\langle \text{if } b \text{ then } S1 \text{ else } S2, s\rangle \rightarrow^{\mathcal{TA}[\![b]\!]+t} s'} \qquad \text{if } \mathcal{B}[\![b]\!]s = \text{tt}$$

$$[if^{ff}] \quad \frac{\langle S_2, s\rangle \rightarrow^{t} s'}{\langle \text{if } b \text{ then } S1 \text{ else } S2, s\rangle \rightarrow^{\mathcal{TA}[\![b]\!]+t} s'} \qquad \text{if } \mathcal{B}[\![b]\!]s = \text{ff}$$

$$[while^{tt}] \quad \frac{\langle S, s\rangle \rightarrow^{t} s'' \qquad \langle \text{while } b \text{ do } S, s''\rangle \rightarrow^{t'} s'}{\langle \text{while } b \text{ do } S, s\rangle \rightarrow^{\mathcal{TB}[\![b]\!]+t+t'} s'} \qquad \text{if } \mathcal{B}[\![b]\!]s = \text{tt}$$

$$[while^{ff}] \quad \langle \text{while } b \text{ do } S, s\rangle \rightarrow^{\mathcal{TB}[\![b]\!]} s \qquad \text{if } \mathcal{B}[\![b]\!]s = \text{ff}$$

Fig. 3. Operational semantics.

We begin by defining the *semantics* of an assertion. As assertions may include (logic) variables, the semantics require an interpretation function to provide the (integer) value of a variable. Given an interpretation, it is convenient to define the states which satisfy an assertion. We will use the notation $\sigma \models^I \varphi$ to denote that state σ satisfies φ in interpretation I, or equivalently that assertion φ is true at state σ, in interpretation I. The definition of $\sigma \models^I \varphi$ is the usual for a first order language. As usual, we are not interested in the particular values associated to variables in a given interpretation I. We are interested in whether or not an assertion is true at all states for all interpretations. This motivates the following definition.

Definition 1 (Validity). $\models \{\varphi\}S\{\psi|t\}$ *if and only if, for every state* σ *and interpretation* I *such that* $\sigma \models \varphi$ *and* $\langle S, \sigma\rangle \rightarrow^{t'} \sigma'$ *we have that* $\sigma' \models \psi$ *and* $\mathcal{A}[\![t]\!]\sigma \geq t'$.

$$\begin{array}{ll}
\mathcal{TA}[\![n]\!] & = C_{CST} \\
\mathcal{TA}[\![x]\!] & = C_{VAR} \\
\mathcal{TA}[\![x[a]]\!] & = \mathcal{TA}[\![a]\!] + C_{ARRAY} \\
\mathcal{TA}[\![a_1 + a_2]\!] & = \mathcal{TA}[\![a_1]\!] + \mathcal{TA}[\![a_2]\!] + C_+ \\
\mathcal{TA}[\![a_1 * a_2]\!] & = \mathcal{TA}[\![a_1]\!] + \mathcal{TA}[\![a_2]\!] + C_* \\
\mathcal{TA}[\![a_1 - a_2]\!] & = \mathcal{TA}[\![a_1]\!] + \mathcal{TA}[\![a_2]\!] + C_- \\
 & \\
\mathcal{TB}[\![\text{true}]\!] & = C_{BOOL} \\
\mathcal{TB}[\![\text{false}]\!] & = C_{BOOL} \\
\mathcal{TB}[\![a_1 = a_2]\!] & = \mathcal{TA}[\![a_1]\!] + \mathcal{TA}[\![a_2]\!] + C_= \\
\mathcal{TB}[\![a_1 \leq a_2]\!] & = \mathcal{TA}[\![a_1]\!] + \mathcal{TA}[\![a_2]\!] + C_\leq \\
\mathcal{TB}[\![\neg b]\!] & = \mathcal{TB}[\![b]\!] + C_\neg \\
\mathcal{TB}[\![b_1 \wedge b_2]\!] & = \mathcal{TB}[\![b_1]\!] + \mathcal{TB}[\![b_2]\!] + C_\wedge \\
\mathcal{TB}[\![b_1 \vee b_2]\!] & = \mathcal{TB}[\![b_1]\!] + \mathcal{TB}[\![b_2]\!] + C_\vee
\end{array}$$

Fig. 4. Cost of arithmetic and boolean expressions.

[Skip] $\quad \{ P \} skip \{ P \mid C_{SKIP} \}$

[Assign] $\quad \{ P[\mathcal{A}[\![a]\!]/x] \}\ x = a\ \{ P \mid \mathcal{TA}[\![a]\!] + C_{ASSIGN\_V} \}$

[Array] $\quad \{ P[\mathcal{A}[\![a_2]\!]/x[\mathcal{A}[\![a_1]\!]]] \}\ x[a_1] = a_2\ \{ P \mid \mathcal{TA}[\![a_1]\!] + \mathcal{TA}[\![a_2]\!] + C_{ASSIGN\_A} \}$

[Seq] $$\frac{\{ P \}\ S_1\ \{ Q \mid t_1 \} \quad \{ Q \}\ S_2\ \{ R \mid t_2 \}}{\{ P \}\ S_1;\ S_2\ \{ R \mid t_1 + t_2 \}}$$

[If] $$\frac{\{ P \wedge \mathcal{B}[\![b]\!] \}\ S_1\ \{ Q \mid t_1 \} \quad \{ P \wedge \neg\mathcal{B}[\![b]\!] \}\ S_2\ \{ Q \mid t_2 \}}{\{ P \}\ \text{if } b \text{ then } S_1 \text{ else } S_2\ \{ Q \mid max(t_1, t_2) + \mathcal{TB}[\![b]\!] \}}$$

[While] $$\frac{\underline{I} \wedge \mathcal{B}[\![b]\!] \Rightarrow \underline{f} \leq \underline{N} \quad \{ \underline{I} \wedge \mathcal{B}[\![b]\!] \wedge \underline{f} = k \}\ S\ \{ \underline{I} \wedge \underline{f} > k \mid \underline{t(k)} \}}{\{ \underline{I} \wedge \underline{f} \geq 0 \}\ \text{while } b \text{ do } S\ \{ I \wedge \neg\mathcal{B}[\![b]\!] \mid \sum_{i=0}^{\underline{N}-1} \underline{t(i)} + (\underline{N}+1) \times \mathcal{TB}[\![b]\!] \}}$$

[Weak] $$\frac{\{P'\}s\{Q'|t'\} \quad P \Rightarrow P' \quad Q' \Rightarrow Q \quad t' \leq t}{\{P\}S\{Q|t\}}$$

Fig. 5. Proof rules for the while language.

If $\models \{\varphi\}S\{\psi|t\}$ we say that the total correctness assertion $\{\varphi\}S\{\psi|t\}$ is *valid.* This approach gives us a worst case analysis for the cost of our program.

We now define a set of proof rules that generate the valid total correctness assertions. The rules of our logic are given in Fig. 5.

In the rules in Fig. 5, the underlined values (f, I and N and $t(k)$) are values provided by an oracle (in an interactive proof system these can be provided by the user; in a non-interactive setting they can be annotated into the program). f is a termination function that must start as a positive value, increase with every iteration, but remain smaller than N. N is therefore the maximum number of iterations. I is the loop invariant, and $t(k)$ is a cost function for the loop body. Given this values, since the *while* loop runs at most N times and, for each iteration k the cost of the loop body is given by $t(k)$ we can take the following upper bound for the while statement: the sum of the cost of all the iterations $\sum_{i=0}^{N-1} t(i)$; plus the sum of evaluating the loop condition, b, each time we enter the while body (at most N); plus one evaluation of b when the condition fails and the loop terminates. This leads to the term $(N+1) * \mathcal{TB}[\![b]\!]$ used in the rule for the *while* loop.

3.1 Verification Condition Generation

In order to implement a verification system based on our logic we define a Verification Conditions (VC) Generator algorithm. The VC function receives a program and a postcondition and returns a set of purely mathematical statements (the verification conditions) needed to guarantee correctness, termination and cost of the program. These verification conditions are then passed to EasyCrypt [6] which attempts to prove them automatically. If it fails, some advise is needed from the user. VC generation relies on the calculation of the weakest precondition (**wpc**). Very briefly, the weakest precondition is the least restrictive requirement needed to guarantee that a postcondition holds after executing a statement. This is a widely used predicate transformer [3,10,18]. The **wpc** function receives a statement and a postcondition and returns a tuple (wp, t), where wp is the weakest precondition of the program and t is an upper bound on the program's cost. The weakest precondition is calculated by a standard algorithm such as the one presented in [2].

4 Amortized Costs

Amortized analysis is a method due to Tarjan for analysing the complexity of a sequence of operations [25]. Instead of reasoning about the worst-case cost of individual operations, amortized analysis is concerned with the worst-case cost of a *sequence* of operations. The advantage of this method is that some operations can be more expensive than others, thus, distributing the cost of expensive operations over the cheaper ones can produce better bounds than analysing worst-case for individual operations.

Let a_i and t_i be, respectively, the amortized costs and the actual costs of each operation. To obtain an amortized analysis it is necessary to define an amortized cost such that

$$\sum_{i=1}^{n} a_i \geq \sum_{i=1}^{n} t_i$$

i.e., for each sequence of operations the total amortized costs is an upper-bound on the total actual costs.

$$\frac{\underline{I} \wedge \mathcal{B}[\![b]\!] \rightarrow \underline{f} \leq \underline{N} \qquad \underline{I} \rightarrow \phi \geq 0 \qquad \{ \underline{I} \wedge \mathcal{B}[\![b]\!] \wedge \} \; S \; \{ \underline{I} \wedge \underline{f} > k \wedge \phi = P_k \mid \underline{a} + \underline{\phi} - P_k \}}{\{ \underline{I} \wedge \underline{f} \geq 0 \wedge \phi = 0 \} \text{ while } b \text{ do } S \; \{ \underline{I} \wedge \neg \mathcal{B}[\![b]\!] \mid \underline{N} \times \underline{a} + (\underline{N} + 1) \times \mathcal{TB}[\![b]\!] \}}$$

Fig. 6. Hoare rule for while statement with amortized costs.

$$\mathbf{wpc}(\text{while } b \text{ do } S, Q) = (\underline{I} \wedge \underline{f} \geq 0 \wedge \underline{\phi} = 0 \, , \; N \times a + (N+1) \times \mathcal{TB}[\![B]\!])$$

$$\begin{aligned}\mathbf{VC}(\text{while } b \text{ do } S, Q) = \; & \{(\underline{I} \wedge \mathcal{B}[\![b]\!] \wedge \underline{f} = \underline{k}) \rightarrow wp_S \wedge a + \phi - P_k \geq t_S\} \cup \\ & \{(\underline{I} \wedge \neg B) \rightarrow Q\} \cup \\ & \{\underline{I} \wedge \mathcal{B}[\![b]\!] \rightarrow \underline{f} \leq \underline{N}\} \cup \\ & \{\underline{I} \rightarrow \phi \geq 0\} \\ & \mathbf{VC}(S, \underline{I} \wedge \underline{f} > k \wedge \underline{\phi} = P_k) \cup \\ & \\ & \text{where } wp_S, t_S = \mathbf{wpc}(S, \underline{I} \wedge \underline{f} > \underline{k} \wedge \underline{\phi} = P_k)\end{aligned}$$

Fig. 7. VCG rules for while statement with amortized costs.

This allows the use of operations with an actual cost that exceeds their amortized cost. Conversely, other operations have a lower actual cost than their amortized cost. Expensive operations can only occur when the difference between the accumulated amortized cost and the accumulated actual cost is enough to cover the extra cost.

There are three methods for amortized analysis: the *aggregate method* (using the above relation directly), the *accounting method* and the *potential method*. Amortized analysis based on the potential method was already used before in type systems which derive upper-bounds on the use of computational resources [14,15,24,26]. Here we use it to prove tighter bounds when the composition of worst-case execution times is overly pessimistic.

We now introduce our modified logic for amortized analysis. To this end, we modify the *while* rule (Fig. 6) to allow deriving more precise bounds.

Similarly to the previous definition f, I, N are all provided by the user. However, this new rule requires additional information from the oracle: an *amortized cost* a and the *potential function* ϕ. a is the amortized cost for one iteration of the while loop and ϕ is a potential function such that knowing that $\langle S, \sigma \rangle \rightarrow^t \sigma_1$ we have that $a + \phi - P_k \geq t$. Here P_k is a logic variable representing the value of ϕ in state σ_1. Given this two new values, and according to the potential method, the cost of the while loop can be defined simply by summing all the amortized costs for the body $N \times a$, plus every evaluation of b, $(N+1) \times \mathcal{TB}[\![b]\!]$.

The VCG is now modified accordingly, as shown in Fig. 7.

$\{n \geq 0 \wedge size == log(n) \wedge \forall i.0 \leq i \wedge i \leq size \rightarrow B[i] == 0\}$

```
i = 0;
while i < n do
  j = 0;
  while B[j] == 1 do
    B[j] = 0;
    j = j + 1
  end;
  B[j] = 1;
  i = i + 1
end
```

$\{n == \sum_{i=0}^{log(n)-1} B[i] \times 2^i | 30n + 30\}$

Fig. 8. Binary counter implementation with annotation.

The new *wpc* rule for *while* returns a new term in the invariants conjunction that stipulates that $\phi = 0$, i.e., the potential function must be zero before the *while* begins. The upper-bound is given by the sum of the amortized cost for every iteration $N \times a$, plus the sum of every $N + 1$ evaluation of b, $(N + 1) \times \mathcal{TB}[\![b]\!]$.

Next we illustrate this new version of the Hoare Logic on two typical applications of amortized analysis: a binary counter and a dynamic array.

4.1 Examples

Binary Counter. In the binary counter algorithm we represent a binary number as an array of zeros and ones. We start with an array with every value at zero and with each iteration we increase the number by one until we reach the desired value. Our implementation can be seen in Fig. 8.

If we were to use a worst-case analysis on this implementation we would get that this algorithm is $\mathcal{O}(n \ \log n)$, meaning we would flip every bit ($\log n$) a total of n times. However, this is not the case, while the first bit ($B[0]$) does flip every iteration, the second bit($B[1]$) flips every other iteration, the third ($B[2]$) every 4th iteration, and so on. We can see a pattern where, each bit $B[i]$ flips every 2^ith iteration. This will mean that, at most, we have $2n$ bit flips, so the algorithm is actually $\mathcal{O}(n)$.

We prove this in our system by defining a constant amortized cost for the external while, and a potential function, denoting the number of ones in the array. By proving that $a + \phi - P_k$ (where P_k is a logic variable defined in our logic rule for the while loop) is an upper-bound on the body of the while we get that $n \times a + (n+1) \times \mathcal{TB}[\![i < n]\!]$ is an upper-bound on the while statement. Since a is constant, our logic derives a $\mathcal{O}(n)$ cost, as expected. The full derivation can be found in the extended version of this paper [23].

$\{n \geq 0\}$

```
size = 1;
x[size];
i = 0;
while i < n do
  if i == size then
    size = size * 2;
    x[size]
  else
    skip
  end;
  x[i] = i;
  i = i + 1
end
```

$\{\forall k.0 \leq k \wedge k \leq n \rightarrow x[k] == k | 2cn + 3n + 8\}$

Fig. 9. Dynamic array implementation with annotation.

Dynamic Array. In the second example we consider a dynamic array where we insert elements and, whenever the array is full, we reallocate memory for a new array, doubling the size of the previous one, as shown in Fig. 9.

We can see that insertion can most of the time be done in constant time, except when there is an overflow. In this case we need to copy all the elements in the array to the new array, which takes linear time in the number of elements on the array. Let us represent the real cost of inserting an element in an array as c. For the amortized analysis we can define a potential function as:

$$\phi = 2c \times (n - \frac{\text{size}}{2})$$

where n=#(elements in array). There are two different scenarios when inserting a new element in the array. Either $n <$ size, in which case, the potential changes by $2c$ and therefore the cost is given by $3c$, or $n =$ size, i.e., the array is full. In this last case the real cost is the cost of copying all the elements to a new array and adding the new element ($c \cdot \text{size} + c$), adding the difference in potential ($2c - 2c \cdot \frac{\text{size}}{2}$), giving us an amortized cost of $3c$. So the cost of the program over n insertions is proved to be $\mathcal{O}(n)$, rather than a worst-case quadratic bound.

5 Exact Costs

The final part of our work consists of extending our language and logic so that we can verify the exact cost of a program instead of an upper-bound like in the previous cases. In this new logic we have that $\models \{\varphi\}S\{\psi|t\}$ if and only if, for all state σ such that $\sigma \models \varphi$ and $\langle S, \sigma\rangle \rightarrow^{t'} \sigma'$ we have that $\sigma' \models \psi$ and $\mathcal{A}[\![t]\!]\sigma = t'$. Notice that this is fairly similar to what was presented in Sect. 3 but now for a

$$[for^{tt}] \quad \frac{\langle S, s[\mathcal{A}[\![a]\!]s]/i\rangle \rightarrow^{t} s'' \qquad \langle \text{for } i = a+1 \text{ to } b \text{ do } S, s''\rangle \rightarrow^{t'} s'}{\langle \text{for } i = a \text{ to } b \text{ do } S, s\rangle \rightarrow^{\mathcal{TA}[\![a<b]\!]+t+t'} s'} \quad \text{if } a < b = \text{tt}$$

$$[for^{ff}] \quad \langle \text{for } i = a \text{ to } b \text{ do } S, s\rangle \rightarrow^{\mathcal{TA}[\![a<b]\!]} s \qquad \text{if } a < b = \text{ff}$$

Fig. 10. Operational semantic of for-loop.

$$\frac{\{P \wedge a \leq i \wedge i \leq b\}\ S\ \{P[i+1/i] \mid t_S\}}{\{P[a/i] \wedge a \leq b\}\ \text{for } i = a \text{ to } b \text{ do } S\ \{P[b+1/i] \mid (b-a) \times t_S\}}$$

$$\{P \wedge b < a\}\ \text{for } i = a \text{ to } b \text{ do } S\ \{P\}$$

Fig. 11. Hoare rule for "for-loop" statement.

$$\frac{\{\ P \wedge b\ \}\ S_1\ \{\ Q \mid t_1\} \qquad \{\ P \wedge \neg b\ \}\ S_2\ \{\ Q \mid t_2\ \} \qquad t_1 == t_2}{\{\ P\ \}\ \text{if } b \text{ then } S_1 \text{ else } S_2\ \{\ Q \mid t_1 + \mathcal{TB}[\![b]\!]\ \}}$$

Fig. 12. Hoare rule for if statement where both branches take exactly the same time to execute.

Hoare triple to be valid the value passed in the cost section needs to represent the exact cost of the program ($\mathcal{A}[\![t]\!]\sigma = t'$). Having this new goal in mind the first thing to consider is that our language needs to be changed. Specifically we need to replace our *while* loop with a well-behaved *for* loop as shown in Fig. 10). The new axiomatic rule for the *for* loop is defined in Fig. 11.

We also modify the rule for conditional statements, by imposing that both branches must execute with the same exact cost. The rule for if is then redefined as shown in Fig. 12. Note that balancing if-branches with, e.g., dummy instructions, is a common technique used in cryptography to eliminate execution time dependencies from branch conditions that may be related to secret data. The new rule is shown in Fig. 12.

Finally, we changed our VCG to reflect these changes in the logic. We now present two examples of our new logic in action. The first example is the Binary Exponentiation algorithm and the second one is the Range Filtering algorithm.

5.1 Examples

Binary Exponentiation. We represent the exponent in binary as an array ($B[\log b]$) of zeros and ones, where $B[0]$ is the least significant bit. We then run through the array and for each position i, if the value is a one we multiply our result by $a^{(2^i)}$ and save this result. So for instance, if we want to compute 3^{10} we first represent $10_{10} = 1010_2$ as an array $[0, 1, 0, 1]$ and the result will be

$\{a \geq 0 \wedge b \geq 0 \wedge b == sum(i, 0, log(b) - 1, 2^B[i])\}$

```
a' = a;
res = 1;
for i = 0 to log(b) - 1 do
  if B[i] == 1 then
    res = res * a'
  else
    res = res * 1
  end
  a' = a' * a'
end
```

$\{res == a^b | 16log(b) + 9\}$

Fig. 13. Binary Exponentiation Implementation with annotations for exact cost.

computed as $3^2 \cdot 3^8$. To prove correctness, we provide the invariant

$$a' = a^{(2^i)} \wedge \text{res} = \sum_{i=0}^{i-1} a^{(2^{B[i]})}$$

Since we're using a for-loop, termination is always guaranteed and therefore doesn't require a proof. In regards to the cost, we successfully prove this program is logarithmic, and it takes exactly $16 \log n + 9$ to run (Fig. 13).

Range Filter. In our last example we implement a simple filter, where, given an array (a) and a range [l..u] we use an auxiliary array (b) to filter if the elements in a are within the range. We provide the invariant $\forall k.(0 \leq k \wedge k < i) \rightarrow (l \leq a[k] \wedge a[k] \leq u \rightarrow \exists j.b[j] = a[i]) \wedge (l > a[k] \wedge a[k] > u \rightarrow \neg(\exists j.b[j] = a[i]))$ in order to prove correctness. Using our VC generator and Easycrypt we prove that, not only is this algorithm correct, the cost we provide of $17n + 22$ is the exact cost of this program. This result allows us to conclude the time it takes to run depends only on the size of the array, and not on its values.

6 Soundness

Proof rules should preserve validity in the sense that, if the premise is valid, then so is the conclusion. When this holds for every rule, we say that the proof systems is *sound*. For the three proposed Hoare logics (cost aware, amortized costs and exact costs) we have shown by induction that every assertion $\{\varphi\}S\{\psi|t\}$ which is the conclusion of a derivation in the proof system, is a valid assertion.

Theorem 1 (Soundness of Hoare Logic). *Let* $\{\varphi\}S\{\psi|t\}$ *be a total correctness assumption. Then* $\vdash \{\varphi\}S\{\psi|t\}$ *implies* $\models \{\varphi\}S\{\psi|t\}$

Proof. The proof follows by induction on the length of the derivation of $\{\phi\}S\{\psi|t\}$ (Fig. 14).

$\{0 \leq l \wedge l < u \wedge n \geq 0\}$

```
j = 0;
for i=0 to n do
  if (l <= a[i] and a[i] <= u)
  then
    b[j] = a[i];
    j = j + 1
  else
    b[j] = b[j];
    j = j + 0
  end
end
```

$\{\forall i.(0 \leq i < n) \rightarrow (l \leq a[i] \leq u \rightarrow \exists j.B[j] = a[i]) \wedge$
$(l > a[i] > u \rightarrow \neg(\exists j.b[j] = a[i]))|17 \ \times n + 22\ \}$

Fig. 14. Array filtering within a range implementation with annotations for exact cost.

We also need to ensure that the verification conditions generator is actually sound with respect to the corresponding Hoare Logic. For this, we also proved, using structural induction on the command Q, a theorem for the three settings (cost aware, amortized costs and exact costs) that states that the VCG algorithm is sound, i.e., the verification conditions generated by the VCG algorithm imply the Hoare triple we wish to prove:

$$\models VGC(\{P\}Q\{R\}) \Rightarrow \vdash \{P\}Q\{R\}$$

This assures that by proving our VCs, we are actually proving the triple. Complete proofs may be found on an extended version of the paper available in [23].

7 Implementation

We have implemented our verification system in OCaml, and all the code and examples are in the GitHub repository

https://github.com/carolinafsilva/time-verification

In Fig. 15 we show the architecture of our tool, with each of the steps that will allow us to meet our goal. Program verification is conducted in three stages:

1. Annotation of the program by the programmer, who specifies the correctness conditions that must be met, as well as the cost upper bound.
2. Implementation of the VCG which, given an annotated program generates a set of goals that need to be proved.

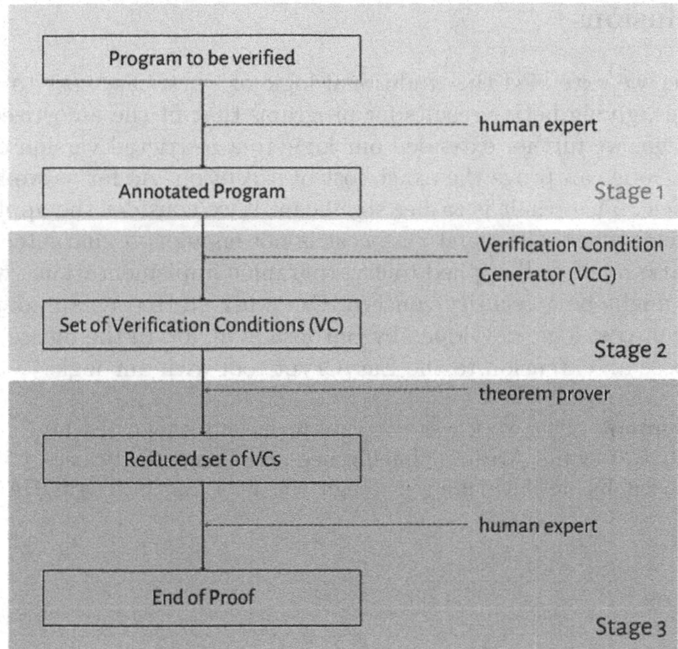

Fig. 15. Architecture of our tool.

3. The proof stage: proof goals are passed to a theorem prover which attempts to prove them automatically. If it fails, some interaction with the user is needed to guide the proof.

Note that in stage three, we discard the proof goals by sending them to an automatic prover (e.g., Easycrypt, why3) for validation. This step might need some assistance from the user since the generated VCs might be too complex to be automatically proved. If all our VCs are validated, we know our program is correct, and we have learned some restrictions on its execution time. Thus, verification is semi-automatic in the sense that in certain situations, the user has to give extra information to the program, either in the form of an oracle that defines some needed parameters or in the proof stage in situations where the proof is interactive.

Programs with loops require additional information to prove correctness, termination, and cost bounds. This information is provided through an oracle. This oracle will request user input when necessary to complete the Verification Condition Generator algorithm. Since information such as invariants is needed multiple times throughout the algorithm, we need to store this information to be easily accessible. To achieve this, we assign a unique identifier to each *while-loop* and create an oracle hashtable to store the oracle information for each loop.

8 Conclusions

In this paper we extended the traditional logic of worst-case cost to use amortized analysis, giving better results for programs that fit the amortized analysis scenario. Then, we further extended our logic to a restricted version of our language, where one can prove the exact cost of execution. As far as we know, this is a novel logic. This result is rather significant if we consider the application to critical systems where the worst-case cost is not enough to guarantee all safety goals. It is also relevant if applied to cryptographic implementations, where timing leakage might be a security concern. Our work started as an adaptation of the EasyCrypt cost logic developed by Barbosa *et al.* [6]. In the future, we would like to propose an extension to the EasyCrypt tool with our logic.

Acknowledgment. This work was partially financially supported by Base Funding UIDB/00027/2020 of the Artificial Intelligence and Computer Science Laboratory – LIACC - funded by national funds through the FCT/MCTES (PIDDAC) and by INESC TEC.

References

1. Albert, E., et al.: Termination and cost analysis with COSTA and its user interfaces. Electron. Notes Theor. Comput. Sci. **258**(1), 109–121 (2009). https://doi.org/10.1016/j.entcs.2009.12.008
2. Almeida, J.B., Frade, M.J., Pinto, J.S., de Sousa, S.M.: Rigorous Software Development - An Introduction to Program Verification. Undergraduate Topics in Computer Science. Springer, London (2011). https://doi.org/10.1007/978-0-85729-018-2
3. Apt, K.R., Olderog, E.: Fifty years of Hoare's logic. Formal Aspects Comput. **31**(6), 751–807 (2019). https://doi.org/10.1007/s00165-019-00501-3
4. Atkey, R.: Amortised resource analysis with separation logic. Log. Methods Comput. Sci. **7**(2) (2011). https://doi.org/10.2168/LMCS-7(2:17)2011
5. Avanzini, M., Lago, U.D.: Automating sized-type inference for complexity analysis. Proc. ACM Program. Lang. **1**(ICFP), 43:1–43:29 (2017). https://doi.org/10.1145/3110287
6. Barbosa, M., Barthe, G., Grégoire, B., Koutsos, A., Strub, P.Y.: Mechanized proofs of adversarial complexity and application to universal composability. In: Proceedings of the 2021 ACM SIGSAC Conference on Computer and Communications Security, CCS 2021, pp. 2541–2563. Association for Computing Machinery, New York (2021). https://doi.org/10.1145/3460120.3484548
7. Brockschmidt, M., Emmes, F., Falke, S., Fuhs, C., Giesl, J.: Alternating runtime and size complexity analysis of integer programs. In: Ábrahám, E., Havelund, K. (eds.) TACAS 2014. LNCS, vol. 8413, pp. 140–155. Springer, Heidelberg (2014). https://doi.org/10.1007/978-3-642-54862-8_10
8. Carbonneaux, Q., Hoffmann, J., Ramananandro, T., Shao, Z.: End-to-end verification of stack-space bounds for C programs. In: O'Boyle, M.F.P., Pingali, K. (eds.) ACM SIGPLAN Conference on Programming Language Design and Implementation, PLDI 2014, Edinburgh, United Kingdom, 09–11 June 2014, pp. 270–281. ACM (2014). https://doi.org/10.1145/2594291.2594301

9. Carbonneaux, Q., Hoffmann, J., Shao, Z.: Compositional certified resource bounds. In: Grove, D., Blackburn, S.M. (eds.) Proceedings of the 36th ACM SIGPLAN Conference on Programming Language Design and Implementation, Portland, OR, USA, 15–17 June 2015, pp. 467–478. ACM (2015). https://doi.org/10.1145/2737924.2737955
10. Dijkstra, E.W.: Guarded commands, non-determinacy and a calculus for the derivation of programs. In: Shooman, M.L., Yeh, R.T. (eds.) Proceedings of the International Conference on Reliable Software 1975, Los Angeles, California, USA, 21–23 April 1975, p. 2. ACM (1975). https://doi.org/10.1145/800027.808417
11. Gulwani, S., Mehra, K.K., Chilimbi, T.: SPEED: precise and efficient static estimation of program computational complexity. In: Proceedings of the 36th Annual ACM SIGPLAN-SIGACT Symposium on Principles of Programming Languages, POPL 2009, pp. 127–139. Association for Computing Machinery, New York (2009). https://doi.org/10.1145/1480881.1480898
12. Haslbeck, M.P.L., Nipkow, T.: Hoare logics for time bounds - a study in meta theory. In: Beyer, D., Huisman, M. (eds.) TACAS 2018, Part I. LNCS, vol. 10805, pp. 155–171. Springer, Cham (2018). https://doi.org/10.1007/978-3-319-89960-2_9
13. Hoffmann, J., Das, A., Weng, S.: Towards automatic resource bound analysis for OCaml. In: Castagna, G., Gordon, A.D. (eds.) Proceedings of the 44th ACM SIGPLAN Symposium on Principles of Programming Languages, POPL 2017, Paris, France, 18–20 January 2017, pp. 359–373. ACM (2017). https://doi.org/10.1145/3009837.3009842
14. Hoffmann, J., Hofmann, M.: Amortized resource analysis with polynomial potential. In: Gordon, A.D. (ed.) ESOP 2010. LNCS, vol. 6012, pp. 287–306. Springer, Heidelberg (2010). https://doi.org/10.1007/978-3-642-11957-6_16
15. Hofmann, M., Jost, S.: Type-based amortised heap-space analysis. In: Sestoft, P. (ed.) ESOP 2006. LNCS, vol. 3924, pp. 22–37. Springer, Heidelberg (2006). https://doi.org/10.1007/11693024_3
16. Kaminski, B.L., Katoen, J., Matheja, C., Olmedo, F.: Weakest precondition reasoning for expected runtimes of randomized algorithms. J. ACM **65**(5), 30:1–30:68 (2018). https://doi.org/10.1145/3208102
17. Leroy, X.: Formal verification of a realistic compiler. Commun. ACM **52**(7), 107–115 (2009). https://doi.org/10.1145/1538788.1538814
18. Loeckx, J., Sieber, K.: The Foundations of Program Verification, 2nd ed. Wiley-Teubner (1987)
19. Nielson, H.R.: A Hoare-like proof system for analysing the computation time of programs. Sci. Comput. Program. **9**(2), 107–136 (1987). https://doi.org/10.1016/0167-6423(87)90029-3
20. Nielson, H.R., Nielson, F.: Semantics with Applications: An Appetizer. Undergraduate Topics in Computer Science. Springer, London (2007). https://doi.org/10.1007/978-1-84628-692-6
21. Radiček, I., Barthe, G., Gaboardi, M., Garg, D., Zuleger, F.: Monadic refinements for relational cost analysis. Proc. ACM Program. Lang. **2**(POPL) (2017). https://doi.org/10.1145/3158124
22. Serrano, A., López-García, P., Hermenegildo, M.V.: Resource usage analysis of logic programs via abstract interpretation using sized types. Theory Pract. Log. Program. **14**, 739–754 (2014). https://doi.org/10.1017/S147106841400057X
23. Silva, A.C., Barbosa, M., Florido, M.: Execution time program verification with tight bounds (2022). Available from Arxiv

24. Simões, H.R., Vasconcelos, P.B., Florido, M., Jost, S., Hammond, K.: Automatic amortised analysis of dynamic memory allocation for lazy functional programs. In: Thiemann, P., Findler, R.B. (eds.) ACM SIGPLAN International Conference on Functional Programming, ICFP 2012, Copenhagen, Denmark, 9–15 September 2012, pp. 165–176. ACM (2012). https://doi.org/10.1145/2364527.2364575
25. Tarjan, R.E.: Amortized computational complexity. SIAM J. Algebraic Discrete Methods **6**(2), 306–318 (1985). https://doi.org/10.1137/0606031
26. Vasconcelos, P., Jost, S., Florido, M., Hammond, K.: Type-based allocation analysis for co-recursion in lazy functional languages. In: Vitek, J. (ed.) ESOP 2015. LNCS, vol. 9032, pp. 787–811. Springer, Heidelberg (2015). https://doi.org/10.1007/978-3-662-46669-8_32

Fluo: A Domain-Specific Language for Experiments in Fluorescence Microscopy (Application Paper)

Birthe van den Berg(✉), Tom Schrijvers, and Peter Dedecker

KU Leuven, Leuven, Belgium
{birthe.vandenberg,tom.schrijvers,peter.dedecker}@kuleuven.be

Abstract. Fluorescence microscopy is a true workhorse in the domain of life sciences, essential for unraveling the inner workings of cells and tissue. It is not only used from day to day in industry, also academia push boundaries in research using and doing fluorescence microscopy. It is in the latter context that software that is sufficiently modular in terms of experiments and hardware is desirable. Existing solutions are too closely tailored to their accompanying hardware setup or too limited in terms of expressivity. We present Fluo: a domain-specific language (DSL) in Haskell for setting up fluorescence microscopy experiments that can be combined and nested freely. It provides domain-specific features such as stage loops and time lapses, and is modular in terms of hardware connections. Fluo has been operational since 2015 at the Nanobiology Lab. It has not only improved researchers' efficiency, but has also given rise to novel research results. For example, performing simultaneous Förster Resonant Energy Transfer (FRET) measurements, a mechanism for tracking energy transfer between a donor-acceptor pair, uses advanced time-lapse experiments and serves as an example use case in the paper. We reflect on the choice of Haskell as a host language and the usability of the DSL.

1 Introduction

Lab technicians often work with microscopes to investigate their biological and (bio-)chemical samples, using, for instance, fluorescence microscopy [27,32,38], where a sample is exposed to light at a certain wavelength and then re-emits light at a different wavelength. This re-emitted light is what fluorescence microscope operators are studying. Fluorescence has advantages over other techniques: (1) its non-invasiveness implies that experiments on cells and tissue can happen in situ and even in vivo; (2) its high selectivity makes it possible to cleanly separate different structures and their interactions (e.g., nucleus versus mitochondria); and (3) the wide availability of fluorescent probes and simple sample preparation provide a low threshold to entry.

Fluorescence microscopy can be used on samples (e.g., human fetal lung Fibroblast cells, often used for vaccine development [6], Fig. 1) with sizes down to micrometers (cells, bacteria) and even beyond the diffraction limit [29]—at

M. Hanus and D. Inclezan (Eds.): PADL 2023, LNCS 13880, pp. 73–82, 2023.
https://doi.org/10.1007/978-3-031-24841-2_5

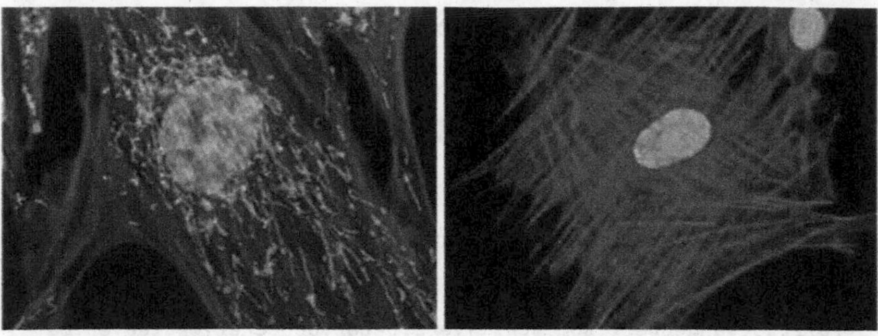

Fig. 1. Human Fetal Lung Fibroblast Cells.

nanometer scale (viruses, proteins, lipids)—using advanced optical approaches, such as Stochastic Optical Reconstruction Microscopy (STORM) [39,40], Photoactivated Localization Microscopy (PALM) [22,28] and Stimulated Emission Depletion (STED) [25]. Fluorescence microscopy is used in many settings: industrial companies such as Galapagos [2], Johnson & Johnson [4] and Roche [10] employ the technique in their day-to-day activities, and companies such as Nikon [8], Olympus [9] and Thermo Fisher Scientific [11] design and sell microscopes for different fluorescence microscopy techniques. Moreover, also in research fluorescence is prevalent, evidenced by the numerous papers in CellPress [1] (impact factor 66.850 in 2022) and Nature [7] (impact factor 69.504 in 2022) on the topic. It goes without saying that research both on novel techniques and on the design of optical paths in microscopes to perform those techniques is desired and impactful in various branches. Accompanying *software packages* are essential to make this research possible and efficient.

Software packages are typically provided by vendors along with their microscopes. Mostly, their software is tailored to their main users and hardware. Notably, these users work with a fixed set of instructions and perform repetitive tasks, such as control tests in the pharmaceutical industry. In contrast, in a research environment, it is not desirable to have software that works only on a specific microscope set-up, has a limited number of options or is too hard-wired to a specific task. Rather, researchers want to have the freedom to conduct custom experiments on custom—often self-assembled—microscopes. In fact, ideally, researchers would have access to a small programming language in which they can formulate custom experiments.

In this work we meet this need: we show how *Fluo*, a domain-specific language, allows its users to write custom programs that represent fluorescence microscopy experiments. These experiments can be combined and nested freely and are modular in terms of hardware configurations, brands and possible add-ons. The language includes domain-specific features such as *stage loops*, in which an acquisition is repeated at different positions on the sample, or *time-lapses*, in which an acquisition is repeated at regular time intervals. Fluo is operational and actively used by the KU Leuven Lab for Nanobiology in their ongoing research. It outperforms existing software packages, such as MicroManager [5], in a research context: it is more expressive as it is a programming language rather than a limited number of options. The domain-specific language was first introduced 8 years ago and has been indispensable in the lab ever since: not only does it improve the efficiency and freedom of researchers, it has also enabled novel experiments. Although Fluo is surprisingly simple and limited from a programmer's point of view, it does have a big impact on the experience and efficiency of its users, who are domain experts.

In this paper, we discuss the architecture of Fluo, along with its most notable features (Sect. 2). We elaborate on a use case: a series of published FRET experiments with advanced irradiation steps (e.g., time-lapses, stage loops) using the domain-specific language (Sect. 3) and reflect (Sect. 4).

2 Overview of Fluo

This section presents Fluo, a domain-specific language (DSL) [16,17,19] in Haskell, that allows users to define their own programs and includes advanced programming constructs tailored to fluorescence microscopy, such as stage loops and time-lapse experiments. Broadly speaking, the set-up consists of three parts: (1) the software, defined in terms of Fluo, (2) the hardware components with accompanying equipment, and (3) the graphical user interface (Fig. 2).

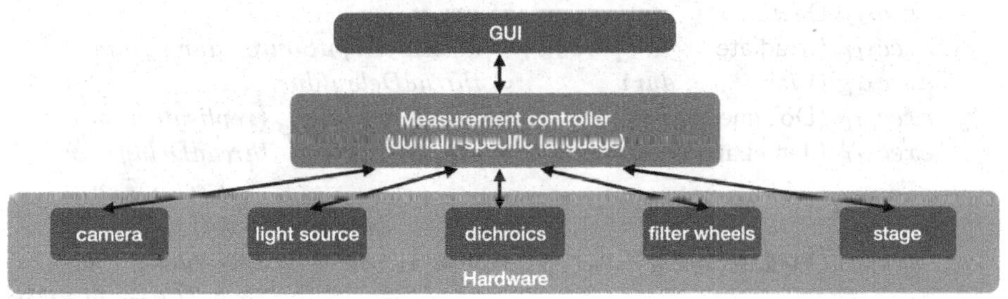

Fig. 2. Generic set-up of the microscope controlled by the DSL.

2.1 Software: Measurement Controller DSL

The measurement controller is the core of the set-up and represents domain-specific knowledge. It is designed as a domain-specific language: a custom-built programming language to formulate experiments in the domain of fluorescence microscopy. It is embedded in Haskell, a widely used *host language* for embedding DSLs because of its rich type system, lazy evaluation and research developments [20,21] on domain-specific languages. In addition, there are already many application areas for which a Haskell DSL has appeared useful: music [30], financial engineering [33] or monadic constraint programming [41]. A domain-specific language has several advantages over other programming languages: its syntax and semantics are tailored to the problem domain, so that its users can understand intuitively what is happening while programming. In fact, the users do not even have to know how to program in the host language, but instead get familiar with the domain-specific features. In what follows we present a simplified syntax and semantics that represent how the embedded DSL works. A program (*Prog*) is a list of *MeasurementElements*, which can recursively contain other programs.

```
type Duration = Double                  data Pos  = Pos {x :: Double,
type Colour   = String                                   y :: Double,
type Name     = String                                   z :: Double}
type Power    = Double                  type Prog = [MeasurementElement]
data MeasurementElement = Detect
                        | Irradiate  Duration (Name, Colour, Power)
                        | Wait       Duration
                        | DoTimes    Int      Prog
                        | TimeLapse  Int      Duration Prog
                        | StageLoop  [Pos]    Prog
```

Using this syntax, we can define programs that can be nested and combined unrestrictedly. An interpreter $exec_{Prog}$ couples the syntax to its actions.

```
exec_Prog prog = foldMap exec_ME prog where
  exec_ME Detect                 = exec_Detect
  exec_ME (Irradiate dur params) = activateLightSource dur params
  exec_ME (Wait      dur)        = threadDelay dur
  exec_ME (DoTimes   n pr)       = mapM exec_Prog (replicate n pr)
  exec_ME (TimeLapse n dur pr)   = mapM (λp →   threadDelay dur
                                                 ≫ exec_Prog p)
                                        (replicate n pr)
  exec_ME (StageLoop poss pr)    = mapM (λpos → setStagePos pos
                                                 ≫ exec_Prog pr) poss
```

Typically, time lapses and stage loops are not available in existing software packages (e.g., MicroManager [5]). Although it might seem straightforward to add them in industrial applications for research purposes, Fluo additionally allows to nest them, and to construct programs in which they can be combined arbitrarily. Furthermore, these novel features also work with different hardware set-ups.

2.2 Hardware

The fluorescence path towards a sample starts at a *light source*, such as a laser or high-power LED, then traverses different *dichroic mirrors*, *filter wheels* and/or a *microscope body*, and finally arrives at the sample with specimen. Depending on their specifications, dichroics let light at a certain wavelength pass through, and reflect the light at another wavelength. In fluorescence microscopy, they are a useful tool to separate the excitation light that goes towards the sample from the emission light, re-emitted by the fluorescent sample and of interest to the microscope user. Filter wheels contain one or more filters that ensure that the light passing through is monochromatic (i.e., of a single wavelength); other wavelengths, which may originate from ambient light, reflections or noise, are filtered out. When the monochromatic light arrives at the sample, the specimen is illuminated. As it is fluorescently labeled, it re-emits light of lower energy (and thus longer wavelength). This re-emitted light again gets filtered by filter wheels and possibly dichroic mirrors. Then, it is captured by a *detector*, usually a camera. The *stage* is the platform on which the sample is placed; it can move along three axes (x, y, z). Often, hardware manufacturers provide a DLL (written in C-code) to interact with their equipment. Where possible, serial communication via COM ports is used to connect to the hardware components.

Fluo interacts with and controls these hardware components. Moreover, it is modular in terms of the hardware brands or possible extensions. For example, Fluo is deployed on an Olympus microscope body (Spectra X, Lumencor) and a Nikon body (ECLIPSE Ti2-E). Multiple brands of stages are supported (Prior, Marzhauser) and recently, a remote stage [12] was added to automatically adapt the stage's z-position to the focusing distance of the sample. To add a different brand of stage (or other equipment), Fluo uses a Haskell type class *Equipment*, which abstracts over different hardware components.

```
instance Equipment Marzhauser where
  motorizedStageName = StageName "Marzhauser"
  supportedStageAxes  = [XAxis, YAxis, ZAxis]
  getStagePos (Marzhauser port) =
    fmap (map read ∘ words ∘ unpack ∘ decodeUtf8)
      (serialWriteAndReadUntilChar port ("?pos\r") '\r' >>=
          λ[x, y, z] → pure (Pos x y z))
  setStagePos (Marzhauser port) (Pos x y z) = ...
```

2.3 Graphical User Interface (GUI)

The graphical user interface for Fluo is written in Igor Pro [3], a scientific software package, developed by WaveMetrics Inc. that is often used for image processing and data acquisition in a (bio)chemical context. The user interface consists of buttons (Fig. 3) that are linked to the underlying DSL syntax and ensure that users, who are domain specialists rather than programmers, can only construct syntactically well-formed programs. Communication between the measurement controller and the GUI is done via a socket through which serialized JSON objects are sent. The GUI then periodically polls to check for available data from the controller.

2.4 An Example Program: From GUI to Hardware

A user can construct a program that uses the Lumencor light source five times to illuminate the sample with a violet colour at a power of 15%, then waits three

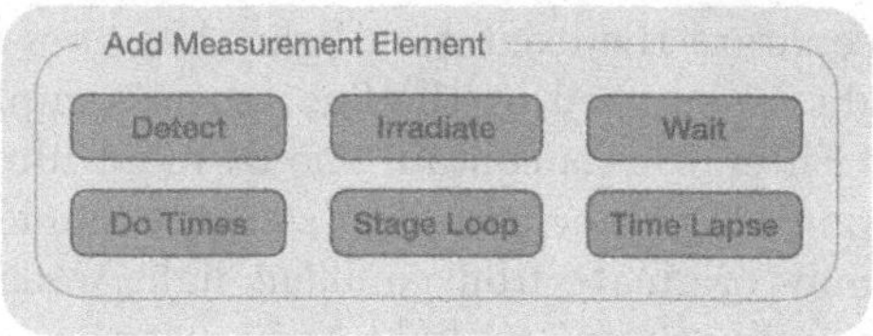

Fig. 3. GUI buttons to construct experiment programs.

seconds and next acquires an image. After five repetitions, it waits another 10 s and acquires a final image.

```
do 5 time(s) in total:
    irradiate 2 s using Lumencor:violet@15;
    wait 3 s
    acquire image(s)
wait 10 s
acquire image(s)
```

Users can see this program being built when using the buttons in the GUI. Although they cannot make syntactic errors using the buttons, they do have to understand program indentation in order to get the program behaviour they expect. This example program translates to the following program in the DSL:

$$\begin{aligned} \mathit{prog} = [&\mathsf{DoTimes}\ 5\ [\mathsf{Irradiate}\ 2\ (\texttt{"Lumencor"}, \texttt{"violet"}, 15), \mathsf{Wait}\ 3, \mathsf{Detect}] \\ &, \mathsf{Wait}\ 10, \mathsf{Detect}] \end{aligned}$$

When these instructions get executed, the DSL communicates with the hardware components required by the program (in this case the Lumencor light source and the filter wheel for violet light).

3 Use Case: FRET Measurements Using Time-Lapses

This section describes an application of the DSL: a study of advanced Förster Resonance Energy Transfer experiments has used Fluo to its full potential.

3.1 Förster Resonance Energy Transfer

Förster Resonance Energy Transfer (FRET) is a type of energy transfer that occurs between two molecules when they are in close proximity (roughly 10 nm). The energy is transferred from the molecule with the higher energy level (the donor) to the molecule with the lower energy level (the acceptor). For that reason, we often speak of FRET pairs, referring to the donor and acceptor. FRET is mostly used to study interactions between molecules in situ and to measure distances or relative orientations between these molecules [35]. There are several ways to quantify the rate of energy transfer between two molecules, the so-called *FRET efficiency*. The approximation used in the Nanobiology Lab is based on the intensity of emission of donor/acceptor [13,26,42,43]. Alternative techniques use donor fluorescence lifetime or spectra, for example. But, the efficiency approximation is not always as accurate as desired: the choice of a FRET pair is a trade-off between enough spectral overlap to ease the FRET efficiency measurements and at the same time minimal spectral overlap to avoid cross-talk. This trade-off becomes even more challenging when measuring multiple FRET pairs simultaneously. Typically, separate measurements are required to do so [31]. One possible mitigation is the choice of two FRET pairs with minimal spectral overlap [15], although these are not readily available. Furthermore, it is possible to share donor or acceptor [24], to do lifetime imaging with non-fluorescent acceptors [14,18,23], or to deploy more advanced techniques based on the orientation or localization of the FRET pairs [34,37].

3.2 Simultaneous FRET Measurements

The Lab for Nanobiology has developed a specialized biosensor[1] for simultaneously measuring multiple FRET pairs in live cells, even if their donor and acceptor spectrally overlap [36]. To make the experiments that led to that biosensor possible, specific time-lapse experiments were required: one wants to track the change in FRET efficiency through time, that is, every x seconds/minutes the biosensor's signal is captured. Each time point consists of three measurements:

1. *Donor excitation - Donor emission* (DD) The correct wavelength of illumination is set for optimal donor excitation; the emission captured is optimal for the donor.
2. *Donor excitation - Acceptor emission* (DA) This is also called the FRET channel; the appropriate wavelength of exposure is set for optimal donor excitation, the emission captured is optimal for the acceptor.
3. *Acceptor excitation - Acceptor emission* (AA) The appropriate wavelength of exposure is set for optimal acceptor excitation, the emission that is captured is optimal for acceptor.

This results in three images (DD, DA, AA) (Fig. 4) for each time point. In addition, a stage loop is nested in time-lapses to capture different positions of the sample at different times.

```
time lapse every 30 s 10 time(s) in total
  stage loop over 5 positions
    acquire image(s)
    irradiate 0.2 s using Lumencor:violet@100
    acquire image(s)
    irradiate 0.2 s using Lumencor:blue@75
    acquire image(s)
```

[1] A biosensor uses a biological component to detect chemical activity.

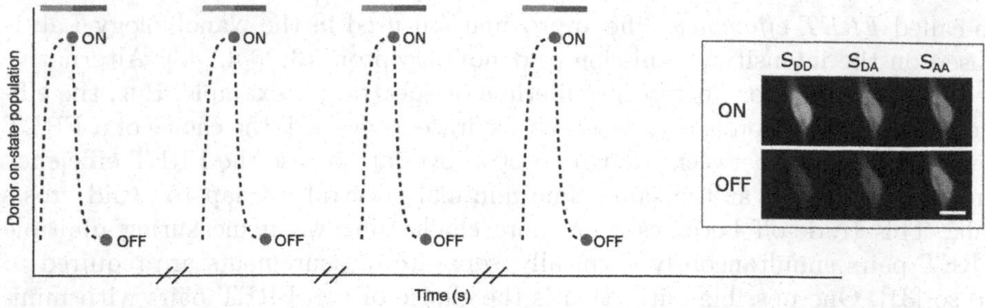

Fig. 4. Irradiation scheme with violet and blue irradiation, and acquisitions (red dots) in on- and off state in the three channels (DD, DA, AA). (Color figure online)

Next, the data is analyzed: for each measurement a background region is determined where no fluorescent cells are located. Fluorescent cells are selected and segmented; these are the regions of interest (ROIs). For each ROI and its background the average intensity timetraces of DD, DA, and AA are calculated, the background is corrected and the final FRET timetrace is calculated by correcting crosstalk (donor and acceptor emission contributions to DA not coming from FRET), filtered based on noise and expected response (e.g., sigmoidal change over time) and converted to FRET efficiency. Finally, this timetrace, averaged with standard deviations, represents the FRET efficiency for each region.

4 Reflection

Since 2015, this DSL has proven its daily use in the KU Leuven Nanobiology Lab. The lab members report that their efficiency of setting up and conducting experiments, which are sometimes very specific to a certain type of microscope, or to a novel configuration, has improved, and that, in addition, new types of experiments have been set up that would not have been possible in a setting with an industrial microscope and accompanying software package. The simultaneous FRET detection using time lapses is a good example thereof.

Compared with manually setting up experiments, Fluo avoids human mistakes, reduces personnel cost for operating the microscope, and improves precision and reproducibility of experiments. Furthermore, the arbitrary nesting of programs and advanced irradiation steps that Fluo allows are a plus compared to industrial software packages. On the one hand, using Haskell as host language has ensured the system to remain robust over time; users have confidence in the code. On the other hand, the development of the DSL was not always straightforward in combination with Haskell's lazy evaluation strategy. For instance, an image that is not needed until later in the process should not be taken lazily. The connection with hardware components has been particularly challenging, mainly because their interfaces are often provided by the manufacturer as C-libraries.

The fact that relatively simple functional programming principles have provided a significant improvement on the end user side is promising for further applications of DSL research.

References

1. Cell Press. https://www.cell.com/
2. Galapagos. https://www.glpg.com/
3. Igor. https://www.wavemetrics.com/
4. Johnson & Johnson. https://www.jnj.com/
5. MicroManager. https://micro-manager.org/
6. MicroscopyU Cell lines. https://www.microscopyu.com/galleries/fluorescence/cells
7. Nature. https://www.nature.com/
8. Nikon. https://www.microscope.healthcare.nikon.com/
9. Olympus. https://www.olympus-lifescience.com/en/microscopes/
10. Roche. https://www.roche.com/
11. Thermo Fisher Scientific. https://www.thermofisher.com/
12. van den Berg, B., Van den Eynde, R., Müller, M., Vandenberg, W., Dedecker, P.: Active focus stabilization for fluorescence microscopy by tracking actively generated infrared spots. bioRxiv (2020). https://doi.org/10.1101/2020.09.22.308197. https://www.biorxiv.org/content/early/2020/09/22/2020.09.22.308197
13. Coullomb, A., et al.: Quanti-fret: a framework for quantitative fret measurements in living cells. Sci. Rep. **10**, 6504 (2020). https://doi.org/10.1038/s41598-020-62924-w
14. Demeautis, C., et al.: Multiplexing PKA and ERK1-2 kinases fret biosensors in living cells using single excitation wavelength dual colour flim. Sci. Rep. **7**, 41026 (2017). https://doi.org/10.1038/srep41026
15. Depry, C., Mehta, S., Zhang, J.: Multiplexed visualization of dynamic signaling networks using genetically encoded fluorescent protein-based biosensors. Pflugers Archiv: Eur. J. Physiol. **465** (2012). https://doi.org/10.1007/s00424-012-1175-y
16. van Deursen, A., Klint, P., Visser, J.: Domain-specific languages: an annotated bibliography. SIGPLAN Not. **35**(6), 26–36 (2000). https://doi.org/10.1145/352029.352035
17. Fowler, M.: Domain Specific Languages, 1st edn. Addison-Wesley Professional (2010)
18. Ganesan, S., Ameer-Beg, S., Ng, T., Vojnovic, B., Wouters, S.: A dark yellow fluorescent protein (YFP)-based resonance energy-accepting chromoprotein (REACh) for Förster resonance energy transfer with GFP. Proc. Natl. Acad. Sci. U.S.A. **103**, 4089–4094 (2006). https://doi.org/10.1073/pnas.0509922103
19. Ghosh, D.: DSL for the uninitiated. Commun. ACM **54**(7), 44–50 (2011). https://doi.org/10.1145/1965724.1965740
20. Gibbons, J.: Free delivery (functional pearl). SIGPLAN Not. **51**(12), 45–50 (2016). https://doi.org/10.1145/3241625.2976005
21. Gibbons, J., Wu, N.: Folding domain-specific languages: deep and shallow embeddings (functional pearl). SIGPLAN Not. **49**(9), 339–347 (2014). https://doi.org/10.1145/2692915.2628138
22. Gould, T.J., Verkhusha, V.V., Hess, S.T.: Imaging biological structures with fluorescence photoactivation localization microscopy. Nat. Protoc. **4**(3), 291–308 (2009)
23. Grant, D., et al.: Multiplexed fret to image multiple signaling events in live cells. Biophys. J. **95**, L69–L71 (2008). https://doi.org/10.1529/biophysj.108.139204
24. Hargett, H., Pan, C.P., Barkley, M.: Two-step fret as a structural tool. J. Am. Chem. Soc. **125**, 7336–7343 (2003). https://doi.org/10.1021/ja034564p
25. Hell, S.W., Wichmann, J.: Breaking the diffraction resolution limit by stimulated emission: stimulated-emission-depletion fluorescence microscopy. Opt. Lett. **19**(11), 780–782 (1994)

26. Hellenkamp, B., et al.: Precision and accuracy of single-molecule fret measurements-a multi-laboratory benchmark study. Nat. Methods **15** (2018). https://doi.org/10.1038/s41592-018-0085-0
27. Herman, B.: Fluorescence microscopy. Garland Science (2020)
28. Hess, S.T., Girirajan, T.P., Mason, M.D.: Ultra-high resolution imaging by fluorescence photoactivation localization microscopy. Biophys. J. **91**(11), 4258–4272 (2006)
29. Huang, B., Bates, M., Zhuang, X.: Super-resolution fluorescence microscopy. Annu. Rev. Biochem. **78**, 993–1016 (2009)
30. Hudak, P., Makucevich, T., Gadde, S., Whong, B.: Haskore music notation – an algebra of music. J. Funct. Program. **6**(3), 465–484 (1996). https://doi.org/10.1017/S0956796800001805
31. Kuchenov, D., Laketa, V., Stein, F., Salopiata, F., Klingmüller, U., Schultz, C.: High-content imaging platform for profiling intracellular signaling network activity in living cells. Cell Chem. Biol. **23** (2016). https://doi.org/10.1016/j.chembiol.2016.11.008
32. Lichtman, J.W., Conchello, J.A.: Fluorescence microscopy. Nat. Methods **2**(12), 910–919 (2005). https://doi.org/10.1038/nmeth817
33. Peyton Jones, S., Eber, J.M., Seward, J.: Composing contracts: an adventure in financial engineering (functional pearl). SIGPLAN Not. **35**(9), 280–292 (2000). https://doi.org/10.1145/357766.351267
34. Piljic, A., Schultz, C.: Simultaneous recording of multiple cellular events by fret. ACS Chem. Biol. **3**, 156–160 (2008). https://doi.org/10.1021/cb700247q
35. Piston, D.W., Kremers, G.J.: Fluorescent protein fret: the good, the bad and the ugly. Trends Biochem. Sci. **32**(9), 407–414 (2007). https://doi.org/10.1016/j.tibs.2007.08.003. https://www.sciencedirect.com/science/article/pii/S0968000407001910
36. Roebroek, T., et al.: Simultaneous readout of multiple fret pairs using photochromism. Nat. Commun. **12**(1), 2005 (2021). https://doi.org/10.1038/s41467-021-22043-0
37. Ross, B., et al.: Single-color, ratiometric biosensors for detecting signaling activities in live cells. eLife **7** (2018). https://doi.org/10.7554/eLife.35458.001
38. Rost, F.W.: Fluorescence Microscopy, vol. 2. Cambridge University Press, Cambridge (1992)
39. Rust, M.J., Bates, M., Zhuang, X.: Stochastic optical reconstruction microscopy (STORM) provides sub-diffraction-limit image resolution. Nat. Methods **3**(10), 793 (2006)
40. Rust, M.J., Bates, M., Zhuang, X.: Sub-diffraction-limit imaging by stochastic optical reconstruction microscopy (STORM). Nat. Methods **3**(10), 793–796 (2006)
41. Schrijvers, T., Stuckey, P., Wadler, P.: Monadic constraint programming. J. Funct. Program. **19**(6), 663–697 (2009). https://doi.org/10.1017/S0956796809990086
42. Zeug, A., Woehler, A., Neher, E., Ponimaskin, E.: Quantitative intensity-based fret approaches-a comparative snapshot. Biophys. J. **103**, 1821–1827 (2012). https://doi.org/10.1016/j.bpj.2012.09.031
43. Zhu, J., et al.: Quantitative fluorescence resonance energy transfer measurements using microarray technology. In: Proceedings of SPIE - The International Society for Optical Engineering (2007). https://doi.org/10.1117/12.741482

Logic Programming

Flexible Job-shop Scheduling for Semiconductor Manufacturing with Hybrid Answer Set Programming (Application Paper)

Ramsha Ali[1], Mohammed M. S. El-Kholany[1,2], and Martin Gebser[1,3](✉)

[1] University of Klagenfurt, Klagenfurt, Austria
{ramsha.ali,mohammed.el-kholany,martin.gebser}@aau.at
[2] Cairo University, Cairo, Egypt
[3] Graz University of Technology, Graz, Austria

Abstract. The complex production processes in modern semiconductor manufacturing involve hundreds of operations on the route of a production lot, so that the period from lot release to completion can stretch over several months. Moreover, high-tech machines performing each of the operations are heterogeneous, may operate on individual wafers, lots or batches of lots in several stages, and require product-specific setups as well as dedicated maintenance procedures. This industrial setting is in sharp contrast to classical job-shop scheduling scenarios, where the production processes and machines are way less diverse and the primary focus is on solving methods for highly combinatorial yet abstract scheduling problems. In this work, we tackle the scheduling of realistic semiconductor manufacturing processes and model their elaborate requirements in hybrid Answer Set Programming, taking advantage of difference logic to incorporate machine processing, setup as well as maintenance times. While existing approaches schedule semiconductor manufacturing processes only locally, by applying greedy heuristics or isolatedly optimizing the allocation of particular machine groups, we study the prospects and limitations of scheduling at large scale.

Keywords: Flexible job-shop scheduling · Answer set programming · Logic programming · Semiconductor manufacturing

1 Introduction

Classical and Flexible Job-shop Scheduling Problems (FJSPs) [5,21], dealing with the allocation of production tasks and the planning of resources such as machines and workers, constitute longstanding combinatorial optimization domains, on which a variety of local and exact search methods have been developed (see [7] for an overview). The importance of optimization in production and supply chain management is increasing

This work was partially funded by KWF project 28472, cms electronics GmbH, FunderMax GmbH, Hirsch Armbänder GmbH, incubed IT GmbH, Infineon Technologies Austria AG, Isovolta AG, Kostwein Holding GmbH, and Privatstiftung Kärntner Sparkasse.

M. Hanus and D. Inclezan (Eds.): PADL 2023, LNCS 13880, pp. 85–95, 2023.
https://doi.org/10.1007/978-3-031-24841-2_6

due to diversified technologies, product customization, quality-of-service targets and ecological footprint [6]. Investigated FJSP optimization objectives include the minimization of makespan, tardiness, flow time and further scheduling criteria [14,20,24].

While abstract scheduling settings come along with a ready problem specification and existing benchmark sets, they do not reflect the complex production processes performed in modern semiconductor manufacturing [8,17]. The latter differ from traditional FJSP in view of long production routes, involving hundreds of operations taking several months to completion in a physical fab, product-specific machine setups, dedicated maintenance procedures as well as stochastic factors like varying process durations, unplanned machine disruptions or reworking steps. Given the ongoing trends towards process digitalization and automation, new-generation semiconductor fabs provide high system integration and at the same time increase the scheduling and control complexity. For modeling modern semiconductor manufacturing processes and facilities, the recent SMT2020 simulation scenario [17] incorporates heterogeneous high-tech machines that may operate on individual wafers, lots or batches of lots in several stages, require product-specific setups as well as dedicated maintenance procedures.

While production management in simulation or physical semiconductor fabs is so far performed locally by means of preconfigured dispatching rules [15] or machine-learned decision making policies [23], exact search and optimization frameworks like Constraint, Integer and Answer Set Programming (ASP) at least in principle enable global approaches to resource allocation and operation scheduling that take the entire range of viable schedules into account and guarantee optimality to the extent feasible. For instance, ASP [18] has been successfully applied for scheduling printing devices [3], specialist teams [19], work shifts [2], course timetables [4], medical treatments [9] and aircraft routes [22]. The hybrid framework of ASP modulo difference logic [16] particularly supports a compact representation and reasoning with quantitative resources like time, which has been exploited in domains such as lab resource [12], train connection [1] and classical job-shop [11] scheduling.

Our work takes advantage of ASP modulo difference logic to model the scheduling of semiconductor manufacturing process, particularly incorporating machine processing, setup as well as maintenance times present in the close-to-reality SMT2020 scenario. While long-term, global optimization of production routes with hundreds of operations is prohibitive in terms of problem size and the need to adapt to unpredictable stochastic factors, our work lays the basis for the future integration of more informed exact scheduling techniques with simulation and reactive decision making methods, i.e., scheduling for some time in advance and rescheduling on demand and/or periodically.

The rest of our paper is organized as follows. In Sect. 2, we introduce the specific requirements of FJSP for semiconductor manufacturing processes. Section 3 presents our hybrid ASP approach, taking advantage of difference logic for handling machine processing, setup and maintenance times. Experimental results on instances reflecting the SMT2020 fab scenario, conclusions and future work are discussed in Sect. 4.

2 Application Scenario

The recent SMT2020 simulation scenario specifies two kinds of semiconductor fab settings, one denoted High-Volume/Low-Mix (HV/LM) with two types of products and the other Low-Volume/High-Mix (LV/HM) with ten kinds of products. Both have in common that, depending on the product type, the production route of each lot includes a large number of operations, i.e., between 300 and 600 of them. Each operation needs some machine in a given tool group to be performed, which may have to be equipped with a product-specific setup before performing the operation. Moreover, periodic maintenance procedures must be applied after processing a certain number of operations or accumulating some amount of processing time on a machine.

Advanced machine capabilities, which are beyond the scope of our prototypical hybrid ASP formulation and left as future work, include the parallel processing of multiple lots in batches and pipelining of operations, where a machine works in several stages and can already proceed with further lots before all stages are completed. Second, we make the simplifying assumption that all machines are idle and each lot can be released at the beginning of the scheduling horizon, while an integration into continuous production or its simulation, respectively, would require the consideration of work in progress, i.e., operations that are currently processed so that the involved lots and machines will become available at some point in the future. Third, we note that the local dispatching rules suggested for the SMT2020 simulation scenario [17] incorporate given lot priorities and thus bias allocation decisions towards low tardiness of high-priority lots, where we instead take the makespan, i.e., the latest completion time over all operations, as optimization objective that is most directly supported by the *clingo*[DL] system [16].

In more detail, a problem instance consists of a set $M = M_1 \cup \cdots \cup M_m$ of *machines*, partitioned into *tool groups* $M_1, \ldots, M_m$, and a set $J = \{j_1, \ldots, j_n\}$ of *jobs*, where

- each $j \in J$ is a sequence $\langle o_{1_j}, \ldots, o_{l_j} \rangle$ of *operations*,
- each operation o_{i_j} is characterized by an associated tool group $M_{i_j} \in \{M_1, \ldots, M_m\}$, a *setup* s_{i_j} and a *processing time* p_{i_j},
- the distinguished setup $s_{i_j} = 0$ is used as a wildcard, i.e., o_{i_j} does not require any specific machine setup, and otherwise equipping a machine with $s_{i_j} \neq 0$ takes some time u_{i_j}, and
- the machines in a tool group M_g may undergo periodic *maintenance* procedures $a_{1_g}, \ldots, a_{h_g}$ such that each a_{f_g} has a duration w_{f_g}, a type $x_{f_g} \in \{\mathit{lots}, \mathit{time}\}$ as well as lower and upper bounds y_{f_g}, z_{f_g} specifying an interval for the number of operations or accumulated processing time at which the maintenance must be applied.

Then, a *schedule* is determined by some sequence $\langle q_{1_m}, \ldots, q_{k_m} \rangle$ of operations and maintenance procedures $\{q_{1_m}, \ldots, q_{k_m}\} \subseteq \{o_{i_j} \in \{o_{1_j}, \ldots, o_{l_j}\} \mid j \in J, m \in M_{i_j}\} \cup \{a_{1_g}, \ldots, a_{h_g} \mid m \in M_g\}$ per machine $m \in M$ such that the following conditions hold:

- Each operation o_{i_j} appears exactly once in the scheduled sequences for machines (still allowing for operations with similar characteristics at separate route positions).
- For each machine $m \in M_g$, some maintenance a_{f_g} and any of the sets $O_{e_m, f_g} = \{q_{d_m} \in \{q_{1_m}, \ldots, q_{e_m}\} \setminus \{a_{1_g}, \ldots, a_{h_g}\} \mid a_{f_g} \notin \{q_{d+1_m}, \ldots, q_{e-1_m}\}\}$ of (non-maintenance) operations performed up to q_{e_m} without applying a_{f_g} in-between, we have that $y_{f_g} \leq |O_{e_m, f_g}|$ (or $y_{f_g} \leq \sum_{o_{i_j} \in O_{e_m, f_g}} p_{i_j}$), in case $q_{e_m} = a_{f_g}$, as well as $|O_{e_m, f_g}| \leq z_{f_g}$ (or $\sum_{o_{i_j} \in O_{e_m, f_g}} p_{i_j} \leq z_{f_g}$) if $x_{f_g} = \mathit{lots}$ (or $x_{f_g} = \mathit{time}$).

This condition expresses that, for any longest subsequence $\langle q_{d_m}, \ldots, q_{e-1_m} \rangle$ of $\langle q_{1_m}, \ldots, q_{k_m} \rangle$ in which the maintenance a_{f_g} is not performed, the number of processed operations or their accumulated processing time (determined by the type $x_{f_g} \in \{lots, time\}$ of a_{f_g} and disregarding any applications of other maintenance procedures than a_{f_g}) must neither exceed the upper bound z_{f_g} nor stay below the lower bound y_{f_g}, where the latter is required only if $q_{e_m} = a_{f_g}$ indicates that the maintenance a_{f_g} is scheduled next in the sequence for machine m.

- The *completion time* c_{i_j} for each operation o_{i_j} with $q_{e_m} = o_{i_j}$, $m \in M_g$ and $t_{i_j} = \max(\{0\} \cup \{u_{i_j} \mid s_{i_j} \notin (\{0\} \cup \{s_{i'_{j'}} \mid 0 < d_m < e_m, \{o_{i'_{j'}}\} = \{o_{i''_{j''}} \in \{q_{d_m}, \ldots, q_{e-1_m}\} \setminus \{a_{1_g}, \ldots, a_{h_g}\} \mid s_{i''_{j''}} \neq 0\}\})\})$ is well-defined inductively by

$$
\begin{aligned}
c_{i_j} = p_{i_j} + \max\big(\{t_{i_j} + \textstyle\sum_{0<d_m, q_{d_m}=o_{i'_{j'}}} c_{i'_{j'}} + \sum_{a_{f_g} \in \{q_{d+1_m}, \ldots, q_{e-1_m}\}} w_{f_g} \mid \\
d_m = \max(\{0\} \cup \{0 < d_m < e_m \mid q_{d_m} \notin \{a_{1_g}, \ldots, a_{h_g}\}\})\} \\
\cup \{c_{i-1_j} \mid 1 < i\}\big).
\end{aligned}
$$

That is, the completion time c_{i_j} of o_{i_j} takes preceding processing, setup and maintenance times for the scheduled sequence on its machine as well as the completion time of the predecessor operation (if any) in its job into account. Hence, a well-defined, finite outcome, which corresponds to the earliest feasible starting time plus the processing time p_{i_j}, is obtained if and only if there are no circular waiting dependencies between operations.

For example, a schedule must not include the sequences $\langle o_{2_2}, o_{1_1} \rangle$ and $\langle o_{2_1}, o_{1_2} \rangle$ for two machines such that the first operations o_{1_1}, o_{1_2} of jobs wait for the out of the row completion of the respective other job's second operation. Given an admissible schedule, the *makespan* $\max\{c_{l_j} \mid j \in J\}$ provides the latest completion time over all jobs' operations, and we take its minimization as optimization objective.

3 Hybrid ASP Approach

As customary in ASP, we represent an FJSP instance in terms of specific facts and a general ASP modulo difference logic encoding, assuming basic familiarity with the input languages of the ASP system *clingo* and its extension *clingo*[DL] (see [13, 16, 18] for detailed descriptions). To begin with, the facts given in Listing 1 constitute a small excerpt of the SMT2020 HV/LM setting, making use of the following input predicates to specify lots, machines and operations:

`route(`*p,r,g,s,t*`)`. The route of any lot of product p includes an operation r requiring the processing time t on some machine in tool group g equipped with setup s. That is, the routes of lots with wafers of the products denoted by `3` and `4` include three operations each, performed by machines in the tool groups `implant_128` and `lithotrack_fe_95` with the setups `su128_1`, `su128_2` and `su450_3`. Note that the constant `0` for the argument s refers to any setup, i.e., a machine needs not be equipped before performing the second or third operation, respectively, along the routes of lots for product `3` and `4`.

```
route(3,1,      implant_128,su128_1,7).
route(3,2,lithotrack_fe_95,       0,8).
route(3,3,      implant_128,su128_2,8).
route(4,1,      implant_128,su128_1,8).
route(4,2,lithotrack_fe_95,su450_3,7).
route(4,3,lithotrack_fe_95,       0,7).
setup(implant_128,su128_1,7). setup(lithotrack_fe_95,su450_3,7).
setup(implant_128,su128_2,8).
pm(      implant_128,     implant_128_mn,lots, 1, 2,5).
pm(      implant_128,     implant_128_qt,lots, 2, 3,3).
pm(lithotrack_fe_95,lithotrack_fe_95_mn,time, 6,10,6).
pm(lithotrack_fe_95,lithotrack_fe_95_qt,time,10,15,5).
tool(implant_128,1..2). tool(lithotrack_fe_95,1).
lot(1,3). lot(1,4).
```

Listing 1. Example FJSP instance

`setup(`g,s,t`)`. Equipping a machine in tool group g with the setup s takes time t, i.e., `7` time units for the setup `su128_1` and `8` for `su128_2`, where `implant_128` is the corresponding tool group, as well as `7` time units for equipping machines of the tool group `lithotrack_fe_95` with `su450_3`.

`pm(`g,a,x,y,z,w`)`. Machines in the tool group g undergo a maintenance procedure a of type $x \in$ {`lots`,`time`} with duration w after processing between y and z operations or accumulating as much processing time, respectively. For example, the maintenance `implant_128_mn` with duration `5` must be applied to machines of the tool group `implant_128` after processing between `1` and `2` operations, and `lithotrack_fe_95_mn` maintenance with duration `6` is required for `lithotrack_fe_95` machines after performing operations whose processing times add up to a sum between `6` and `10`.

`tool(`g,m`)`. The tool group g includes a machine m, i.e., two machines belong to the tool group `implant_128` and one to `lithotrack_fe_95`.

`lot(`l,p`)`. The operations for a lot l with wafers of the product p need to be scheduled, where we have one lot for product `3` and another for product `4` in Listing 1.

Figure 1 shows an optimal schedule for the instance in Listing 1 in terms of the resulting makespan 54, where operations are indicated by the numbers l, p, r of their lots, products and route positions, and machine, setup and maintenance labels are abbreviated for readability. The first operations of the lots for product `3` and `4` can be released after equipping the two machines in tool group `implant_128` with the required setup `su128_1`. Then, both lots need to be processed on by the single machine in tool group `lithotrack_fe_95`, where it turns out as advantageous to proceed with the operation requiring the setup `su450_3`. In this way, the setup time can be spent while the first operation on the route is still processed, while we would otherwise need to postpone the setup change to the completion of another operation not requiring any specific setup. Note that our FJSP formulation does not include opportunistic setup changes before operations that do not require the specific setup, considering that stochastic factors

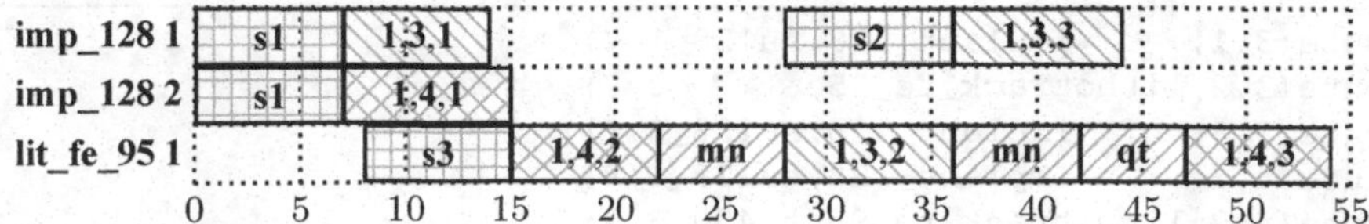

Fig. 1. Optimal schedule for the example FJSP instance in Listing 1

necessitate frequent adaptions to new circumstances and make the precise anticipation of future machine assignments impossible in practice. Further focussing on the maintenances applied to the machine in tool group `lithotrack_fe_95`, the maintenance procedure `lithotrack_fe_95_qt` is required once and `lithotrack_fe_95_mn` twice in view of their upper bounds on accumulated processing time. Such maintenances delay the release of operations before which they are applied, particularly when both are necessary in a row before the third operation of the lot for product `4`, while distributing the operations performed on the two machines in tool group `implant_128` avoids the need of scheduling the maintenance procedures `implant_128_mn` and `implant_128_qt`.

The main parts of our general ASP modulo difference logic encoding for FJSP are shown in Listing 2.[1] For a convenient representation of operations to schedule, the rule in line 3 maps given lot identifiers `L`, products `P`, operation numbers `R`, and setups `S` to a tuple `(L,P,R,S)` accompanied by the tool group `G` and the processing time `T` for performing the operation. A fixed machine assignment based on the lexicographic order of operations can be determined by means of the rule in lines 4–5, provided that the value `1` declared for the constant `f` in line 2 is not overridden using the `--const` option of *clingo*[DL]. For example, the fixed machine assignment derived for the facts in Listing 1 and the tool group `implant_128` is represented by atoms `assign((1,3,1,su128_1),implant_128,7,1)`, `assign((1,3,3,su128_2),implant_128,8,2)` and `assign((1,4,1,su128_1),implant_128,8,1)`, whose last arguments identify a machine to perform each operation. If the fixed assignment is deactivated by changing the value for constant `f`, the (choice) rule in line 6 enables the flexible assignment of some machine in the tool group required for an operation.

The next block of rules from line 9 to line 14 determines the sequence of operations on each machine. To this end, `share/5` atoms derived by the rule in line 9 provide pairs of distinct operations `O` and `O'` assigned to the same machine. The two rules in line 10 and 11 make sure that either `order(O,O',T)` or `order(O',O,T')` is derived to signal that `O` is performed before `O'` or vice versa, where `T` or `T'` is the processing time of the respective operation coming first. Dependencies due to the chosen machine-wise operation sequences and successor operations along the routes of lots are taken together by `later/3` atoms obtained from the rules in lines 12–14, providing the basis to propagate conditions on the release times of operations by difference logic constraints.

Due to space limitations, we do not include the rules for deriving further atoms in Listing 2, but explain the meaning of their predicates in the following:

[1] Our encoding as well as test instances are available online at: https://github.com/prosysscience/FJSP-SMT2020.

```
% fixed/flexible machine assignment
#const f = 1.
op((L,P,R,S),G,T) :- lot(L,P), route(P,R,G,S,T).
assign(O,G,T,M) :- op(O,G,T), I = #count{O': op(O',G,T'), O'<O},
                   tool(G,N), not tool(G,N+1), M = I\N+1, f = 1.
{assign(O,G,T,M): tool(G,M)} = 1 :- op(O,G,T).

% sequence of operations per machine
share(G,O,O',T,T') :- assign(O,G,T,M), assign(O',G,T',M), O<O'.
{order(O,O',T)} :- share(G,O,O',T,T').
 order(O',O,T') :- share(G,O,O',T,T'), not order(O,O',T').
later(O,O',T) :- order(O,O',T).
later((L,P,R-1,S),(L,P,R,S'),T) :- op((L,P,R-1,S),G,T),
                                   op((L,P,R,S'),G',T').

% [rules handling setup/maintenance times skipped for brevity]

% constraints for maintenance type 'lots'
:- pm(G,A,lots,X,Y,W), op(O,G,T), not nofirst(O),
   not maintain(G,A,O).
:- pm(G,A,lots,X,Y,W), op(O,G,T),
   #count{O': op(O',G,T'), not ignore(G,A,O,O')} > Y.
:- pm(G,A,lots,X,Y,W), maintain(G,A,O), nolast(G,A,O),
   #count{O': op(O',G,T'), not ignore(G,A,O,O')} < X.

% difference logic constraints on operations' starting times
&diff{0-O } <= -T :- setuptime(O,T).
&diff{O-O'} <= -T :- later(O,O',T).
&diff{O-O'} <= -T - T' :- order(O,O',T), pmsetuptime(O',T').
&diff{(L,P,R,S)-makespan} <= -T :- op((L,P,R,S),G,T),
                                   not op((L,P,R+1,_),_,_).
```

Listing 2. Hybrid FJSP encoding

`maintain(`g,a,o`)`. A maintenance procedure a is applied before performing the operation o on some machine in the tool group g. Such atoms are obtained from choice rules, where the approach to link maintenance procedures to operations before which they take place makes use of the fact that maintenance could be entirely avoided without upper bounds on the number of lots or the processing time at which the maintenance must be applied. Hence, scheduling some maintenance is exclusively needed to create the capacity for processing further operations afterwards.

`nofirst(`o`)`. The operation o is not the first in the sequence of processing on its machine, as signaled by some atom of the form `order(`o'`,`o`,`t'`)`.

`nolast(`g,a,o`)`. The operation o is not the last operation on its machine in tool group g before which the maintenance a is applied, i.e., `order(`o`,`o'`,`t`)` as well as `maintain(`g`,`a`,`o'`)` yield some later operation o' such that a is applied before.

`ignore(`g,a,o,o'`)`. The operation o' is not taken into account in checking the lower and upper bounds for maintenance a when applied before operation o. That is, either o' is performed on a different machine than o in the tool group g, o' comes before o in the sequence of processing on the same machine, or the maintenance a is repeated before some operation in-between o and o' (including $o' \neq o$ itself).

`setuptime(`o,t`)`. Equipping the machine performing operation o with the required setup takes t time units, where `setuptime(`o`,0)` is always the case and a positive time t applies in addition when the setup for o cannot be reused from a preceding operation on the same machine.

`pmsetuptime(`o,t`)`. The total time taken for maintenance (possible several procedures) and setup before operation o is at least t. That is, any maintenance durations and possibly the setup time for o are added up as gap needed between preceding operations on the same machine and o.

Based on the above predicates, the constraints from line 19 to line 24 handle the lower and upper bounds for maintenance procedures in terms of processed `lots`. (The constraints dealing with maintenance procedures considering accumulated processing `time` are analogous and skipped in Listing 2 for brevity.) To begin with, the constraint in lines 19–20 requires that maintenances are associated with the first operation performed on each machine. Note that such initial maintenance is asserted merely for checking the lower bound of operations to perform before repeating the maintenance, while its duration is disregarded and does not delay the release of the first operation on a machine. The second constraint in lines 21–22 addresses upper bounds restricting the number of processed operations after which a maintenance must be repeated. Here it is sufficient to assert that not more than the upper bound for a maintenance a many atoms `ignore(`g`,`a`,`o`,`o'`)` are underived regarding operations o' performed by machines in tool group g. Such operations o' are performed no earlier than o on the same machine without applying the maintenance a in-between (thus including o itself), so that the upper bound must not be exceeded no matter whether a is actually scheduled before o. The latter is different with the constraint on the lower bound for a in lines 23–24, where o must not be the last operation on its machine before which the maintenance a is applied.

Finally, the difference logic constraints from line 27 to line 31 propagate waiting dependencies along machine-wise operation sequences and the routes of lots, here using tuples of the form `(L,P,R,S)` introduced by the rule in line 3 as integer variables representing the release times of operations. In particular, the difference logic constraint asserted in line 27 expresses that the setup time (which can be zero) for an operation must at least be spent before the operation can be released. Line 28 states that later operations on a machine or the route of a lot must wait for their predecessors to complete. Relative to predecessor operations on the same machine, the time for performing maintenance procedures and possibly changing the setup is in line 29 included as additional gap delaying the release of a later operation. The difference logic constraint in lines 30–31 then introduces the integer variable `makespan` as upper bound on the completion times of lots by investigating the last operations on their routes. This allows for activating makespan minimization as optimization objective by supplying `--minimize-variable=makespan` as an option to *clingo*[DL]. In fact, taking the

Table 1. Results for FJSP instances with fixed and flexible machine assignment

O × M	Fixed machine assignment				Flexible machine assignment			
	Makespan	Time	Conflicts	Constraints	Makespan	Time	Conflicts	Constraints
10 × 2	120	<1	44	2249	120	<1	44	2258
15 × 3	169	<1	7910	8980	169	<1	16562	12916
20 × 3	224	TO	6091128	23751	204	TO	5869538	33682
20 × 4	120	<1	88	4569	105	15	162085	45364
25 × 4	171	<1	16303	15862	160	TO	2656812	89214
30 × 4	219	405	2626786	32872	227	TO	2868459	157258
30 × 5	191	93	1027764	26460	169	TO	2702824	159548
30 × 6	120	<1	128	6828	153	TO	2250537	158243
35 × 6	142	<1	925	11526	123	TO	2947938	181867
40 × 6	192	18	189240	30657	210	TO	1711929	387488
45 × 6	226	TO	3553153	49191	300	TO	1226970	554625

makespan as comparably simple objective has the pragmatic advantage that we can directly use the built-in linear search for a minimum value by *clingo*[DL], while more fine-grained criteria like lexicographical makespan [10] or tardiness [1, 12] minimization require reification of integer variables' values by regular atoms in order to express the objective function(s) in terms of the optimization statements available in ASP [13].

4 Experimental Evaluation and Conclusion

We extracted test instances for FJSP scheduling from the STM2020 scenario by considering a part of the available tool groups along with their setups and maintenances (see footnote 1). The numbers of operations to schedule and machines to perform them are given in the **OxM** column in Table 1, and we ran *clingo*[DL] (v1.4.0) for makespan minimization up to 600 s, aborted runs indicated by 'TO', on an Intel® Core^{TM}i7-8650U CPU Dell Latitude 5590 machine. Our comparison includes the fixed and flexible machine assignment options described in the previous section, for which we report the obtained makespan, solving time, conflicts and constraints at termination or abort in case of 'TO'.

With the fixed machine assignment, we observe that the transition from trivial runs taking less than a second to aborts is sharp when increasing the number of operations, as the number of search conflicts goes up. A similar behavior is encountered even earlier, i.e., on smaller instances, with the flexible machine assignment, where the up to one order of magnitude greater number of constraints also yields a significantly increased problem size. This is certainly a reason why the flexible machine assignment sometimes leads to schedules of larger makespan in the given time than fixed machine assignment, even though a flexible schedule of the same or strictly smaller makespan always exists.

Our FJSP encoding for semiconductor manufacturing reflects work in progress. As future work, we target the improvement of its space and search efficiency, addition of yet missing capabilities like batch processing and pipelining of operations, decomposition techniques making better tradeoffs than either fixed or fully flexible machine assignment, and favorable integration with simulation and reactive decision making methods.

References

1. Abels, D., Jordi, J., Ostrowski, M., Schaub, T., Toletti, A., Wanko, P.: Train scheduling with hybrid answer set programming. Theory Pract. Logic Program. **21**(3), 317–347 (2021)
2. Abseher, M., Gebser, M., Musliu, N., Schaub, T., Woltran, S.: Shift design with answer set programming. Fundamenta Informaticae **147**(1), 1–25 (2016)
3. Balduccini, M.: Industrial-size scheduling with ASP+CP. In: Delgrande, J.P., Faber, W. (eds.) LPNMR 2011. LNCS (LNAI), vol. 6645, pp. 284–296. Springer, Heidelberg (2011). https://doi.org/10.1007/978-3-642-20895-9_33
4. Banbara, M., et al.: teaspoon: solving the curriculum-based course timetabling problems with answer set programming. Ann. Oper. Res. **275**(1), 3–37 (2019)
5. Brucker, P., Schlie, R.: Job-shop scheduling with multi-purpose machines. Computing **45**(4), 369–375 (1990)
6. Ceylan, Z., Tozan, H., Bulkan, S.: A coordinated scheduling problem for the supply chain in a flexible job shop machine environment. Oper. Res. **21**(2), 875–900 (2021). https://doi.org/10.1007/s12351-020-00615-0
7. Chaudhry, I., Khan, A.: A research survey: review of flexible job shop scheduling techniques. Int. Trans. Oper. Res. **23**(3), 551–591 (2015)
8. Da Col, G., Teppan, E.C.: Industrial size job shop scheduling tackled by present day CP solvers. In: Schiex, T., de Givry, S. (eds.) CP 2019. LNCS, vol. 11802, pp. 144–160. Springer, Cham (2019). https://doi.org/10.1007/978-3-030-30048-7_9
9. Dodaro, C., Galatà, G., Grioni, A., Maratea, M., Mochi, M., Porro, I.: An ASP-based solution to the chemotherapy treatment scheduling problem. Theory Pract. Logic Program. **21**(6), 835–851 (2021)
10. Eiter, T., Geibinger, T., Musliu, N., Oetsch, J., Skocovský, P., Stepanova, D.: Answer-set programming for lexicographical makespan optimisation in parallel machine scheduling. In: Proceedings of the Eighteenth International Conference on Principles of Knowledge Representation and Reasoning (KR 2021), pp. 280–290. AAAI Press (2021)
11. El-Kholany, M.M.S., Gebser, M., Schekotihin, K.: Problem decomposition and multi-shot ASP solving for job-shop scheduling. Theory Pract. Logic Program. **22**(4), 623–639 (2022)
12. Francescutto, G., Schekotihin, K., El-Kholany, M.M.S.: Solving a multi-resource partial-ordering flexible variant of the job-shop scheduling problem with hybrid ASP. In: Faber, W., Friedrich, G., Gebser, M., Morak, M. (eds.) JELIA 2021. LNCS (LNAI), vol. 12678, pp. 313–328. Springer, Cham (2021). https://doi.org/10.1007/978-3-030-75775-5_21
13. Gebser, M., et al.: Potassco user guide (2019). https://potassco.org
14. Hassanzadeh, A., Rasti-Barzoki, M., Khosroshahi, H.: Two new meta-heuristics for a bi-objective supply chain scheduling problem in flow-shop environment. Appl. Soft Comput. **49**, 335–351 (2016)
15. Holthaus, O.: Efficient dispatching rules for scheduling in a job shop. Int. J. Prod. Econ. **48**(1), 87–105 (1997)
16. Janhunen, T., Kaminski, R., Ostrowski, M., Schaub, T., Schellhorn, S., Wanko, P.: Clingo goes linear constraints over reals and integers. Theory Pract. Logic Program. **17**(5–6), 872–888 (2017)
17. Kopp, D., Hassoun, M., Kalir, A., Mönch, L.: SMT2020–a semiconductor manufacturing testbed. IEEE Trans. Semiconductor Manuf. **33**(4), 522–531 (2020)

18. Lifschitz, V.: Answer Set Programming. Springer, Heidelberg (2019). https://doi.org/10.1007/978-3-030-24658-7
19. Ricca, F., et al.: Team-building with answer set programming in the Gioia-Tauro seaport. Theory Pract. Logic Program. **12**(3), 361–381 (2012)
20. Sahraeian, R., Rohaninejad, M., Fadavi, M.: A new model for integrated lot sizing and scheduling in flexible job shop problem. J. Ind. Syst. Eng. **10**(3), 72–91 (2017)
21. Taillard, E.: Benchmarks for basic scheduling problems. Eur. J. Oper. Res. **64**(2), 278–285 (1993)
22. Tassel, P., Rbaia, M.: A multi-shot ASP encoding for the aircraft routing and maintenance planning problem. In: Faber, W., Friedrich, G., Gebser, M., Morak, M. (eds.) JELIA 2021. LNCS (LNAI), vol. 12678, pp. 442–457. Springer, Cham (2021). https://doi.org/10.1007/978-3-030-75775-5_30
23. Waschneck, B., et al.: Optimization of global production scheduling with deep reinforcement learning. Procedia CIRP **72**, 1264–1269 (2018)
24. Xing, L., Chen, Y., Wang, P., Zhao, Q., Xiong, J.: A knowledge-based ant colony optimization for flexible job shop scheduling problems. Appl. Soft Comput. **10**(3), 888–896 (2010)

Integrating ASP-Based Incremental Reasoning in the Videogame Development Workflow (Application Paper)

Denise Angilica(✉), Giovambattista Ianni, Francesco Pacenza, and Jessica Zangari

Department of Mathematics and Computer Science, University of Calabria, Rende, CS 87036, Italy
{denise.angilica,giovambattista.ianni,francesco.pacenza, jessica.zangari}@unical.it

Abstract. Challenging fields like real-time videogames constitute an ideal, reproducible and controllable ground for researching and experimenting on the new developments of incremental reasoners for Answer Set Programming (ASP). On the other hand, declarative methods show potential in cutting down development costs in commercial videogames: nonetheless, fulfilling the strict time requirements of this type of stream reasoning-like application is still an unsurpassed obstacle. Incremental reasoning techniques might help in overcoming this latter. In this work we report about the integration of an incremental ASP engine in a framework conceived for adding declarative decision-making modules in the typical videogame development workflow. Namely, the two systems are *Incremental-DLV2*, a recently introduced multi-shot incremental solver based on the ASP semantics, and *ThinkEngine*, a tool for developing declarative modules working in the context of the Unity game engine. After describing the features of both systems, we give an example showing how to program a declarative-based videogame character. We discuss how we adapted the architecture of *ThinkEngine* for accommodating incremental reasoning, and report about experiments showing the impact in performance after the introduction of *Incremental-DLV2*.

Keywords: Declarative logic · Answer set programming · Knowledge representation · Games and videogames · Stream reasoning

1 Introduction

The interest of the videogame industry in AI research is not new, both whether we are talking of inductive/machine learning-based techniques or knowledge-based, deductive techniques [25]. In this respect, declarative methods show potential benefits, like enabling the possibility of specifying parts of the game logic in a few lines of high-level statements. Possible applications range from defining the general game logic, to describing non-player characters, programming tactic and/or strategic credible AI behaviors, to expressing path planning desiderata, non-player resource management policies and so on.

M. Hanus and D. Inclezan (Eds.): PADL 2023, LNCS 13880, pp. 96–106, 2023.
https://doi.org/10.1007/978-3-031-24841-2_7

Many examples of the usage of declarative languages in the industrial videogame realm exist, starting from the pioneer F.E.A.R. game [24], which used STRIPS-based planning [23]. Other remarkable examples are the games Halo [5] and Black & White [4]. If we look at videogames from the basic research perspective, it must be noted the longstanding interest in using (video-)games as a controllable and reproducible setting in which to face open issues: one might cite the GDL [20], VGDL [29] and Ludocore [31] languages adopted for declaratively describing General Game Playing [22]. The Planning Domain Definition Language (PDDL) found natural usage in the videogame realm [10,27,28]; among its sister languages, we will herein focus particularly on Answer Set Programming (ASP), the known declarative paradigm with a tradition in modeling planning problems, robotics, computational biology as well as other industrial applications [18]. ASP does not come last in its experimental usage in videogames: it has been used to various extents, e.g., for declaratively generating level maps [30] and artificial architectural buildings [7]; it has been used as an alternative for specifying general game playing [33], for defining artificial players [13] in the Angry Birds AI competition [26], and for modelling resource production in real-time strategy games [32], to cite a few. Despite this potential, however, performance and integration shortcomings limited so far the widespread adoption of declarative methods in professional videogames: ASP makes no exception in this respect.

Two aspects are of concern: *i)* integration, i.e., the ease of wiring declarative modules with other standard parts of the game implementation, and *ii)* performance in real-time contexts. Concerning the first issue, we build on our recent proposal of *ThinkEngine* [1,9], a tool working in the known Unity game engine [2], which allows to wire declaratively-programmed *Brains* to videogame implementations. We will concentrate here on performance issues: it must be considered that much research has been done in the last years on incremental ASP engines capable of working on fast paced and repeated decision-making tasks [11,14,19,21]. These contributions enlarged the range of applications of ASP to highly demanding settings like stream reasoning [15,17]. It is thus appealing to explore the possibility of integrating these new reasoners in the videogame development workflow, although this requires some architectural changes. We particularly focus on *Incremental-DLV2*, an incremental ASP reasoner [14]. *Incremental-DLV2* is especially conceived for transparently solving repeated reasoning tasks, by reusing previously done computational effort and caching so-called *overgrounded programs*, with no need of manually adding procedural directives controlling the incremental evaluation.

In this paper, after presenting the *ThinkEngine* system and the *Incremental-DLV2* reasoner, we focus on the integration of the latter into the former, thus potentially increasing the pace of repeated reasoning tasks. We briefly describe the declarative implementation of an AI in a representative game. We then report about experiments on the new *ThinkEngine* version, compared with its older version which makes use of a standard, non-incremental, ASP solver.

2 The *ThinkEngine*

ThinkEngine [9] is a system allowing to integrate declarative-based reasoning modules in a videogame or any other kind of software developed in the known Unity game engine [2]. The main components of the framework are the so called *Brains*. Brains can be attached at will to game characters, they can drive parts of the game logic, and can be used in general for delivering AI at the tactical or strategic level within the game at hand.

One can have *planner brains* or *reactive brains*, which respectively make different type of decisions: deliberative ones (i.e., *plans*), which, in the terminology of our *ThinkEngine*, are sequences of actions to be executed in a programmable order, or *reactive* decisions which can have an immediate impact on the game scene. Plans work in the spirit of the Goal Oriented Action Planning methodology (GOAP), a popular way of deploying academic planning in the realm of videogames since its introduction in F.E.A.R. [24]. Multiple planner brains can be prioritised to control the same character, thus introducing a form of multiple behavior programming. Moreover, as game environments are subject to fast changes, one can program the appropriate plan aborting logic.

ThinkEngine has been developed having the integration of declarative ASP modules in mind, but other types of automated reasoning can be in principle wired (e.g., PDDL), by writing the appropriate glue code. ASP specifications are composed of set of rules, hard and soft constraints, by means of which it is possible to express qualitative and quantitative statements. In general, a set of input values F (called *facts*), describing the current state of the world, are fed together with an ASP specification S to a *solver*. Solvers in turn produce sets of outputs $AS(S \cup F)$ called *answer sets*. Answer sets contain the result of a decision-making process in terms of logical assertions which, depending on the application domain at hand, might encode actions to be made, employee shifts to be scheduled, protein sequences, and so on.

ThinkEngine is integrated in the Unity game engine both at design time and at run time. At *design time*, one can add, wire and program brains in the game editor. A brain can be wired to sensor inputs, and one can decide where a brain acts on the game by connecting *actuators* or defining *plan actions*. Sensor values can be aggregated within a time window according to a policy of choice (like max, min, average, oldest or newest value), while *Triggers* can be defined to program when brain reasoning activities must take place at run time. Brains can be embedded in reusable objects, called *prefabs* in the Unity terminology. At run time, most of the game runs in the usual single-threaded game loop [3], while brain reasoning tasks are offloaded to separate threads. An information passing layer allows communication between brains and the main game loop. In this latter a sensor update cycle periodically refreshes sensor readings. Inputs (sensor readings) and outputs (actuator values or plans) of brains are bidirectionally converted to/from the ASP-Core2 format [12], according to a properly defined mapping discipline between the game world data and ASP logic assertions.

3 The *Incremental-DLV2* Reasoner

Incremental-DLV2 [14] represents the evolution of the standard, non-incremental ASP system *DLV2* [8] towards multi-shot reasoning, i.e., it allows the repeated execution of a reasoning task over iterative sequences of varying inputs (called *shots*). Given a fixed ASP specification S composed of logic rules, and a set of facts F_i, expressing inputs for the current shot i, *Incremental-DLV2* computes the answer sets $AS(S \cup F_i)$ of S over F_i. Then, *Incremental-DLV2* maintains itself alive while waiting for another set of facts F_{i+1} for the next shot $i+1$. The key feature of *Incremental-DLV2*, as suggested by its name, lies in the capability of performing the evaluation of S across the shots in an incremental way, by relying on the so called overgrounding technique [21]. More in detail, when evaluating P over F_i at a shot i, *Incremental-DLV2* performs two consecutive steps: grounding (also called instantiation) and solving (also called answer set search). Instantiating a program consists in generating, rule by rule, substitutions of variables with constants, thus obtaining an equivalent propositional program; ASP *grounders* are geared towards considering only relevant substitutions to mitigate the theoretical exponential blow-up [16] in the number of instantiated rules. Compared with standard solvers, *Incremental-DLV2* caches and maintains across the shots a monotonically increasing ground program G, which helps in performing instantiations incrementally. Only possibly new information appearing at a new shot is taken into account when adding new rules. Finally, in the solving step, *Incremental-DLV2* applies SAT-inspired algorithms on G to compute the corresponding answer sets, using its integrated solver.

To comply with the multi-shot paradigm, when started, *Incremental-DLV2* works on a stay online basis, and waits for commands. We illustrate next how the system works over a simple dynamic setting based on the vertex-covering problem. The reader is referred to the official ASP-Core2 documentation for more detail on syntax and semantics [12]. In particular, we recall that rules in the form $a_1|\dots|a_m$:- $b_1,\dots,b_n$ stand for the logical implication $a_1 \vee \dots \vee a_m \leftarrow b_1,\dots,b_n$. If $a_1,\dots,a_m$ are missing then the rule denotes a hard constraint, while the operator :~ denotes a weak (soft) constraint. Search spaces are defined with *disjunctions* or *choice rules* with their appropriate syntax.

Given an undirected graph $G = (V, E)$, the problem requires to select a set $C \subseteq V$ such that all edges are covered (i.e., for every edge $(a, b) \in E$ either $a \in C$ or $b \in C$). If the graph G is specified by means of facts over predicates *node* and *edge*, then the specification S_{VC} below encodes the problem in ASP:

$$
\begin{array}{ll}
r: & inC(X) \mid outC(X) \text{ :- } node(X). \\
s: & \text{:- } edge(X,Y),\ \texttt{not}\ inC(X),\ \texttt{not}\ inC(Y).
\end{array}
$$

In our dynamic version of the problem, we have to reason over a graph whose structure changes over time (i.e., at each shot, nodes/edges can be added/removed). We describe below the behaviour of the system across three possible shots where input facts change. After loading S_{VC}, one can send the first shot F_1 to *Incremental-DLV2* which may be: $node(1..3).\ edge(1,2).\ edge(2,3).\ edge(1,3)$. In response to F_1, the system produces 5 answer sets, in which C corresponds to

{1, 2}, {1, 3}, {2, 3}, {1, 2, 3}, respectively. Internally, *Incremental-DLV2* accumulates the overgrounded program G_{VC} reported below.

$$
\begin{array}{ll}
r_1 & inC(1) \mid outC(1) \text{ :- } \cancel{node(1)}. \\
r_2 & inC(2) \mid outC(2) \text{ :- } \cancel{node(2)}. \\
r_3 & inC(3) \mid outC(3) \text{ :- } \cancel{node(3)}. \\
s_1 & \text{:- } \cancel{edge(1,2)},\ inC(1),\ inC(2). \\
s_2 & \text{:- } \cancel{edge(2,3)},\ inC(2),\ inC(3). \\
s_3 & \text{:- } \cancel{edge(1,3)},\ inC(1),\ inC(3).
\end{array}
$$

Crossed out atoms represent occurred simplifications due to information known to be certainly true stemming from input facts at shot 1. Suppose now that in the shot 2, inputs F_2 is loaded and contains the facts: $node(1..3)$. $edge(1,2)$. $edge(1,3)$. $edge(2,3)$. $node(4..5)$. $edge(4,5)$. $edge(1,4)$. $edge(1,5)$. $inC(4)$. At this point, the input graph is thus composed by two additional nodes connected to each other and connected also to node 1; moreover, node 4 is already known to be in C. When asking the computation of $AS(S_{VC} \cup F_2)$, thanks to the overgrounding-based instantiation strategy, the system only generates the additional ground rules describing the new additions, and adds them to G_{VC}:

$$
\begin{array}{ll}
r_4 & inC(4) \mid outC(4) \text{ :- } \cancel{node(4)}. \\
r_5 & inC(5) \mid outC(5) \text{ :- } \cancel{node(5)}. \\
s_4 & \text{:- } \cancel{edge(4,5)},\ \cancel{inC(4)},\ inC(5). \\
s_5 & \text{:- } \cancel{edge(1,5)},\ inC(1),\ inC(5). \\
s_6 & \text{:- } \cancel{edge(1,4)},\ inC(1),\ \cancel{inC(4)}.
\end{array}
$$

Notably, simplifications made at shot 1 remain valid since all facts in shot 1 are also facts of shot 2. All possible coverings must thus include node 4, hence there are 7 solutions (i.e., answer sets), in which C is {4, 1, 3}, {4, 1, 2}, {4, 1, 2, 3}, {4, 1, 3, 5}, {4, 1, 2, 5}, {4, 1, 2, 3, 5}, {4, 2, 3, 5}, respectively.

Finally, suppose the system is input with F_3: $node(1..3)$. $edge(1,2)$. $edge(2,3)$. $edge(1,3)$. $node(4..5)$. $inC(4)$. $edge(1,4)$. $edge(4,5)$. The graph is updated removing the edge between nodes 1 and 5. Now, no facts are unseen in previous shots: hence, no additional ground rules are generated and no grounding effort is needed; the only update in G_{VC} consists of the desimplification of $edge(1,5)$ in the rule s_5 since it cannot be longer assumed as a certain information. In turn, the possible solutions are the same 7 ones computed at shot 2 with the addition of {4, 2, 3}. It is worth noting that the savings in computational time, obtained by properly reusing rules generated at previous shots, occurs with no need of knowing anything about possible incoming input facts in advance. Moreover, the management of the incremental computation is completely automatic and transparent to the user, who is not required to define a priori what is fixed and what is volatile in a specification.

4 How to Program Brains: An Example

In order to give an idea of how AI declarative modules can be integrated within applications developed in Unity we herein illustrate how we approached the Pac-Man game. In this classic game the main character moves in a 2D labyrinth in which it has to eat all the appearing pellets while avoiding four moving ghosts. Energizers can also be eaten to give a temporary power up mode where ghosts can be chased and killed by the main character. The game starts with a labyrinth full of pellets and with the main character and the ghosts in their initial position.

We implemented an AI controlling the main character composed of three planner brains. Each planning brain is attached to a trigger function, which can be used to define under which condition a brain should compute a plan. Competing planner brains follow a general prioritization rule: the availability of a plan of higher priority makes the execution of lesser priority plans to be aborted.

The introduction of multiple brains controlling the main character is meant to cope with the rapid changes of the game world in Pac-Man. The idea is to be faster yet less accurate with a simple AI working in the early seconds of a game, while allowing a "smarter" AI more time: this latter takes control as soon as it starts to produce decisions. Moreover, a dedicate AI is triggered when it is needed to manage emergency situations. More in detail:

- a "faster" AI (Planner3) produces a single action at each iteration and has priority 3 (lower priority values come first). In order to make this planner empirically faster, this ASP specification measures distances using the lesser accurate *taxicab metric* [6], ignoring the existence of walls.
- a more sophisticated AI (Planner2) produces plans each containing two actions and has priority 2. The distance between cells is computed considering the actual configuration of the grid and finding an actual path between each pair of free cells. Yet, accuracy comes at the cost of lower reasoning speed. As soon as Planner2 has a plan ready, its better priority make it to supersede Planner3. Planner3 is then permanently disabled by falsifying its trigger condition.
- an emergency AI (Planner1) makes the Pac-Man run away from too close ghosts and has priority 1. Planner1 is not triggered in power up mode. If triggered, it does not however produce plans unless ghosts are closer to the Pac-Man of a given threshold.

We report here some snippet from the Planner2 AI. The presented AI. expects input information like Pac-Man and ghosts positions, coordinates of free cells and of pellets and energizers. Actions that this AI can plan correspond to the possible movements *right, left, up, down*, and are encoded with propositional assertions like `next(step,dir)`.

```
%GUESSING FOR EACH ACTION OF THE PLAN WHICH DIRECTION TO PURSUE
{next(S,Move) : move(Move)}=1 :- planStep(S).

% COMPUTING THE DISTANCE BETWEEN COUPLE OF GRID CELLS
step(0,1). step(1,0).
adj(X1,Y1,X2,Y2):- tile(X1,Y1), tile(X2,Y2), step(DX,DY), X2=X1+DX, Y2=Y1+DY.
adj(X1,Y1,X2,Y2):- tile(X1,Y1), tile(X2,Y2), step(DX,DY), X2=X1-DX, Y2=Y1-DY.
dist(X1,Y1,X2,Y2,1) :- tile(X1,Y1), adj(X1,Y1,X2,Y2).
dist(X1,Y1,X3,Y3,Dp):- dist(X1,Y1,X2,Y2,D),adj(X2,Y2,X3,Y3),D=Dp-1,dist(Dp).
nonMinDist(X1,Y1,X2,Y2,D1):-dist(X1,Y1,X2,Y2,D1),dist(X1,Y1,X2,Y2,D2),D1>D2.
min_dist(X1,Y1,X2,Y2,D):-not nonMinDist(X1,Y1,X2,Y2,D),dist(X1,Y1,X2,Y2,D).
```

```
%IN ORDER TO WRITE COMPACT RULES, MOVES ARE CONVERTED AS FOLLOWS
%(EXAMPLE FOR "left" MOVE)
move(S,-1,0):-next(S,"left").

%DERIVING THE FIRST CELL REACHED AFTER MOVING ACCORDING TO THE STEP CHOICE
%WE DON'T WANT TO MOVE TOWARDS A WALL
nextCell(S,X+Dx,Y+Dy) :- pacman(S-1,X, Y), move(S,Dx,Dy), tile(X+Dx,Y+Dy).
moveOk(S) :- nextCell(S,X,Y).
:- not moveOk(S), planStep(S).

%DERIVING ALL THE CELL THAT WILL BE REACHED IN EACH STEP
%EACH STEP OF THE PLAN IS MEANT TO REACH THE NEAREST INTERSECTION
reach(S,0,X,Y):- nextCell(S,X,Y).
reach(S,N+1,X+Dx,Y+Dy) :- reach(S,N,X,Y), not inters(X,Y), move(S,Dx,Dy),
    tile(X+Dx,Y+Dy), maxN(MN), N<=MN.

%THE PAC-MAN SHOULD NOT CROSS PATH WITH A GHOST IF NOT IN POWER-UP MODE
:~ not powerup(S), ghost(S-1,X,Y,_), reach(S,X,Y), planStep(S). [1@5, S,X,Y]

%THE PAC-MAN SHOULD EAT ENERGIZERS IF NOT IN POWER-UP MODE
:~ energizer(S-1,X,Y), not reach(S,X,Y), not powerup(S). [1@4, S,X,Y]

%THE PAC-MAN SHOULD STAY NEAR GHOSTS IF IN POWER-UP MODE
:~ distPacmanNextGhost(S,MD,_), powerup(S). [MD@3, MD,S]

%THE PAC-MAN SHOULD EAT AS MANY PELLETS AS POSSIBLE
:~ pellet(S-1,X,Y), not reach(S,X,Y),  planStep(S). [1@2, S,X,Y]
%THE PAC-MAN SHOULD AVOID LOOPING RIGHT-LEFT OR UP-DOWN
:~ next(S-1,X), next(S,Y), opposite(X,Y). [1@1, X,Y,S]
%PLAN GENERATION
applyAction(S,"MovePacman"):-planStep(S).
actionArgument(S,"Direction", X):-next(S,X),planStep(S).
actionArgument(S,"X", X):-pacman(S-1,X,_),planStep(S).
actionArgument(S,"Y", Y):-pacman(S-1,_,Y),planStep(S).
```

The rules computing the distances between cells exploit most of the potential of incremental grounding: these ground rules are instantiated only in the first shot and remain available until the last one. Weak constraints attribute a cost to each possible decision and drive choices towards directions with minimal cost like, e.g., staying away from ghosts.

5 Integration and Experiments

ThinkEngine can easily embed different reasoners. We thus accommodated *Incremental-DLV2* in the existing architecture by properly dealing with the fact that it is an online solver keeping an internal state updated across the shots.

To assess the impact of the new integrated version of *ThinkEngine*, we compared the run time performance of some significant games selected from our showcase[1], using both *DLV2* and *Incremental-DLV2*. In each game, namely Pac-Man, Space Invaders and Frogger, the main player is controlled by one or more concurrent brains. In particular, Frogger is equipped with a reactive brain while the other two games are provided with three competing planner brains each.

[1] https://github.com/DeMaCS-UNICAL/ThinkEngine-Showcase.

Table 1. Characteristics of brains.

Game	Brain	Main operation	Avg #Sensors	#Actions	#Const
Pac-Man	Planner1	Path finding	3505	2	~1K
	Planner2	Path finding	4061		
	Planner3	Taxicab distance	4104	1	
Space Invaders	Planner1	Move away from missile	8	10	<1K
	Planner2	Motion planning	120		~1G
	Planner3	Motion planning	58		
Frogger	Reactive	Obstacle avoidance	156	1	<1K

Table 2. Statistics of incremental and standard ASP reasoners.

Brain	#Iter.	Reasoner	Avg time (s)	Max RAM (MB)
Pac-Man Planner1	49	*DLV2*	0.352	16.22
		Incr-DLV2	0.107	112.43
Pac-Man Planner2	56	*DLV2*	0.633	38.91
		Incr-DLV2	0.296	320.02
Pac-Man Planner3	73	*DLV2*	0.202	14.29
		Incr-DLV2	0.032	42.45
Frogger	86	*DLV2*	0.140	4.98
		Incr-DLV2	0.019	58.07
Space Invaders Planner1	42	*DLV2*	0.139	4.20
		Incr-DLV2	0.001	5.82
Space Invaders Planner2	76	*DLV2*	0.960	18.43
	3	*Incr-DLV2*	11.291	2791.23
Space Invaders Planner3	50	*DLV2*	0.897	30.22
	3	*Incr-DLV2*	13.305	1521.12

All brains are associated to a proper ASP specification of varying features in terms of the type of operation performed, number of sensor readings and length of generated plans (this latter applies to planner brains only). Moreover, it is important to note that the range of sensor readings has an impact on the number of constant values introduced in the grounding step (last column of Table 1).

These features are summarized in Table 1. The tested brains include both simple or more complex AIs which are triggered depending on the game situation. For instance, an AI in charge of managing an emergency situation has to be really fast: this means either *i)* a brain should use simple operations (like basing decisions on the taxicab distance instead of looking for complex paths) or *ii)* it should have a really low number of sensor readings, or *iii)* it should produce a very short plan, or a combination thereof.

Table 2 shows details about the performance of both *Incremental-DLV2* and *DLV2* when fed in input with the respective brain specifications on a series of input facts taken from real game executions. Tests were performed on a desktop machine equipped with an Intel CPU i7-11800H with 32 GB of RAM. It can be seen that for Pac-Man, Frogger and one instance of Space Invaders, the decision time is significantly reduced in the incremental execution with respect to the standard one. This comes at the cost of a higher (but still reasonable) memory consumption. In terms of playability, of course, the lower the time required to synthesize a plan, the higher the possibility it to be executed. Indeed, if too much time is required in order to compute a plan then, very likely, the plan itself will be aborted due to the changes occurred to the world game. Also, faster computation times imply an higher number of decisions per second, thus extending the range of real-time games where *ThinkEngine* can be in principle adopted.

With regards to the other two brains for Space Invaders, instead, we could tests respectively only two and three shots of the incremental evaluation since the next shot required so much RAM (more than 15 GB) that the process was killed by the operating system. It must be considered that sensors readings in the Space Invaders domain have a wide range, since the game grid has been modelled to a very fine-grained level. This produces over a billion of different constant symbols, and an accordingly large (over)grounded program. The results suggests that incremental techniques pay off in domains that imply small to average ground program size, while they require further optimization to compete in settings where sensor reading values are denser. Nonetheless, one should observe that it is a good practice to use declarative methods on qualitative information, thus abstracting away fine-grained values. This calls for introducing better abstraction methodologies when wiring declarative brains on environments where usually information is expressed in floating point precision. ASP specifications, inputs and detailed results of our experiments can be found on GitHub[2].

Acknowledgements. This work has been partially supported by the Italian MIUR Ministry and the Presidency of the Council of Ministers under the project "Declarative Reasoning over Streams" under the "PRIN" 2017 call (Project 2017M9C25L_001) and under Italian Ministry of Economic Development (MISE) under the PON project "MAP4ID - Multipurpose Analytics Platform 4 Industrial Data", N. F/190138/01-03/X44.

References

1. Thinkengine on github. https://github.com/DeMaCS-UNICAL/ThinkEngine
2. Unity 3d game engine. https://unity3d.com/unity
3. Unity, order of execution for event functions. https://docs.unity3d.com/Manual/ExecutionOrder.html
4. Black & White (2001). https://www.ea.com/games/black-and-white
5. Halo (2001). https://www.xbox.com/en-US/games/halo
6. Taxicab norm distance. In: Sammut, C., Webb, G.I. (eds.) Encyclopedia of Machine Learning and Data Mining, p. 1232. Springer, Heidelberg (2017). https://doi.org/10.1007/978-0-387-30164-8_812

[2] https://github.com/DeMaCS-UNICAL/ThinkEngine-PADL-Experiments.

7. van Aanholt, L., Bidarra, R.: Declarative procedural generation of architecture with semantic architectural profiles. In: CoG (2020)
8. Alviano, M., et al.: The ASP system DLV2. In: Balduccini, M., Janhunen, T. (eds.) LPNMR 2017. LNCS (LNAI), vol. 10377, pp. 215–221. Springer, Cham (2017). https://doi.org/10.1007/978-3-319-61660-5_19
9. Angilica, D., Ianni, G., Pacenza, F.: Declarative AI design in unity using answer set programming. In: CoG, pp. 417–424. IEEE (2022)
10. Bartheye, O., Jacopin, E.: A real-time pddl-based planning component for video games. In: AIIDE. The AAAI Press (2009)
11. Beck, H., Eiter, T., Folie, C.: Ticker: a system for incremental ASP-based stream reasoning. Theory Pract. Log. Program. **17**(5–6), 744–763 (2017)
12. Calimeri, F., et al.: ASP-Core-2 input language format. Theory Pract. Log. Program. **20**(2), 294–309 (2020)
13. Calimeri, F., et al.: Angry-hex: an artificial player for angry birds based on declarative knowledge bases. IEEE Trans. Comput. Intell. AI Games **8**(2), 128–139 (2016)
14. Calimeri, F., Ianni, G., Pacenza, F., Perri, S., Zangari, J.: ASP-based multi-shot reasoning via DLV2 with incremental grounding. In: PPDP, pp. 2:1–2:9. ACM (2022)
15. Calimeri, F., Manna, M., Mastria, E., Morelli, M.C., Perri, S., Zangari, J.: I-dlv-sr: a stream reasoning system based on I-DLV. Theory Pract. Log. Program. **21**(5), 610–628 (2021)
16. Calimeri, F., Perri, S., Zangari, J.: Optimizing answer set computation via heuristic-based decomposition. Theory Pract. Log. Program. **19**(4), 603–628 (2019)
17. Dodaro, C., Eiter, T., Ogris, P., Schekotihin, K.: Managing caching strategies for stream reasoning with reinforcement learning. Theory Pract. Log. Program. **20**(5), 625–640 (2020)
18. Erdem, E., Gelfond, M., Leone, N.: Applications of answer set programming. AI Mag. **37**(3), 53–68 (2016)
19. Gebser, M., Kaminski, R., Kaufmann, B., Schaub, T.: Multi-shot ASP solving with clingo. Theory Pract. Log. Program. **19**(1), 27–82 (2019)
20. Genesereth, M.R., Love, N., Pell, B.: General game playing: overview of the AAAI competition. AI Mag. **26**(2), 62–72 (2005)
21. Ianni, G., Pacenza, F., Zangari, J.: Incremental maintenance of overgrounded logic programs with tailored simplifications. Theory Pract. Log. Program. **20**(5), 719–734 (2020)
22. Liebana, D.P., et al.: General video game AI: competition, challenges and opportunities. In: AAAI (2016)
23. Nilsson, N.: STRIPS planning systems. In: Artificial Intelligence: A New Synthesis, pp. 373–400 (1998)
24. Orkin, J.: Three states and a plan: the AI of fear. In: Game developers conference. vol. 2006, p. 4. CMP Game Group SanJose, California (2006)
25. Pfau, J., Smeddinck, J.D., Malaka, R.: The case for usable AI: what industry professionals make of academic AI in video games. In: CHI PLAY (Companion), pp. 330–334. ACM (2020)
26. Renz, J., Ge, X., Gould, S., Zhang, P.: The angry birds AI competition. AI Mag. **36**(2), 85–87 (2015)
27. Robertson, J., Young, R.M.: The general mediation engine. In: Experimental AI in Games: Papers from the 2014 AIIDE Workshop. AAAI Technical Report WS-14-16, vol. 10, no. 3, pp. 65–66 (2014)
28. Robertson, J., Young, R.M.: Automated gameplay generation from declarative world representations. In: AIIDE, pp. 72–78. AAAI Press (2015)

29. Schaul, T.: A video game description language for model-based or interactive learning. In: CIG, pp. 1–8. IEEE (2013)
30. Smith, A.M., Mateas, M.: Answer set programming for procedural content generation: a design space approach. IEEE Trans. Comput. Intell. AI Games **3**(3), 187–200 (2011)
31. Smith, A.M., Nelson, M.J., Mateas, M.: LUDOCORE: a logical game engine for modeling videogames. In: CIG, pp. 91–98. IEEE (2010)
32. Stanescu, M., Certický, M.: Predicting opponent's production in real-time strategy games with answer set programming. IEEE Trans. Comput. Intell. AI Games **8**(1), 89–94 (2016)
33. Thielscher, M.: Answer set programming for single-player games in general game playing. In: Hill, P.M., Warren, D.S. (eds.) ICLP 2009. LNCS, vol. 5649, pp. 327–341. Springer, Heidelberg (2009). https://doi.org/10.1007/978-3-642-02846-5_28

Dynamic Slicing of Reaction Systems Based on Assertions and Monitors

Linda Brodo[1], Roberto Bruni[2], and Moreno Falaschi[3](✉)

[1] Dipartimento di Scienze economiche e aziendali, Università di Sassari, Sassari, Italy
brodo@uniss.it

[2] Dipartimento di Informatica, Università di Pisa, Pisa, Italy
bruni@di.unipi.it

[3] Dipartimento di Ingegneria dell'Informazione e Scienze Matematiche, Università di Siena, Siena, Italy
moreno.falaschi@unisi.it

Abstract. Reaction Systems (RSs) are a successful computational framework inspired by biological systems. RSs can involve a large number of reactions and entities, which makes it difficult the debugging of quite long computations that traverses complex states. Slicing is a technique which is useful for simplifying a debugging process, by selecting a portion of the program containing the faulty code. We define the first dynamic slicer for RSs and show that it can help the user to detect the origins of bugs and also to inspect more closely the behavior of the model, highlighting the parts that need more investigation. Our slicer allows to select part of a (faulty) state and then eliminates from the computation the information which is not relevant for the selected items. We show how to use the slicer on some biological models in RSs. Our dynamic backward slicer is based on an SOS semantics of RSs. In order to automatize the slicing process we describe how to use monitors for identifying the states which violate a safety specification and hence are suitable to start the slicing. We have integrated our slicer in BioResolve, a prototype implementation in Prolog which provides also other features such as computations represented as colored graphs, and verification of properties.

Keywords: Reaction systems · SOS semantics · Program slicing · Assertions · Monitors · Natural computation

1 Introduction

Reaction Systems (RSs) [13] are a computational framework inspired by systems of living cells. Their constituents are a finite set of entities and a finite set of reactions over entities. RSs have shown to be a general computational model whose application ranges from the modeling of biological phenomena [7,9,11,17] to molecular chemistry [25].

Research supported by Università degli Studi di Sassari *Fondi di Ateneo per la ricerca 2020*, by MIUR Project *Programma di Sviluppo - Dipartimenti di Eccellenza 2018–2022*, by MIUR PRIN Project 201784YSZ5 *ASPRA–Analysis of Program Analyses*, and by the INdAM - GNCS Project CUP_E55F2200027001 *Proprietà qualitative e quantitative di sistemi reversibili.*

M. Hanus and D. Inclezan (Eds.): PADL 2023, LNCS 13880, pp. 107–124, 2023.
https://doi.org/10.1007/978-3-031-24841-2_8

The classical semantics of RSs is defined as a reduction system whose states are sets of entities (coming from an external context or produced at the previous step).

Several tools are available to simulate RSs or verify that certain properties are met. However, writing reactions is an error-prone activity and inspecting their execution can be difficult even for medium-sized RSs. If some mistake is done at the specification level and some inexplicable result is observed during the simulation, then a manual inspection of the computation is often necessary in order to understand the nature of the problem. In fact, we are not aware of any debugging systems for RSs.

The aim of this paper is to propose an automatic technique to ease the debugging of RSs. The idea is to trigger the slicing of the computation when certain events are detected, so to highlight and focus the attention on the entities and reactions that are directly responsible for the unexpected event. Slicing was introduced in pioneering works by Mark Weiser [27] as a static technique. Then it was extended by introducing the so called dynamic program slicing [21], which supports the debugging process by selecting a portion of the program containing the faulty code. Dynamic program slicing has been applied to several programming paradigms (see [26] for a survey).

In order to trigger the slicing we adapt the monitoring framework in [2] to the setting of RSs. To this aim we build on a Labelled Transition System (LTS) semantics of RSs and introduce a flexible modal logic of assertions over transition labels that can be used to express safety conditions. Formulas are then used to synthesize suitable monitors for the LTS semantics of RSs: they inspect the labels as the computation progresses and when a violation is detected they trigger the slicing of the trace from the faulty state.

The slicing framework follows three main steps as in [18] (there in the quite different setting of Concurrent Constraint Programming, which is a monotonic language). First the dynamics of RSs is extended to an enriched semantics that considers states with their computation history. Second, we take a partial computation which is considered faulty by the user, or which is determined automatically as faulty by the monitor. The fault is identified by marking a set of entities in the last state reached. The third step is an automatic algorithm that removes from the history the information not relevant to derive the entities selected in the second step.

We consider here two possible slicing algorithms: the first situation is when we have a computation independent from the context, the second is when the computation is context dependent, which means a computation driven by entities provided by the environment. In both cases, we show that the sliced computation is a simplification of the original one, which highlights the entities, contexts and reactions that are essential to produce the marked entities. We have developed a prototype implementation in Prolog, freely available online. Here the use of a declarative approach is useful to guarantee the correctness of the implementation, which closely mirrors the theoretical definitions.

Organization. In Sect. 2 we summarize the basics of RSs. In Sect. 3 we recall a more convenient process algebra for RSs [14,15]. In Sect. 4 we define the slicing technique for RSs. In Sect. 5 we show a biological example to illustrate our framework and implemented tools. In Sect. 6 we present a specialized assertion language and we introduce monitors. Monitors can be derived from safety formulas in modal logic to mark automatically the entities and start the slicing. We discuss some related work in Sect. 7 and future work in Sect. 8, together with concluding remarks.

2 Reaction Systems

The theory of Reaction Systems (RSs) [13] was born in the field of Natural Computing to model the qualitative behavior of biochemical reactions in living cells. We recall here the main concepts as introduced in the classical set theoretic version. In the following, we use the term *entities* to denote generic molecular substances (e.g., atoms, ions, molecules) that may be present in the states of a biochemical system.

Let S be a (finite) set of entities. A reaction in S is a triple $a = (R, I, P)$, where $R, I, P \subseteq S$ are finite, non empty sets and $R \cap I = \emptyset$. The sets R, I, P are the sets of *reactants*, *inhibitors*, and *products*, respectively. All reactants have to be present in the current state for the reaction to take place. The presence of any of the inhibitors blocks the reaction. Products are the outcome of the reaction, to be released in the next state. We denote with $rac(S)$ the set of all reactions over S. Given $W \subseteq S$, the result of $a = (R, I, P) \in rac(S)$ on W, denoted by $res_a(W)$, is defined as follows:

$$res_a(W) \triangleq \begin{cases} P & \text{if } en_a(W) \\ \emptyset & \text{otherwise} \end{cases} \qquad en_a(W) \triangleq R \subseteq W \wedge I \cap W = \emptyset$$

where $en_a(W)$ is called the *enabling predicate*.

A Reaction System is a pair $\mathcal{A} = (S, A)$ where S is the set of entities, and $A \subseteq rac(S)$ is a finite set of reactions over S. Given $W \subseteq S$, the result of the reactions A on W, denoted $res_A(W)$, is just the lifting of res_a, i.e., $res_A(W) \triangleq \cup_{a \in A} res_a(W)$.

Since living cells are seen as open systems that react to environmental stimuli, the behavior of a RS is formalized in terms of an *interactive process*. Let $\mathcal{A} = (S, A)$ be a RS and let $n \geq 0$. An n-steps *interactive process* in $\mathcal{A}$ is a pair $\pi = (\gamma, \delta)$ s.t. $\gamma = \{C_i\}_{i \in [0,n]}$ is the *context sequence* and $\delta = \{D_i\}_{i \in [0,n]}$ is the *result sequence*, where $C_i, D_i \subseteq S$ for any $i \in [0, n]$, $D_0 = \emptyset$, and $D_{i+1} = res_A(D_i \cup C_i)$ for any $i \in [0, n-1]$. The context sequence γ represents the environment, while the result sequence δ is entirely determined by γ and A. We call $\tau = W_0, \ldots, W_n$ the *state sequence*, with $W_i \triangleq C_i \cup D_i$ for any $i \in [0, n]$. Note that each state W_i in τ is the union of the context C_i at step i and the result set $D_i = res_A(W_{i-1})$ from step $i - 1$. Note also that the result of a computation step does not depend on the order of application of the reactions.

Example 1. We consider a toy RS defined as $\mathcal{A} \triangleq (S, A)$ where $S \triangleq \{\mathsf{a}, \mathsf{b}, \mathsf{c}\}$, and the set of reactions $A \triangleq \{a_1\}$ only contains the reaction $a_1 \triangleq (\{\mathsf{a}, \mathsf{b}\}, \{\mathsf{c}\}, \{\mathsf{b}\})$, more concisely written as $(\mathsf{ab}, \mathsf{c}, \mathsf{b})$. Then, we consider a 4−*steps* interactive process $\pi \triangleq (\gamma, \delta)$, where $\gamma \triangleq \{C_0, C_1, C_2, C_3\}$, with $C_0 \triangleq \{\mathsf{a}, \mathsf{b}\}$, $C_1 \triangleq \{\mathsf{a}\}$, $C_2 \triangleq \{\mathsf{c}\}$, and $C_3 \triangleq \{\mathsf{c}\}$; and $\delta \triangleq \{D_0, D_1, D_2, D_3\}$, with $D_0 \triangleq \emptyset$, $D_1 \triangleq \{\mathsf{b}\}$, $D_2 \triangleq \{\mathsf{b}\}$, and $D_3 \triangleq \emptyset$. Then, the resulting state sequence is $\tau = W_0, W_1, W_2, W_3 = \{\mathsf{a}, \mathsf{b}\}, \{\mathsf{a}, \mathsf{b}\}, \{\mathsf{b}, \mathsf{c}\}, \{\mathsf{c}\}$. In fact, it is easy to check that, e.g., $W_0 = C_0$, $D_1 = res_A(W_0) = res_A(\{\mathsf{a}, \mathsf{b}\}) = \{\mathsf{b}\}$ because $en_a(W_0)$, and $W_1 = C_1 \cup D_1 = \{\mathsf{a}\} \cup \{\mathsf{b}\} = \{\mathsf{a}, \mathsf{b}\}$.

3 SOS Rules for Reaction Systems

In order to define our automatic slicing technique, we find it convenient to exploit the algebraic syntax for RSs introduced in [15]. Inspired by process algebras such as CCS [22], simple SOS inference rules define the behavior of each operator. This induces a LTS semantics for RSs, where states are terms of the algebra, each transition corresponds to a step of the RS and transition labels retain some information on the entities needed to perform each step. Transition labels and SOS rules will allow us to pair RSs with monitors and to easily enrich state information with histories, respectively.

Definition 1 (RS processes). *Let S be a set of entities. A* RS process P *is any term defined by the following grammar:*

$$\mathsf{P} := [\mathsf{M}] \qquad \mathsf{M} := (R, I, P) \mid D \mid \mathsf{K} \mid \mathsf{M}|\mathsf{M} \qquad \mathsf{K} ::= \mathbf{0} \mid X \mid C.\mathsf{K} \mid \mathsf{K}+\mathsf{K} \mid \mathsf{rec}\, X.\, \mathsf{K}$$

where $R, I, P \subseteq S$ are non empty sets of entities, $C, D \subseteq S$ are possibly empty set of entities, and X is a process variable.

While in principle reactions, entities and context processes could be seen as components to be handled separately (and in the implementation is more convenient to do so), we prefer to mix them together to provide a compositional account of their interactions. A RS process P embeds a *mixture* process M obtained as the parallel composition of some reactions (R, I, P), some set of current entities D (possibly the empty set), and some *context* K. We write $\prod_{i\in I} \mathsf{M}_i$ for the parallel composition of all M_i with $i \in I$.

A context process K is a possibly non-deterministic and recursive system: the nil context $\mathbf{0}$ stops the computation; the prefixed context $C.\mathsf{K}$ makes the entities C available to the reactions, and then leaves K be the context offered at the next step; the non-deterministic choice $\mathsf{K}_1 + \mathsf{K}_2$ allows the context to behave as either K_1 or K_2; X is a process variable, and $\mathsf{rec}\ X.\ \mathsf{K}$ is the usual recursive operator of process algebras. Choice and recursion can combine in-breadth (different evolutions) and in-depth (evolutions of different length, possibly infinite) analysis. The ability to compose contexts in parallel is useful, e.g., to handle different entities with different strategies or to reduce combinatorial explosion at the specification level. For example, the context $\mathsf{rec}\ X.\ \mathsf{a}.X + \emptyset.X | \mathsf{rec}\ X.\ \mathsf{b}.X + \emptyset.X$ can recursively offer any combination of a and b.

We say that P and P' are structurally equivalent, written $\mathsf{P} \equiv \mathsf{P}'$, when they denote the same term up to the laws of commutative monoids (unit, associativity and commutativity) for parallel composition $\cdot|\cdot$, with $\emptyset$ as the unit, and the laws of idempotent and commutative monoids for choice $\cdot + \cdot$, with $\mathbf{0}$ as the unit. We also assume $D_1|D_2 \equiv D_1 \cup D_2$ for any $D_1, D_2 \subseteq S$.

Definition 2 (RSs as RS processes). *Let $\mathcal{A} = (S, A)$ be a RS, and $\pi = (\gamma, \delta)$ an n-step interactive process in $\mathcal{A}$, with $\gamma = \{C_i\}_{i\in[0,n]}$ and $\delta = \{D_i\}_{i\in[0,n]}$. For any step $i \in [0, n]$, the corresponding RS process $[\![\mathcal{A}, \pi]\!]_i$ is defined as follows:*

$$[\![\mathcal{A}, \pi]\!]_i \triangleq \left[\prod_{a\in A} a \mid D_i \mid \mathsf{K}_{\gamma^i} \right]$$

$$\frac{}{D \xrightarrow{\langle(D,\emptyset)\rhd\emptyset,\emptyset,\emptyset\rangle} \emptyset}\ (\mathit{Ent}) \qquad \frac{}{C.\mathsf{K} \xrightarrow{\langle(\emptyset,C)\rhd\emptyset,\emptyset,\emptyset\rangle} \mathsf{K}}\ (\mathit{Cxt})$$

$$\frac{\mathsf{K}_1 \xrightarrow{\ell} \mathsf{K}_1'}{\mathsf{K}_1 + \mathsf{K}_2 \xrightarrow{\ell} \mathsf{K}_1'}\ (\mathit{Suml}) \qquad \frac{\mathsf{K}_2 \xrightarrow{\ell} \mathsf{K}_2'}{\mathsf{K}_1 + \mathsf{K}_2 \xrightarrow{\ell} \mathsf{K}_2'}\ (\mathit{Sumr}) \qquad \frac{\mathsf{K}[^{\mathsf{rec}\ X.\mathsf{K}}/_{X}] \xrightarrow{\ell} \mathsf{K}'}{\mathsf{rec}\ X.\ \mathsf{K} \xrightarrow{\ell} \mathsf{K}'}\ (\mathit{Rec})$$

$$\frac{}{(R,I,P) \xrightarrow{\langle(\emptyset,\emptyset)\rhd R,I,P\rangle} (R,I,P) \mid P}\ (\mathit{Pro}) \qquad \frac{J \subseteq I \quad Q \subseteq R \quad J \cup Q \neq \emptyset}{(R,I,P) \xrightarrow{\langle(\emptyset,\emptyset)\rhd J,Q,\emptyset\rangle} (R,I,P)}\ (\mathit{Inh})$$

$$\frac{\mathsf{M}_1 \xrightarrow{\ell_1} \mathsf{M}_1' \quad \mathsf{M}_2 \xrightarrow{\ell_2} \mathsf{M}_2' \quad \ell_1 \frown \ell_2}{\mathsf{M}_1 \mid \mathsf{M}_2 \xrightarrow{\ell_1 \cup \ell_2} \mathsf{M}_1' \mid \mathsf{M}_2'}\ (\mathit{Par}) \qquad \frac{\mathsf{M} \xrightarrow{\langle(D,C)\rhd R,I,P\rangle} \mathsf{M}' \quad R \subseteq D \cup C}{[\mathsf{M}] \xrightarrow{\langle(D,C)\rhd R,I,P\rangle} [\mathsf{M}']}\ (\mathit{Sys})$$

Fig. 1. SOS semantics of the RS processes.

where the context $\mathsf{K}_{\gamma^i} \triangleq C_i.C_{i+1}.\cdots.C_n.\mathbf{0}$ *is the serialization of the entities offered by* γ^i *(the shifting of* γ *at the i-th step). We write* $[\![\mathcal{A},\pi]\!]$ *as a shorthand for* $[\![\mathcal{A},\pi]\!]_0$.

Example 2. The encoding of the RS $\mathcal{A} = (S, A)$, in Example 1, is as follows:

$$\mathsf{P} \triangleq [\![\mathcal{A},\pi]\!] = [\![(\{\mathsf{a},\mathsf{b},\mathsf{c}\},\{(\mathsf{ab},\mathsf{c},\mathsf{b})\}),\pi]\!] = [(\mathsf{ab},\mathsf{c},\mathsf{b}) \mid \emptyset \mid \mathsf{K}_\gamma] \equiv [(\mathsf{ab},\mathsf{c},\mathsf{b}) \mid \mathsf{K}_\gamma]$$

where $\mathsf{K}_\gamma = \{\mathsf{a},\mathsf{b}\}.\{\mathsf{a}\}.\{\mathsf{c}\}.\{\mathsf{c}\}.\mathbf{0}$, written more concisely as $\mathsf{ab.a.c.c}.\mathbf{0}$. Note that $D_0 = \emptyset$ is inessential and can be discarded thanks to structural congruence.

A transition label ℓ is a tuple $\langle(D,C) \rhd R,I,P\rangle$ with $D,C,R,I,P \subseteq S$. The sets D,C record the entities currently in the system; the set R records entities whose presence is assumed (either acting as reactants or as inhibitors); the set I records entities whose absence is assumed (either acting as inhibitors or as missing reactants); the set P records the products of enabled reactions.

The operational semantics of RS processes is defined by the SOS rules in Fig. 1. The process $\mathbf{0}$ has no transition. The rule (*Ent*) makes available the entities in the (possibly empty) set D, then reduces to $\emptyset$. As a special instance of (*Ent*), $\emptyset \xrightarrow{\langle(\emptyset,\emptyset)\rhd\emptyset,\emptyset,\emptyset\rangle} \emptyset$. The rule (*Cxt*) says that a prefixed context process $C.\mathsf{K}$ makes available the entities in the set C and then reduces to K. The rule (*Rec*) is the classical rule for recursion. The rules (*Suml*) and (*Sumr*) select a move of either the left or the right component, resp., discarding the other process. The rule (*Pro*) executes the reaction (R,I,P) (its reactants, inhibitors, and products are recorded in the label), which remains available at the next step together with P. The rule (*Inh*) applies when the reaction (R,I,P) should not be executed; it records in the label the possible causes for which the reaction is disabled: possibly some inhibiting entities ($J \subseteq I$) are present or some reactants ($Q \subseteq R$) are missing, with $J \cup Q \neq \emptyset$, as at least one cause is needed. The rule (*Par*) puts two processes in parallel by pooling their labels and joining all the set components of the labels. The sanity check $\ell_1 \frown \ell_2$ is required to guarantee that reactants and inhibitors are consistent (see definition below, where we let $W_i \triangleq D_i \cup C_i$):

$$\langle(D_1,C_1) \rhd R_1,I_1,P_1\rangle \frown \langle(D_2,C_2) \rhd R_2,I_2,P_2\rangle \triangleq (W_1 \cup W_2 \cup R_1 \cup R_2) \cap (I_1 \cup I_2) = \emptyset$$

In the conclusion of rule (*Par*) we write $\ell_1 \cup \ell_2$ for the component-wise union of labels (see definition below, where the notation $X_{1,2} \triangleq X_1 \cup X_2$ is used):

$$\langle (D_1, C_1) \rhd R_1, I_1, P_1 \rangle \cup \langle (D_2, C_2) \rhd R_2, I_2, P_2 \rangle \triangleq \langle (D_{1,2}, C_{1,2}) \rhd R_{1,2}, I_{1,2}, P_{1,2} \rangle$$

Finally, the rule (*Sys*) requires that all the processes of the systems have been considered, and also checks that all the needed reactants are actually available in the system ($R \subseteq D \cup C$). In fact this constraint can only be met on top of all processes. The check that inhibitors are absent ($I \cap (D \cup C) = \emptyset$) is embedded in rule (*Par*).

Example 3. Let us consider the RS process $\mathsf{P}_0 \triangleq [(\mathsf{ab}, \mathsf{c}, \mathsf{b}) \mid \mathsf{ab.a.c.c.}\mathbf{0}]$ from Example 2. The process P_0, and its next state P_1 have a unique outgoing transition:

$$[(\mathsf{ab}, \mathsf{c}, \mathsf{b}) \mid \mathsf{ab.a.c.c.}\mathbf{0}] \xrightarrow{\langle (\emptyset, \mathsf{ab}) \rhd \mathsf{ab}, \mathsf{c}, \mathsf{b} \rangle} [(\mathsf{ab}, \mathsf{c}, \mathsf{b}) \mid \mathsf{b} \mid \mathsf{a.c.c.}\mathbf{0}] \xrightarrow{\langle \mathsf{b}, \mathsf{a} \rhd \mathsf{ab}, \mathsf{c}, \mathsf{b} \rangle} [(\mathsf{ab}, \mathsf{c}, \mathsf{b}) \mid \mathsf{b} \mid \mathsf{c.c.}\mathbf{0}]$$

The process $\mathsf{P}_2 = [(\mathsf{ab}, \mathsf{c}, \mathsf{b}) \mid \mathsf{b} \mid \mathsf{c.c.}\mathbf{0}]$ has three outgoing transitions, sharing the same target process $\mathsf{P}_3 \triangleq [(\mathsf{ab}, \mathsf{c}, \mathsf{b}) \mid \mathsf{c.}\mathbf{0}]$, each providing a different justification why reaction $(\mathsf{ab}, \mathsf{c}, \mathsf{b})$ is not enabled:

1. $\mathsf{P}_2 \xrightarrow{\langle (\mathsf{b}, \mathsf{c}) \rhd \mathsf{c}, \emptyset, \emptyset \rangle} \mathsf{P}_3$ where the presence of c inhibited the reaction;
2. $\mathsf{P}_2 \xrightarrow{\langle (\mathsf{b}, \mathsf{c}) \rhd \emptyset, \mathsf{a}, \emptyset \rangle} \mathsf{P}_3$ where the absence of a inhibited the reaction.
3. $\mathsf{P}_2 \xrightarrow{\langle (\mathsf{b}, \mathsf{c}) \rhd \mathsf{c}, \mathsf{a}, \emptyset \rangle} \mathsf{P}_3$ where both the presence of c and the absence of a inhibited the reaction; this label is thus more informative than the previous two.

In the following we assume transitions $\mathsf{P} \xrightarrow{\langle (D,C) \rhd R, I, P \rangle} \mathsf{P}'$ guarantee that any instance of the rule (*Inh*) is applied in a way that maximizes the sets J and Q (see [15]). The following theorem from [15] shows how the set-theoretic dynamics of a RS matches the SOS semantics of its RS process.

Theorem 1. *Let $\mathcal{A} = (S, A)$ be a RS, and $\pi = (\gamma, \delta)$ an n-step interactive process in $\mathcal{A}$ with $\gamma = \{C_i\}_{i \in [0,n]}$, $\delta = \{D_i\}_{i \in [0,n]}$, and let $\mathsf{P}_i \triangleq [\![\mathcal{A}, \pi]\!]_i$ for any $i \in [0, n]$. Then:*

1. *$\forall i \in [0, n-1]$, if $\mathsf{P}_i \xrightarrow{\langle (D,C) \rhd R, I, P \rangle} \mathsf{P}$ then $D = D_i$, $C = C_i$, $P = D_{i+1}$ and $\mathsf{P} \equiv \mathsf{P}_{i+1}$;*
2. *$\forall i \in [0, n-1]$, there exists $R, I \subseteq S$ such that $\mathsf{P}_i \xrightarrow{\langle (D_i, C_i) \rhd R, I, D_{i+1} \rangle} \mathsf{P}_{i+1}$.*

3.1 SOS Rules for the Slicing Computation

The slicing computation, presented in the next section, needs a slightly different state configuration that includes the whole past state sequence $W_i = C_i \cup D_i$. To do that, we enrich the state configuration $[\mathsf{M}]$ by prefixing it with the list of the previous result and context sets, written $(\mathsf{D}, \mathsf{C})[\mathsf{M}]$, where (D, C) stands for the list $(D_0, C_0), \ldots, (D_n, C_n)$. The formal Definition 1 is thus updated to carry on the history of the computation.

Definition 3 (RS processes with history). *Let $\mathcal{A} = (S, A)$ be a RS, and $\pi = (\gamma, \delta)$ an n-step interactive process in $\mathcal{A}$, with $\gamma = \{C_i\}_{i\in[0,n]}$ and $\delta = \{D_i\}_{i\in[0,n]}$. For any step $i \in [0, n]$, the corresponding new process configuration $[\![\mathcal{A}, \pi]\!]_i$ is defined as follows:*

$$[\![\mathcal{A}, \pi]\!]_i \triangleq (\mathsf{D}, \mathsf{C})[\mathsf{M}]$$

where $(\mathsf{D}, \mathsf{C}) \equiv (D_0, C_0), \ldots, (D_{i-1}, C_{i-1})$, *and,* $[\mathsf{M}] = [\prod_{a\in A} a \mid D_i \mid \mathsf{K}_{\gamma^i}]$.

The next step consists in enriching the operational semantics to deal with the history. We only have to modify the (Sys) inference rule in Fig. 1 as follows:

$$\frac{\mathsf{M} \xrightarrow{\langle (D,C)\rhd R,I,P\rangle} \mathsf{M}' \quad R \subseteq D \cup C}{(\mathsf{D}, \mathsf{C})[\mathsf{M}] \xrightarrow{\langle (D,C)\rhd R,I,P\rangle} (\mathsf{D}, \mathsf{C})::(D, C)[\mathsf{M}']} \ (\mathit{HistSys})$$

where, given the history $(\mathsf{D}, \mathsf{C}) = (D_0, C_0), \ldots, (D_{i-1}, C_{i-1})$, we let the notation $(\mathsf{D}, \mathsf{C})::(D, C)$ stand for the history $(D_0, C_0), \ldots, (D_{i-1}, C_{i-1}), (D, C)$.

4 Slicing RS Computations

Dynamic slicing is a technique that helps the user to debug her program by simplifying a partial execution trace, thus depurating it from parts which are irrelevant to finding the bug. It can also help to highlight parts of the programs which have been wrongly ignored by the execution. Our slicing technique consists of three main steps.

Enriched Semantics (Step S1). The slicing process requires some extra information from the execution of the processes. More precisely, (1) at each operational step we need to highlight the reactions that have been applied; and (2) we need to determine the part of the context which adds to the previous state the entities which are necessary to produce the marked entities in the following state. For solving (1) and (2), in Sect. 3 we have introduced an enriched semantics that records computation sequences. We need to keep track of the state sequence of the computation for the slicing process, by keeping separated the produced entities D_i in a computation step from the context C_i.

Marking the State (Step S2). Let us suppose that the final configuration in a partial computation is (D_m, C_m). The user selects a subset $D_{sliced} \subseteq D_m$ that may explain the (wrong) behavior of the program. In Sect. 6 we describe an assertion language and monitors to automatize the selection.

Trace Slice (Step S3). Starting from the the pair (D_{sliced}, C_m) denoting the user's marking, we define a backward slicing step. Roughly, this step allows us to eliminate from the execution trace all the information not related to D_{sliced}. Starting from this sliced final state and proceeding backwards we can compute for each computation step the information which is relevant to produce the marked elements in the final state.

Marking Algorithms. In the following we assume that the reactions are numbered consecutively by positive integer numbers, and denote the j-th reaction in the RS by the notation r_j. Let us explain how the slicing Algorithm 1 works.

Input: - a trace $(D_0, C_0) \xrightarrow{N_1} \cdots \xrightarrow{N_m} (D_m, C_m)$
- a marking $D_{sliced} \subseteq D_m$

Output: a sliced trace $(D'_0, C'_0) \xrightarrow{N'_1} \cdots \xrightarrow{N'_m} (D_{sliced}, C_m)$

begin
  let $D'_m = D_{sliced}$
  for $i = m$ *to* 1 **do**
    let $D'_{i-1} = \emptyset \wedge C'_{i-1} = \emptyset \wedge N'_i = \emptyset$
    for $j \in N_i$ *where* $r_j = (R_j, I_j, P_j)$, *such that* $(D'_i \cap P_j \neq \emptyset)$ **do**
      let $N'_i = N'_i \cup \{j\}$
      let $D'_{i-1} = D'_{i-1} \cup R_j$
      if $\neg en_{r_j}(D_{i-1})$, **then** $C'_{i-1} = C'_{i-1} \cup (R_j \backslash D_{i-1})$
    end
  end
end

Algorithm 1: Trace Slicer for context dependent computations

Input: - a trace $D_0 \xrightarrow{N_1} \cdots \xrightarrow{N_m} D_m$
- a marking $D_{sliced} \subseteq D_m$

Output: a sliced trace $D'_0 \xrightarrow{N'_1} \cdots \xrightarrow{N'_m} D_{sliced}$

begin
  let $D'_m = D_{sliced}$
  for $i = m$ *to* 1 **do**
    let $D'_{i-1} = \emptyset \wedge N'_i = \emptyset$
    for $j \in N_i$ *where* $r_j = (R_j, I_j, P_j)$, *such that* $D'_i \cap P_j \neq \emptyset$ **do**
      let $N'_i = N'_i \cup \{j\}$
      let $D'_{i-1} = D'_{i-1} \cup R_j$
    end
  end
end

Algorithm 2: Trace Slicer for context independent computations

As a matter of notation, please notice that each history $(D_0, C_0)::\cdots::(D_m, C_m)$ computed in m steps by Definition 3, defines a *trace* $(D_0, C_0) \xrightarrow{N_1} \cdots \xrightarrow{N_m} (D_m, C_m)$ on which we perform the slicing computation, where N_i is the set of reactions applied in the $i - th$ computation step. Here each reaction is simply represented by its numeric position in the list of reactions, i.e., $N_i = \{j \mid en_{r_j}(D_{i-1} \cup C_{i-1})\}$ for any $i \in [1, m]$. Abusing the notation, in the following we write $r_j \in N$ whenever $j \in N$.

Our algorithm returns a sliced trace which contains only the (usually rather small) subset of the entities which are necessary for deriving the marked entities. Let us now describe it informally. Let us consider the more complex case of context dependent computations. First of all the user has to indicate the subset D_{sliced} of the entities in the last state of the computation D_m that she wants to mark. Then the backward slicing process can start. Now let us consider the iteration i of the slicer. Marking the relevant information in previous state (D_{i-1}, C_{i-1}) requires analyzing the rules which have been applied at step $i - 1$. So, if $r_j \in N_{i-1}$ and $r_j = (R_j, I_j, P_j)$ then we need to check if r_j produces at least one entity which is marked in the next state. If this is the case, then j is added to the set of marked rules. Then it is necessary to check if the context C_{i-1} was essential for applying rule r_j, or if all necessary entities were already included in D_{i-1}. Thus it is necessary to compute the entities in C_{i-1} which are missing in D_{i-1} in order for rule r_j to be enabled, and those entities are marked in (added to) context C'_{i-1}. The elements in R_j are added to the marked entities in D'_{i-1}. For the computations which are context independent the part of transformation which is related to the context can clearly be eliminated (see Algorithm 2).

The following proposition states what is kept or removed by the slicing algorithm at each step.

Proposition 1. *Let* $(D_0, C_0) \xrightarrow{N_1} \cdots \xrightarrow{N_m} (D_m, C_m)$ *be a context dependent computation and let* $(D'_0, C'_0) \xrightarrow{N'_1} \cdots \xrightarrow{N'_m} (D_{sliced}, C_m)$ *be the sliced trace corresponding to a given marking* D_{sliced}. *Then,*
1) $\forall i \in [1, m]$, $r_j = (R_j, I_j, P_j) \in N'_i$ *iff* $r_j \in N_i$ *and there exists* $e \in D'_i$ *s.t.* $e \in P_j$.
2) $\forall i \in [0, m-1]$, $e \in D'_i$ *iff there exists* $r_j = (R_j, I_j, P_j) \in N'_{i+1}$ *such that* $e \in R_j$.
3) $\forall i \in [0, m-1]$, $e \in C'_i$ *iff there exists* $r_k = (R_k, I_k, P_k) \in N'_{i+1}$ *such that* $e \in C_i \cap R_k \wedge e \notin D_i$.

Proof (Sketch). The proof is by induction on the number n of computation steps.
Base case $n = 1$)

Let us first prove property (1).

Let the marked subset of D_1 be the set $D'_1 \subseteq D_1$ as defined in line 2 of Algorithm 1. Then, by line 5, for a reaction $r_j = (R_j, I_j, P_j) \in N_0$, we have that:
$r_j \in N'_0$ (by line 6) iff $D'_1 \cap P_j = \emptyset$ (by line 5) iff there exists $e \in D'_1$ s.t. $e \in P_j$. Thus, property (1) holds.

Let us now prove property (2).

In Algorithm 1, D'_0 is initialised to the empty set in line 4. Thus, we have that an entity $e \in D'_0$ iff (by line 7) $e \in R_j$ and (by line 5) $r_j = (R_j, I_j, P_j) \in N_1$ and $D'_1 \cap P_j \neq \emptyset$ iff $e \in R_j$ and (by line 6) $r_j = (R_j, I_j, P_j) \in N'_1$ and $D'_1 \cap P_j \neq \emptyset$ iff (by Property (1)) $e \in R_j$ and $r_j = (R_j, I_j, P_j) \in N'_1$. Thus the property holds.

Let us now prove property (3).

By line 4, $C'_0 = \emptyset$. Then a new entity e can be added to C'_0 only by line 8 and only if the condition in line 5 holds. Hence, by line 5, $k \in N_0$, with $r_k = (R_k, I_k, P_k)$, hence (1) $e \in R_k \subseteq D_0 \cup C_0$. By line 8 it must hold $e \in R_k \backslash D_0 \subseteq (D_0 \cup C_0) \backslash D_0$ (as r_k is enabled) $\subseteq C_0$, thus: (2) $e \in C_0$. By line 8 $e \in R_k \backslash D_0$, and (3) $e \notin D_0$. By (1) and (2) we get $e \in C_0 \cap R_k$ and together with (3) we get that the property holds.
Inductive case $n > 1$)

We start by considering the step $(D_{n-1}, C_{n-1}) \xrightarrow{N_n} (D_n, C_n)$.

In the first iteration of the **for** statement in line 5 in Algorithm 1, thus with $i = n-1$, we can show that properties (1–3) hold, in a way completely similar to the proof of the base case. Thus we obtain that property (1) holds for $i = n$, and properties (2) and (3) hold for $i = n - 1$.

Now we consider the computation $(D_0, C_0) \xrightarrow{N_1} \cdots \xrightarrow{N_{n-1}} (D_{n-1}, C_{n-1})$ w.r.t. the marking D'_{n-1} computed in previous step (see line 7 of Algorithm 1). By line 5 the set R_j in line 7 is a subset of D_{n-1}, and hence D'_{n-1} is a subset of D_{n-1} and determines a marking D_{sliced} for the trace $(D_0, C_0) \xrightarrow{N_1} \cdots \xrightarrow{N_{n-1}} (D_{n-1}, C_{n-1})$, then the property follows by the inductive hypothesis, as the number of steps is $n - 1$. □

Proposition 1 addresses context independent computations when contexts are empty.

5 Implementation

In this section we show how to check a biological model by our slicing methodology. The implementation is available on-line[1], with a small manual to use it. It extends the tool BioReSolve[2], which already provided a friendly environment for simulation, analysis and verification of RSs. The tool has been developed and tested under SWI-Prolog and exploits a few library predicates for efficiency reasons. DCG Grammar rules are used to ease the writing of RS specifications. Its features include the possibility to simulate single traces or generate the whole graph of the LTS (where user defined predicates can be used to color each node depending on its content so to improve readability), to verify modal logic formulas, to check bisimulation based equivalences between different RSs, to deal with quantitative aspects of RSs, such as delays and duration for produced entities and linear handling of concentration levels (see [15,16] for details).

5.1 A Computation with the Interpreter of Reaction Systems

For a computation which does not use assertions the interpreter gives the user some choices: (1) whether she wants to make a context independent computation; (2) the possibility to specify the maximum number m of computation steps.

The interpreter will show the corresponding trace $(D_0, C_0) \xrightarrow{N_1} \cdots \xrightarrow{N_m} (D_m, C_m)$, emphasizing the elements D_i, C_i, N_i. Then the user has to provide the entities that she wants to mark in D_m. Finally, the interpreter will compute the corresponding sliced computation and present it. Let us see one example for illustrating our tool.

Example 4. We consider a RS defined in [12], to model a network for gene regulation. These networks represent the interactions among genes regulating the activation of specific cell functions. The RS models a fragment of the network for controlling the process of differentiation of T helper (Th) lymphocytes, which play a fundamental role in the immune system. We introduced one wrong reaction in the RS model that can be found in our tool website[3] and performed some experiments. For instance we made a context free computation, starting from the initial state containing only the entity `ifngammah`. The computation, limited to 6 steps, produced the following sequence of states.

```
[ [ifngammah], [ifngammarh], [stat1h], [socs1,socs2], [tbeth],
  [ifngammah,socs1,tbeth], [ifngammah,ifngammarm,socs1,tbeth] ]
```

Now we wanted to focus on the molecule `tbeth` in the last computation state, and hence we used our slicer, marking `tbeth`. The outcome was the following:

```
[ [ifngammah], [ifngammarh], [stat1h], [socs1], [tbeth],
  [socs1,tbeth], [tbeth] ]
```

[1] http://www.di.unipi.it/~bruni/LTSRS/slicingBioReSolve.zip.
[2] http://www.di.unipi.it/~bruni/LTSRS/.
[3] http://www.di.unipi.it/~bruni/LTSRS/wrongspec.pl.

with the following sequence of reaction numbers applied in the steps:
`[ [5], [8], [19], [18], [27,29], [18,27] ].`

The sliced sequence can now easily be interpreted. Clearly `[tbeth]` was produced as a result of the application of four reaction rules in a sequence, those with numbers `[5],[8],[19],[18]`. Then reaction `[27]` reintroduced `tbeth` in each following step. So, the sliced sequence produced a much simpler trace, with an easy interpretation. It is now immediate to see that `[tbeth]` was introduced by rule `[18]`, rewriting entity `[socs1]`, which is recognized as an error, because `[tbeth]` should be introduced by `[stat1h]`. Thus the user can now correct reaction `[18]` for `[stat1h]`.

6 A Logical Framework for the Slicing Algorithm

In this section we try to further automate the slicing process. A user can describe a property to be checked along a computation. A computation is stopped when a state S does not satisfy the property. Then slicing starts from S, which is automatically marked.

To specify properties we reuse the simple assertion language introduced in [15], which is tailored on the labels of the LTS generated by SOS semantics in Sect. 3. Then, we rely on a fragment of the recursive extension of the Hennessy Milner Logic [2], called sHML, to formally express properties to be verified along a RS process execution. We exploit the monitor technique [1] to check each state of the RS process execution with respect to the required property. To this aim, we apply the translation from sHML formula to monitors given in [2]. Some modifications are required as in the original proposal sHML logic works on action names, whereas we work with our assertions.

6.1 The Assertion Language

The labels of our LTS carry on a large amount of information about the activity executed during each transition, our assertion is a formula that predicates on those labels. Hereafter, we assume that the context can be non-deterministic.

Example 5. Here follows an example of some properties which we may verify:
Has the reaction $(\mathsf{ab}, \mathsf{c}, \mathsf{b})$ been applied?
Has the entity a played both as reactant and as product?

Definition 4 (Assertion Language). *Given a set of entities S, assertions* F *on S are built from the following syntax, where $E \subseteq S$ and $Pos \in \{\mathcal{D}, \mathcal{C}, \mathcal{W}, \mathcal{R}, \mathcal{I}, \mathcal{P}\}$:*

$$\mathsf{F} ::= \mathbf{tt} \mid E \subseteq Pos \mid ? \in Pos \mid \mathsf{F} \vee \mathsf{F} \mid \mathsf{F} \wedge \mathsf{F} \mid \neg \mathsf{F}$$

Pos distinguishes different positions in the labels: $\mathcal{D}$ stands for entities produced in the previous transition, $\mathcal{C}$ for entities provided by the context, $\mathcal{W}$ for their union, $\mathcal{R}$ for reactants, $\mathcal{I}$ for inhibitors, and $\mathcal{P}$ for products. An assertion $E \subseteq Pos$, checks the membership of a subset of entities E in a given Pos, $? \in Pos$ is a test of non-emptiness of Pos, $\mathsf{F}_1 \vee \mathsf{F}_2$ denotes a disjunction, $\mathsf{F}_1 \wedge \mathsf{F}_2$ is a conjunction, $\neg\mathsf{F}$ is a negation.

$$\frac{}{\mathsf{F}.m \xrightarrow{\mathsf{F}} m}\ (Pro) \quad \frac{}{\mathbf{yes} \otimes m \xrightarrow{\tau} m}\ (Ver1) \quad \frac{}{\mathbf{no}_s \otimes m \xrightarrow{\tau} \mathbf{no}_s}\ (Ver2) \quad \frac{m \xrightarrow{\tau} m'}{m \otimes n \xrightarrow{\tau} m' \otimes n}\ (Par1)$$

$$\frac{m_1 \xrightarrow{\mathsf{F}} m_1'}{m_1 + m_2 \xrightarrow{\mathsf{F}} m_1'}\ (Sum) \quad \frac{m[^{\mathsf{rec}\,X.m}/_X] \xrightarrow{\mathsf{F}} m'}{\mathsf{rec}\ X.\ m \xrightarrow{\mathsf{F}} m'}\ (Rec) \quad \frac{m \xrightarrow{\mathsf{F}_1} m' \qquad n \xrightarrow{\mathsf{F}_2} n'}{m \otimes n \xrightarrow{\mathsf{F}_1 \wedge \mathsf{F}_2} m' \otimes n'}\ (Par2)$$

Fig. 2. SOS semantics of the monitors.

Definition 5 (Satisfaction of Assertion). *Let* P *be a RSs process, let* $\ell = \langle (D, C) \rhd R, I, P \rangle$ *be a transition label, and* F *be an assertion. We write* $\ell \models \mathsf{F}$ *(read as the transition label* ℓ *satisfies the assertion* F*) if and only if the following hold:*

$$\begin{array}{llll} \ell \models E \subseteq Pos & \mathit{iff}\ E \subseteq \mathsf{select}(\ell, Pos) & \ell \models \mathsf{F}_1 \vee \mathsf{F}_2 & \mathit{iff}\ \ell \models \mathsf{F}_1 \vee \ell \models \mathsf{F}_2 \\ \ell \models ? \in Pos & \mathit{iff}\ \mathsf{select}(\ell, Pos) \neq \emptyset & \ell \models \mathsf{F}_1 \wedge \mathsf{F}_2 & \mathit{iff}\ \ell \models \mathsf{F}_1 \wedge \ell \models \mathsf{F}_2 \\ & & \ell \models \neg \mathsf{F} & \mathit{iff}\ \ell \not\models \mathsf{F} \end{array}$$

$$\mathsf{select}(\langle (D, C) \rhd R, I, P \rangle, Pos) \triangleq \begin{cases} D & \mathit{if}\ Pos = \mathcal{D}, \quad C\ \mathit{if}\ Pos = \mathcal{C} \\ D \cup C & \mathit{if}\ Pos = \mathcal{W}, \quad R\ \mathit{if}\ Pos = \mathcal{R} \\ I & \mathit{if}\ Pos = \mathcal{I}, \quad\ P\ \mathit{if}\ Pos = \mathcal{P} \end{cases}$$

Example 6. Some assertions matching the queries listed in Example 5 are:

– $\mathsf{F}_1 \triangleq \{\mathsf{ab}\} \subseteq \mathcal{R} \wedge \{\mathsf{c}\} \subseteq \mathcal{I}$, while $\mathsf{F}_1' \triangleq \neg \mathsf{F}_1$ is verified if (ab, c, b) is not applied,
– $\mathsf{F}_2 \triangleq \{\mathsf{a}\} \subseteq \mathcal{R} \wedge \{\mathsf{a}\} \subseteq \mathcal{P}$.

6.2 Monitors

Differently from [1], in our context monitors check if transition labels satisfy a given property. A process monitor stops when a verdict is reached, thus we omit **0**. The $\mathbf{no}_s$ verdict is equipped with a set of entities, $s \subseteq S$, used as markers for the slicing.

Definition 6. *A monitor process is defined by the grammar:*

$$m, n \in \mathsf{Mon} ::= \mathbf{no}_s \mid \mathbf{yes} \mid \mathsf{F}.m \mid m + n \mid m \otimes n \mid \mathsf{rec}\ X.m \mid X$$

where X *comes from a countably infinite set of monitor variables, and the set* $s \subseteq S$.

The syntax of a monitor is similar to that of a context process: actions are replaced by properties to be verified by the process action. A 'verdict' can be **yes** or $\mathbf{no}_s$ for acceptance or rejection respectively. Sum $m + n$ is used to provide monitors with different behaviors, while $m \otimes n$ is used to compose monitors in parallel.

The semantics is in Fig. 2, where the symmetric rules are omitted. The set of transition labels is composed by the set of the formulas of the assertion language in Definition 4 plus a special silent action τ. The verdicts do nothing.

Table 1. Monitored systems.

$$\frac{p \xrightarrow{\ell} p' \quad m \xrightarrow{\mathsf{F}} m' \quad \ell \models \mathsf{F}}{m \lhd p \xrightarrow{(\ell,\mathsf{F})} m' \lhd p'}\ (\mathit{Exec}) \qquad \frac{m \xrightarrow{\tau} m' \quad p}{m \lhd p \xrightarrow{\tau} m' \lhd p}\ (\tau)$$

Table 2. Syntax and semantics for the cHML.

Syntax $\phi, \psi \in \text{cHML} ::= \mathbf{tt} \mid \mathbf{ff}_s \mid [\mathsf{F}].\phi \mid \phi \wedge \psi \mid \mathsf{max}X.\phi \mid X$

Semantics

$[\![\mathbf{tt}, \rho]\!] \triangleq \mathcal{P}$ $\qquad [\![\mathbf{ff}_s, \rho]\!] \triangleq \emptyset$

$[\![[\mathsf{F}].\phi, \rho]\!] \triangleq \{p \mid \forall q.\, p \xrightarrow{\ell} q, \ell \models \mathsf{F} \text{ and } q \in [\![\phi, \rho]\!]\}$ $\qquad [\![\phi \wedge \psi, \rho]\!] \triangleq [\![\phi, \rho]\!] \cap [\![\psi, \rho]\!]$

$[\![\mathsf{max}\, X.\phi, \rho]\!] \triangleq \bigcup\{\mathsf{P} \mid \mathsf{P} \subseteq [\![\phi, \rho[X \mapsto \mathsf{P}]]\!]\}$ $\qquad [\![X, \rho]\!] \triangleq \rho(X)$

where ρ is a set of formula definitions, and $\mathcal{P}$ is the whole set of processes.

Table 3. Rules for deriving a process monitor.

$\mathsf{m}(\mathbf{ff}_s) \triangleq \mathbf{no}_s$	$\mathsf{m}([\mathsf{F}].\phi) \triangleq \mathsf{F}.\mathsf{m}(\phi) + \neg\mathsf{F}.\mathbf{yes}$	$\mathsf{m}(\mathsf{max}\, X.\phi) \triangleq \mathsf{rec}\, x.\mathsf{m}(\phi)$
$\mathsf{m}(\mathbf{tt}) \triangleq \mathbf{yes}$	$\mathsf{m}(\phi \wedge \psi) \triangleq \mathsf{m}(\phi) \otimes \mathsf{m}(\psi)$	$\mathsf{m}(X) \triangleq x$

Definition 7. *A monitored system is a monitor $m \in \mathsf{Mon}$ and a process $p \in \mathsf{Proc}$ that run side-by-side, denoted $m \lhd p$. The behavior of a monitored system is defined by the derivation rules in Table 1.*

If a monitored system $m \lhd p$ reaches a verdict like $\mathbf{no}_s \lhd q$, then a violation is detected and q is the state where the slicing starts by marking the entities in the set s.

Monitorability. Typically we are interested in verifying if certain assertions are met along a process execution. Writing the corresponding monitors is an error prone task. Following [2], we prefer to write property specifications as formulas in a fragment of the (recursive) Hennessy Milner logic, called sHML. The syntax and semantics of sHML are reported in Table 2. For example, given a certain assertion F (see Definition 4) we can write the sHML formula $\mathsf{max}\, X.([\mathsf{F}].X \wedge [\neg\mathsf{F}].\mathbf{ff})$ to specify that "the computation should exhibit transition labels satisfying F and stops as soon as a violation is detected". Of course, other properties can be required, e.g. that F_1 and F_2 are satisfied in alternation. Following the monitor synthesis in [3], and recalled in Table 3, we obtain that, after some logical simplifications, a monitor implementing the previous formula is: $\mathsf{m} = \mathsf{rec}\, X.(\neg\mathsf{F}.X + \mathsf{F}.\mathbf{no})$. While it may seem that monitors closely resemble sHML formulas, we argue that the box modality $[\mathsf{F}].\phi$ is much more convenient to write and to manage than the sum $\mathsf{F}.\mathsf{m}(\phi) + \neg\mathsf{F}.\mathbf{yes}$.

Example 7. Let $\mathsf{P}_0 \triangleq [(\mathsf{ab}, \mathsf{c}, \mathsf{b}) \mid \mathsf{ab}.\mathsf{a}.\mathsf{c}.\mathsf{c}.\mathbf{0}]$ be the process in Example 3. We want to study a computation where all the visited states satisfy the following formula: $\mathsf{F} \triangleq \{\mathsf{b}\} \subseteq \mathcal{R} \wedge \{\mathsf{c}\} \not\subseteq \mathcal{C}$. Then we need a sHML formula of the format described above: $\phi \triangleq \mathsf{max}\, X.([\mathsf{F}].X \wedge [\neg\mathsf{F}]\mathbf{ff}_{\{\mathsf{c}\}})$, where the idea is to flag the presence of the entity c as a fault to understand why it has been produced. The corresponding process monitor is thus: $\mathsf{m} \triangleq \text{rec } X.(\mathsf{F}.X + \neg\mathsf{F}.\mathbf{no}_{\{\mathsf{c}\}})$. The execution of the monitored process $\mathsf{m} \lhd \mathsf{P}_0$ proceeds by applying the rules in Table 1.

$$\frac{\mathsf{P}_0 \xrightarrow{\langle(\emptyset,\mathsf{ba})\rhd\mathsf{ab},\mathsf{c},\mathsf{b}\rangle} \mathsf{P}_1 \qquad \mathsf{m} \xrightarrow{\mathsf{F}} \mathsf{m} \qquad \langle(\emptyset,\mathsf{ba}) \rhd \mathsf{ab},\mathsf{c},\mathsf{b}\rangle \models \mathsf{F}}{\mathsf{m} \lhd \mathsf{P}_0 \xrightarrow{(\langle(\emptyset,\mathsf{ba})\rhd\mathsf{ab},\mathsf{c},\mathsf{b}\rangle,\mathsf{F})} \mathsf{m} \lhd \mathsf{P}_1} \;(\mathit{Exec})$$

where $\mathsf{P}_1 \triangleq [(\mathsf{ab},\mathsf{c},\mathsf{b}) \mid \mathsf{b} \mid \mathsf{a.c.c.}\mathbf{0}]$. The next step is derived by using rule (Exec) again:

$$\frac{\mathsf{P}_1 \xrightarrow{\langle(\mathsf{b},\mathsf{a})\rhd\mathsf{ab},\mathsf{c},\mathsf{b}\rangle} \mathsf{P}_2 \qquad \mathsf{m} \xrightarrow{\mathsf{F}} \mathsf{m} \qquad \langle(\mathsf{b},\mathsf{a}) \rhd \mathsf{ab},\mathsf{c},\mathsf{b}\rangle \models \mathsf{F}}{\mathsf{m} \lhd \mathsf{P}_1 \xrightarrow{(\langle(\mathsf{b},\mathsf{a})\rhd\mathsf{ab},\mathsf{c},\mathsf{b}\rangle,\mathsf{F})} \mathsf{m} \lhd \mathsf{P}_2} \;(\mathit{Exec})$$

where $\mathsf{P}_2 \triangleq [(\mathsf{ab},\mathsf{c},\mathsf{b}) \mid \mathsf{b} \mid \mathsf{c.c.}\mathbf{0}]$. The computation ends after applying the (Exec) rule:

$$\frac{\mathsf{P}_2 \xrightarrow{\langle(\mathsf{b},\mathsf{c})\rhd\mathsf{bc},\emptyset,\emptyset\rangle} \mathsf{P}_3 \qquad \mathsf{m} \xrightarrow{\neg\mathsf{F}} \mathbf{no}_{\{\mathsf{c}\}} \qquad \langle(\mathsf{b},\mathsf{c}) \rhd \mathsf{bc},\emptyset,\emptyset\rangle \models \neg\mathsf{F}}{\mathsf{m} \lhd \mathsf{P}_2 \xrightarrow{(\langle(\mathsf{b},\mathsf{c})\rhd\mathsf{bc},\emptyset,\emptyset\rangle,\neg\mathsf{F})} \mathbf{no}_{\{\mathsf{c}\}} \lhd \mathsf{P}_3} \;(\mathit{Exec})$$

where $\mathsf{P}_3 \triangleq [(\mathsf{ab},\mathsf{c},\mathsf{b}) \mid \mathsf{c.}\mathbf{0}]$. The computation stops and P_3 is a starting point for backward slicing on $\{\mathsf{c}\}$.

Example 8. Given the assertion $\mathsf{F} = \{\texttt{ifngammah}, \texttt{tbeth}\} \not\subseteq \mathcal{W}$, the formula $\phi \triangleq \max X.([\mathsf{F}].X \wedge [\neg\mathsf{F}]\mathbf{ff}_{\{\mathsf{tbeth}\}})$ would automatize the slicing in Example 4, by selecting `tbeth` with the monitor, if it is found present.

Example 9. We consider here a complex and very large biological example in which there is a continuous interaction with the environment, represented by the sequence of the provided contexts. This example is due to [23], where a RS was introduced to replicate one of the experiments in [20] related to a dynamical model of ErbB receptor signal transduction in human mammary epithelial cells. The non-receptor tyrosine kinase Src and receptor tyrosine kinase epidermal growth factor receptor (EGFR/ErbB1) have been established as collaborators in cellular signaling and their combined disregulation plays key roles in human cancers, including breast cancer. The RS contains 6,720 reactions and a sequence of 1,000 contexts.

We installed a monitor looking for the introduction of an EFGR lysosome, which is essential for a bifurcation in the pathway of our experiment. We discovered that the lysosome is introduced after 11 computation steps. Then we marked the lysosome in the last state of the computation of length 11, and computed the slicing.

The resulting sequence and contexts are still rather large, but notably smaller. Moreover, in the sliced sequence the biologist can identify other entities that can be used for further slicing simplifications. We report here an excerpt of the sliced sequence. The first and last three states of the *sliced computation* are:

```
[[alpha_ir,alpha_qr,alpha_sr,ap2,arf,cas,clathrin,egfr_free,grb2,
  hip1r,ilk,myosin,pak,pdk1,pi4k,pi5k,pip2_45,pten,rho,rhok,rin],
 ...
 [egfr_egfr_egf_mvb,egfr_egfr_egf_pm,egfr_ub,escrt_i,rab5],
 [alix,egfr_egfr_egf_mvb,eps15,escrt_iii,rab7],
 [egfr_egfr_egf_lysosome] ]
```

The last states of the original computation consisted of about 100 different molecules each, which made the trace pretty unreadable, while the slice gives a much simpler sequence, which can be easier to inspect, and can be used for further slicing derivations. We show the sequence of sliced contexts, which emphasizes that only a small number of the entities provided by the contexts are really useful/necessary for the sliced computation:

```
[[alpha_il,alpha_ql,alpha_sl,calm,cdc42,egf,egfr_contr,erk,fak,
  gak,pa,pip2_34,pip3_345,pp2a,ras,src,stress],
 [alpha_sl,cdc42,egf,fak,gak,pa,pip2_34,pip3_345,rho,src,stress],
 [cdc42,egfr_contr,fak,gak,pip3_345,rho,src,stress],
 [cdc42,egfr_contr,fak,pip3_345,ras,src],
 [cdc42,egf,egfr_contr,pip3_345,src],
 [egf,fak,pip3_345,src],
 [egf,egfr_contr,pip3_345,src],
 [egf],
 [],
 [] ]
```

Each context in the original computation adds at every step 30 entities, most of which after the first computation step become irrelevant, while only 17 were useful for the first step. In the last two computation steps the context does not contribute at all to the computation of the marked entity. Concerning the applied reactions, in the first step 36 of them are useful, and this number decreases heavily after another two steps. In the last three steps of the computation the relevant reactions are just the following ones:

```
[1558,1574,2456,3276,5573], [91,1556,3008,3279,5599], [1554]
```

Thus, it is possible to know exactly, within the 6,720 reactions, the few ones which are relevant to compute the marked entities, as well as the relevant entities introduced by the contexts. Then it is possible to proceed to identify a bug in the model following the same strategy exemplified in Example 4 In practice the biologist analyses the simplified sliced sequence by using her knowledge of the intended model, or she can express the properties to be monitored automatically. We also note that the slicer even if it is a prototype in a declarative language can compute the sliced sequences rather efficiently.

We performed our experiments on a Mac Powerbook Pro with OSX 11.7, 2,6 GHz intel i7 6 core, with 16 GB ram. The slicing for the small Example 4 was executed in 9 ms, while the big Example 9 took 6000 ms. Considering the number of reactions (about 300 times more) and the fact that the context is absent in the first example and it is pretty large in the second one we can notice that the computation time increases linearly with the number of reactions. This is a result that we might expect also theoretically by an analysis of the algorithm, which depends directly on the number of reactions and the size of the context. We are working on an optimization of the implementation by recording in the history also the list of the reactions applied in the computation.

Our implementation introduces several novel features not covered in the literature and has been designed as a tool for verification, for slicing, as well as for rapid prototyping extensions of RSs, not just for their simulations. For ordinary interactive RSs there are already some performant simulators, such as `brsim`, written in Haskell [8] or such as `HERESY` which is a GPU-based simulator of RSs, written using CUDA [23]. Using Prolog has had the advantage of a more flexible, safe and rapid implementation.

7 Related Work

Dynamic program slicing has been applied to several programming paradigms, for instance to imperative programming [21], functional programming [24], Term Rewriting [4], functional logic programming [5,6], and Constraint Concurrent Programming [19]. RSs use term rewriting and sets manipulation for its basic computation mechanism. Thus, [4,6] have a similarity with our work in the adoption of a backward style of computation of the slicer and [6] uses assertions to stop the computation and to start the slicing process. However our framework and the one in [4,6] are quite different. [6] considers the Maude language, and [4] is oriented towards functional computations. In the language that they consider there are not inhibitors, which introduce a kind of non-monotonic behaviour in the rewriting rule, nor a notion of interactive context. We tried to use the framework in [4,6] on some examples, but we were losing too much information, and the resulting simplified computations were not informative. We have instead defined a framework which is totally specialised and suitable for RSs, we use different logics specific for RSs and we are the first to consider monitors for checking the verification process in a slicer.

8 Conclusions and Future Work

We have presented the first framework for dynamic slicing of RSs. The slicer computes automatically a simplified computation, with the information relevant to derive the marked one. We can deal with computations with or without a context. Monitors help to identify states which violate a property and to automate marking of such states, using a flexible language to specify assertions over computations. We have described our prototype implementation in Prolog, and have applied our tool to some bioinformatics examples.

We plan to add forward slicing to our interpreter, and improve its interface, besides generalising it to some quantitative extensions of RSs for modelling speeds and delays in reactions and discrete concentration levels that are handled linearly by reactions [16]. We can then study the relation and application of our techniques to a framework to model biological systems in rewriting logic which has several similarities with Reaction Systems [10].

Acknowledgment. We thank the anonymous reviewers for their comments and suggestions that helped us to improve our paper.

References

1. Aceto, L., Achilleos, A., Francalanza, A., Ingófsdóttir, A., Kjartansson, S.: Determinizing monitors for HML with recursion. J. Log. Algebr. Methods Program. **111**, 100515 (2020). https://doi.org/10.1016/j.jlamp.2019.100515
2. Aceto, L., Achilleos, A., Francalanza, A., Ingófsdóttir, A., Lehtinen, K.: The best a monitor can do. In: Proc. CSL 2021. LIPIcs, vol. 183, pp. 7:1–7:23 (2021). https://doi.org/10.4230/LIPIcs.CSL.2021.7

3. Aceto, L., Achilleos, A., Francalanza, A., Ingólfsdóttir, A., Lehtinen, K.: An operational guide to monitorability with applications to regular properties. Softw. Syst. Model. **20**(2), 335–361 (2021). https://doi.org/10.1007/s10270-020-00860-z
4. Alpuente, M., Ballis, D., Espert, J., Romero, D.: Backward trace slicing for rewriting logic theories. In: Bjørner, N., Sofronie-Stokkermans, V. (eds.) CADE 2011. LNCS (LNAI), vol. 6803, pp. 34–48. Springer, Heidelberg (2011). https://doi.org/10.1007/978-3-642-22438-6_5
5. Alpuente, M., Ballis, D., Frechina, F., Romero, D.: Using conditional trace slicing for improving Maude programs. Sci. Comput. Program. **80**, 385–415 (2014). https://doi.org/10.1016/j.scico.2013.09.018
6. Alpuente, M., Ballis, D., Frechina, F., Sapiña, J.: Debugging Maude programs via runtime assertion checking and trace slicing. J. Log. Algebr. Meth. Program. **85**, 707–736 (2016). https://doi.org/10.1016/j.jlamp.2016.03.001
7. Azimi, S.: Steady states of constrained reaction systems. Theor. Comput. Sci. **701**, 20–26 (2017). https://doi.org/10.1016/j.tcs.2017.03.047
8. Azimi, S., Gratie, C., Ivanov, S., Petre, I.: Dependency graphs and mass conservation in reaction systems. Theor. Comput. Sci. **598**, 23–39 (2015). https://doi.org/10.1016/j.tcs.2015.02.014
9. Azimi, S., Iancu, B., Petre, I.: Reaction system models for the heat shock response. Fund. Inform. **131**(3–4), 299–312 (2014). https://doi.org/10.3233/FI-2014-1016
10. Baggi, M., Ballis, D., Falaschi, M.: Quantitative pathway logic for computational biology. In: Degano, P., Gorrieri, R. (eds.) CMSB 2009. LNCS, vol. 5688, pp. 68–82. Springer, Heidelberg (2009). https://doi.org/10.1007/978-3-642-03845-7_5
11. Barbuti, R., Gori, R., Levi, F., Milazzo, P.: Investigating dynamic causalities in reaction systems. Theor. Comput. Sci. **623**, 114–145 (2016). https://doi.org/10.1016/j.tcs.2015.11.041
12. Barbuti, R., Gori, R., Milazzo, P.: Encoding Boolean networks into reaction systems for investigating causal dependencies in gene regulation. Theor. Comput. Sci. **881**, 3–24 (2021). https://doi.org/10.1016/j.tcs.2020.07.031
13. Brijder, R., Ehrenfeucht, A., Main, M., Rozenberg, G.: A tour of reaction systems. Int. J. Found. Comput. Sci. **22**(07), 1499–1517 (2011). https://doi.org/10.1142/S0129054111008842
14. Brodo, L., Bruni, R., Falaschi, M.: Enhancing reaction systems: a process algebraic approach. In: Alvim, M.S., Chatzikokolakis, K., Olarte, C., Valencia, F. (eds.) The Art of Modelling Computational Systems: A Journey from Logic and Concurrency to Security and Privacy. LNCS, vol. 11760, pp. 68–85. Springer, Cham (2019). https://doi.org/10.1007/978-3-030-31175-9_5
15. Brodo, L., Bruni, R., Falaschi, M.: A logical and graphical framework for reaction systems. Theoret. Comput. Sci. **875**, 1–27 (2021). https://doi.org/10.1016/j.tcs.2021.03.024
16. Brodo, L., Bruni, R., Falaschi, M., Gori, R., Levi, F., Milazzo, P.: Exploiting modularity of SOS semantics to define quantitative extensions of reaction systems. In: Aranha, C., Martín-Vide, C., Vega-Rodríguez, M.A. (eds.) TPNC 2021. LNCS, vol. 13082, pp. 15–32. Springer, Cham (2021). https://doi.org/10.1007/978-3-030-90425-8_2
17. Corolli, L., Maj, C., Marinia, F., Besozzi, D., Mauri, G.: An excursion in reaction systems: from computer science to biology. Theor. Comput. Sci. **454**, 95–108 (2012). https://doi.org/10.1016/j.tcs.2012.04.003
18. Falaschi, M., Gabbrielli, M., Olarte, C., Palamidessi, C.: Slicing concurrent constraint programs. In: Hermenegildo, M.V., Lopez-Garcia, P. (eds.) LOPSTR 2016. LNCS, vol. 10184, pp. 76–93. Springer, Cham (2017). https://doi.org/10.1007/978-3-319-63139-4_5
19. Falaschi, M., Gabbrielli, M., Olarte, C., Palamidessi, C.: Dynamic slicing for concurrent constraint languages. Fundam. Informaticae **177**(3–4), 331–357 (2020). https://doi.org/10.3233/FI-2020-1992

20. Helikar, T., et al.: A comprehensive, multi-scale dynamical model of ErbB receptor signal transduction in human mammary epithelial cells. PLoS ONE **8**(4), 1–9 (2013). https://doi.org/10.1371/journal.pone.0061757
21. Korel, B., Laski, J.: Dynamic program slicing. Inf. Process. Lett. **29**(3), 155–163 (1988). https://doi.org/10.1016/0020-0190(88)90054-3
22. Milner, R. (ed.): A Calculus of Communicating Systems. LNCS, vol. 92, pp. 138–157. Springer, Heidelberg (1980). https://doi.org/10.1007/3-540-10235-3
23. Nobile, M.S., et al.: Efficient simulation of reaction systems on graphics processing units. Fundam. Informaticae **154**(1–4), 307–321 (2017). https://doi.org/10.3233/FI-2017-1568
24. Ochoa, C., Silva, J., Vidal, G.: Dynamic slicing of lazy functional programs based on redex trails. Higher Order Symbol. Comput. **21**(1–2), 147–192 (2008). https://doi.org/10.1007/s10990-008-9023-7
25. Okubo, F., Yokomori, T.: The computational capability of chemical reaction automata. Nat. Comput. **15**(2), 215–224 (2015). https://doi.org/10.1007/s11047-015-9504-7
26. Silva, J.: A vocabulary of program slicing-based techniques. ACM Comput. Surv. **44**(3), 1–41 (2012). https://doi.org/10.1145/2187671.2187674
27. Weiser, M.: Program slicing. IEEE Trans. Softw. Eng. **10**(4), 352–357 (1984). https://doi.org/10.1109/TSE.1984.5010248

Multiple Query Satisfiability of Constrained Horn Clauses

Emanuele De Angelis[1(✉)], Fabio Fioravanti[2(✉)], Alberto Pettorossi[1,3(✉)], and Maurizio Proietti[1(✉)]

[1] IASI-CNR, Rome, Italy
{emanuele.deangelis,maurizio.proietti}@iasi.cnr.it
[2] DEc, University 'G. d'Annunzio', Chieti-Pescara, Italy
fabio.fioravanti@unich.it
[3] DICII, University of Rome 'Tor Vergata', Rome, Italy
pettorossi@info.uniroma2.it

Abstract. We address the problem of checking the satisfiability of a set of constrained Horn clauses (CHCs) possibly including more than one query. We propose a transformation technique that takes as input a set of CHCs, including a set of queries, and returns as output a new set of CHCs, such that the transformed CHCs are satisfiable if and only if so are the original ones, and the transformed CHCs incorporate in each new query suitable information coming from the other ones so that the CHC satisfiability algorithm is able to exploit the relationships among all queries. We show that our proposed technique is effective on a non trivial benchmark of sets of CHCs that encode many verification problems for programs manipulating algebraic data types such as lists and trees.

1 Introduction

Constrained Horn Clauses (CHCs) have been advocated as a logical formalism very well suited for automatic program verification [3,7]. Indeed, many verification problems can be reduced to problems of checking satisfiability of CHCs [14], and several effective CHC *solvers* are currently available as back-end tools for program verification purposes [4,8,17,19,20].

Following the CHC-based verification approach, a program is translated into a set of definite CHCs (that is, clauses whose head is different from *false*), which capture the semantics of the program, together with a set of queries (that is, clauses whose head is *false*), which specify the program properties to be verified. Very often the CHC translation of the verification problem generates several queries. In particular, this is the case when the program includes several functions, each one having its own contract (that is, a pair of a pre-condition and a post-condition).

CHC solvers try to show the satisfiability of a set of CHCs of the form: $P \cup \{\mathit{false} \leftarrow G_1, \ldots, \mathit{false} \leftarrow G_n\}$, where P is a set of definite CHCs, and $\mathit{false} \leftarrow G_1, \ldots, \mathit{false} \leftarrow G_n$ are queries, by trying to show in a separate way the satisfiability of each set $P \cup \{\mathit{false} \leftarrow G_i\}$, for $i = 1, \ldots, n$. This approach may

M. Hanus and D. Inclezan (Eds.): PADL 2023, LNCS 13880, pp. 125–143, 2023.
https://doi.org/10.1007/978-3-031-24841-2_9

not always be effective, as the solver may not be able to exploit, possibly mutual, dependencies among the various queries. There is a simple way of combining all queries into one (as done, for instance, in the CHC solver competition [9]): we introduce a new predicate f, and from the above mentioned set of CHCs we get $P \cup \{\mathit{false} \leftarrow f, f \leftarrow G_1, \ldots, f \leftarrow G_n\}$. However, also in this case, existing solvers will handle each query separately and then combine the results.

In this paper we propose, instead, a technique that, given a set $P \cup \{\mathit{false} \leftarrow G_1, \ldots, \mathit{false} \leftarrow G_n\}$ of CHCs, derives an *equisatisfiable* set $P' \cup \{\mathit{false} \leftarrow G'_1, \ldots, \mathit{false} \leftarrow G'_n\}$, for whose satisfiability proof a CHC solver may exploit the mutual interactions among the n satisfiability proofs, one for each query.

Our technique builds upon the transformation approach for verifying contracts that we presented in previous work and implemented in the VeriCaT tool [11]. The algorithm of VeriCaT takes as input a set of CHCs that manipulate algebraic data types (ADTs) such as lists and trees, and a set of CHCs defining contracts by means of catamorphisms [22], and returns as output a set of CHCs without ADT variables such that the original set is satisfiable if the new set is satisfiable. For CHCs without ADT variables state-of-the-art solvers are more effective in proving satisfiability, and hence validity of contracts.

The objective of the transformation algorithm presented in this paper, which we call $\mathcal{T}_{mq}$, is not to eliminate ADT variables, rather, it is to incorporate into the clauses relative to a particular query some additional constraints that are derived from other queries. These additional constraints are often very beneficial to the CHC solvers when trying to check the satisfiability of a given set of clauses, thereby enhancing their ability to verify program properties. Algorithm $\mathcal{T}_{mq}$ is both sound and complete, that is, the transformed clauses are satisfiable if and only if so are the original ones. The completeness of $\mathcal{T}_{mq}$ is very important because if a property does *not* hold, it allows us to infer the unsatisfiability of the original clauses. Thus, whenever the solver shows the unsatisfiability of the transformed clauses, we deduce the invalidity of the property to be verified.

2 Preliminary Notions

We consider constrained Horn clauses that are defined in a many-sorted first order language $\mathcal{L}$ with equality (=) whose constraints are expressed using linear integer arithmetic (*LIA*) and boolean (*Bool*) expressions. A *constraint* is a quantifier-free formula c, where the *LIA* constraints may occur as subexpressions of boolean constraints, according to the SMT approach [2]:

$$\begin{aligned} c \ ::= \ & d \mid B \mid \mathit{true} \mid \mathit{false} \mid \sim c \mid c_1 \,\&\, c_2 \mid c_1 \vee c_2 \mid c_1 \Rightarrow c_2 \mid c_1 = c_2 \mid \\ & \mathit{ite}(c, c_1, c_2) \mid t = \mathit{ite}(c, t_1, t_2) \\ d \ ::= \ & t_1 = t_2 \mid t_1 < t_2 \mid t_1 \leq t_2 \mid t_1 \geq t_2 \mid t_1 > t_2 \end{aligned}$$

where: (i) B is a boolean variable, (ii) $\sim$, $\&$, $\vee$, and $\Rightarrow$ denote negation, conjunction, disjunction, and implication, respectively, (iii) the ternary function *ite* denotes the if-then-else operator, and (iv) t is a *LIA* term of the form $a_0 + a_1 X_1 + \cdots + a_n X_n$ with integer coefficients $a_0, \ldots, a_n$ and variables

$X_1, ..., X_n$. The equality symbol will be used both for integers and booleans. We will often write $B = true$ (or $B = false$) as B (or $\sim B$). The theory of *LIA* and boolean constraints will be denoted by $LIA \cup Bool$. The integer and boolean sorts are said to be *basic sorts.* A recursively defined sort (such as the sort of lists and trees) is said to be an *algebraic data type* (ADT, for short).

An *atom* is a formula of the form $p(t_1, \ldots, t_m)$, where p is a predicate symbol not occurring in $LIA \cup Bool$, and $t_1, \ldots, t_m$ are first order terms in $\mathcal{L}$. A *constrained Horn clause* (CHC), or simply, a *clause*, is an implication of the form $H \leftarrow c, G$. The conclusion H, called the *head*, is either an atom or *false*, and the premise, called the *body*, is the conjunction of a constraint c and a conjunction G of zero or more atoms. A clause is said to be a *query* if its head is *false*, and a *definite clause*, otherwise. Without loss of generality, we assume that every atom of the body of a clause has distinct variables (of any sort) as arguments. The set of all variables occurring in an expression e is denoted by $vars(e)$. By $bvars(e)$ (or $adt\text{-}vars(e)$) we denote the set of variables in e whose sort is a basic sort (or an ADT sort). The *universal closure* of a formula φ is denoted by $\forall(\varphi)$.

Let $\mathbb{D}$ be the usual interpretation for the symbols of theory $LIA \cup Bool$. By $M(P)$ we denote the *least* $\mathbb{D}$-model of a set P of definite clauses [18].

Now, in order to characterize the class of queries that can be handled using our transformation technique, we introduce (see Definition 1 below) a class of recursive schemata defined by CHCs [11]. That class is related to those of *morphisms*, *catamorphisms*, and *paramorphisms* considered in functional programming [16,22]. We will not introduce a new terminology here and we will refer to our schemata as *generalized catamorphisms*, or *catamorphisms*, for short.

Let f be a predicate symbol with $m+n$ arguments (for $m \geq 0$ and $n \geq 0$) whose sorts are $\alpha_1, \ldots, \alpha_m, \beta_1, \ldots, \beta_n$, respectively. We say that f is a *functional predicate* from $\alpha_1 \times \ldots \times \alpha_m$ to $\beta_1 \times \ldots \times \beta_n$, with respect to a set P of definite clauses, if $M(P) \models \forall X,Y,Z.\ f(X,Y) \wedge f(X,Z) \rightarrow (Y = Z)$, where X is an m-tuple of distinct variables, and Y and Z are n-tuples of distinct variables. Given the atom $f(X, Y)$, we say that X and Y are the tuples of the *input* and *output* variables of f, respectively. Predicate f is said to be *total* if $M(P) \models \forall X \exists Y.\ f(X, Y)$. In what follows, a 'total, functional predicate' f from a tuple α of sorts to a tuple β of sorts will be called a 'total function' and denoted by $f \in [\alpha \rightarrow \beta]$ (the set P of clauses that define f will be understood from the context).

Definition 1 (Generalized Catamorphisms). *A* generalized list catamorphism, *shown in Fig.* 1 (A), *is a total function* $h \in [\sigma \times list(\beta) \rightarrow \varrho]$, *where*: (*i*) σ, β, ϱ, *and* τ *are (products of) basic sorts,* (*ii*) $list(\beta)$ *is the sort of lists of elements of sort* β, (*iii*) $base1$ *is a total function in* $[\sigma \rightarrow \varrho]$, (*iv*) f *is a catamorphism in* $[\sigma \times list(\beta) \rightarrow \tau]$ *and* (*v*) $combine1$ *is a total function in* $[\sigma \times \beta \times \varrho \times \tau \rightarrow \varrho]$. *Similarly, a* generalized tree catamorphism *is a total function* $t \in [\sigma \times tree(\beta) \rightarrow \varrho]$ *defined as shown in Fig.* 1 (B).

Note that the above definition is recursive, that is, the predicates f and g are defined by instances of schemata (A) and (B), respectively. Examples of catamorphisms will be shown in the following section.

$$\text{(A)}\quad \begin{array}{l} h(X,[\,],Y) \leftarrow \mathit{base1}(X,Y).\\ h(X,[H|T],Y) \leftarrow\\ \quad f(X,T,Rf),\\ \quad h(X,T,R),\\ \quad \mathit{combine1}(X,H,R,Rf,Y). \end{array} \qquad \text{(B)}\quad \begin{array}{l} t(X,\mathit{leaf},Y) \leftarrow \mathit{base2}(X,Y).\\ t(X,\mathit{node}(L,N,R),Y) \leftarrow\\ \quad g(X,L,RLg),\ \ g(X,R,RRg),\\ \quad t(X,L,RL),\ \ t(X,R,RR),\\ \quad \mathit{combine2}(X,N,RL,RR,RLg,RRg,Y). \end{array}$$

Fig. 1. (A) Generalized list catamorphism. (B) Generalized tree catamorphism.

3 An Introductory Example

Let us consider the clauses of Fig. 2, which are the result of translating an iterative program for Insertion Sort. (Details on how this translation can be performed are outside the scope of this paper.) The clauses in Fig. 2 will be called *program clauses* and the predicates defined by those clauses will be called *program predicates*.

We have that *ins_sort*(Xs, Ys, Zs) holds if Zs is the ordered list of integers (here and in what follows, the order is with respect to $\leq$) obtained by inserting every element of the list Ys in the proper position of the ordered list Xs. Thus, the result of sorting a list Ys is the list Zs such that *ins_sort*($[\,]$, Ys, Zs) holds. The predicate *ins_sort* depends (see clause 2) on the predicates *empty_list* and *ord_ins*. We have that *empty_list*(L) holds if the list L is empty, and *ord_ins*(Y, $Xs1$, $Xs2$, Ys, Zs) holds if: (i) the concatenation of the lists $Xs1$ and $Xs2$ is ordered, (ii) Y is greater than or equal to the last element of $Xs1$, and (iii) *ins_sort*(Xs', Ys, Zs) holds, where Xs' is the concatenation of $Xs1$ and the ordered list obtained by inserting (according to the $\leq$ order) Y into $Xs2$. As usual, *append*(Xs, Ys, Zs) holds if Zs is the concatenation of the lists Xs and Ys, and *snoc*(Xs, Y, Zs) holds if *append*(Xs, $[Y]$, Zs) holds.

It is not immediate to see why the above properties for *ins_sort* and *ord_ins* are valid. This is also due to the fact that the predicates *ins_sort* and *ord_ins* are mutually recursive. The goal of this paper is to present a technique based on CHC transformations that allows us to automatically prove properties expressed by possibly mutually recursive predicates, using a CHC solver.

Now we will present the clauses that formalize the properties we want to show for our Insertion Sort example. This set of clauses is made out of two subsets: a set of definite CHCs that define the catamorphisms used in the queries, and a set of queries that specify the program properties to be shown.

In Fig. 3 we present the definite CHCs defining the three catamorphisms we will use. They are: (i) *ordered*, which takes as input a list Ls and returns a boolean B such that if Ls is ordered, then $B = \mathit{true}$, otherwise $B = \mathit{false}$; (ii) *first*, which takes as input a list Ls and returns a boolean B and an element F such that if Ls is empty, then $B = \mathit{false}$ and $F = 0$ (this value for F is an arbitrary integer and will not be used elsewhere), otherwise $B = \mathit{true}$ and F is the head of Ls; and (iii) *last*, which is analogous to *first*, except that it returns the last element L, if any, instead of the first element.

The predicates of these catamorphisms are called *property predicates*.

1. $ins\_sort(Xs, [\,], Xs).$
2. $ins\_sort(Xs, [Y|Ys], S) \leftarrow$
   $empty\_list(L),$
   $ord\_ins(Y, L, Xs, Ys, S).$
3. $append([\,], Ys, Ys).$
4. $append([X|Xs], Ys, [X|Zs]) \leftarrow$
   $append(Xs, Ys, Zs).$
5. $snoc([\,], X, [X]).$
6. $snoc([H|T], X, [H|TX]) \leftarrow$
   $snoc(T, X, TX).$
7. $empty\_list([\,]).$
8. $ord\_ins(Y, Xs1, [\,], Ys, Zs) \leftarrow$
   $snoc(Xs1, Y, Xs1Y),$
   $ins\_sort(Xs1Y, Ys, Zs).$
9. $ord\_ins(Y, Xs1, [X|Xs2], Ys, Zs) \leftarrow$
   $Y \leq X,\ snoc(Xs1, Y, Xs1Y),$
   $snoc(Xs1Y, X, Xs1YX),$
   $append(Xs1YX, Xs2, Xs),$
   $ins\_sort(Xs, Ys, Zs).$
10. $ord\_ins(Y, Xs1, [X|Xs2], Ys, Zs) \leftarrow$
   $Y > X,\ snoc(Xs1, X, Xs1X),$
   $ord\_ins(Y, Xs1X, Xs2, Ys, Zs).$

Fig. 2. Program clauses for Insertion Sort.

11. $ordered([\,], B) \leftarrow B.$
12. $ordered([H|T], B) \leftarrow$
   $B = (B1 \Rightarrow (H \leq F \ \&\ B2)),$
   $first(T, B1, F), ordered(T, B2).$
13. $first([\,], B, F) \leftarrow {\sim}B \ \&\ F{=}0.$
14. $first([H|T], B, F) \leftarrow B \ \&\ F{=}H.$
15. $last([\,], B, L) \leftarrow {\sim}B \ \&\ L{=}0.$
16. $last([H|T], B, L) \leftarrow$
   $B \ \&\ L{=}ite(B1, L1, H), last(T, B1, L1).$

Fig. 3. Property clauses for Insertion Sort: the catamorphisms.

Now we present the set of queries that specify the program properties. We assume that: (i) at most one property is specified for each program predicate, and (ii) the property related to the program predicate p is expressed as an implication of the form: $p(\ldots),\ cata_1(\ldots),\ \ldots,\ cata_n(\ldots) \rightarrow d$, where the $cata_i(\ldots)$'s are catamorphisms and d is a constraint. This implication can be expressed as a query by adding ${\sim}d$ to its premise and changing its conclusion to *false*.

For our Insertion Sort example, in Fig. 4 we specify four properties by introducing a query for each of the four program predicates *ins_sort*, *ord_ins*, *append*, and *snoc*.

(1) For *ins_sort*, query $q1$ states that given a list of integers Ys, if Xs is ordered and $ins\_sort(Xs, Ys, Zs)$ holds, then Zs is ordered.

(2) For *ord_ins*, query $q2$ states that if the concatenation of the lists $Xs1$ and $Xs2$ is ordered (that is, $Xs1$ and $Xs2$ are ordered, and the last element of $Xs1$ is less than or equal to the first element of $Xs2$), Y is greater than or equal to the last element of $Xs1$, and $ord\_ins(Y, Xs1, Xs2, Ys, Zs)$ holds, then Zs is ordered.

(3) For *append*, query $q3$ states that if $Xs1$ and $Xs2$ are ordered, and the last element of $Xs1$, if any, is less than or equal to the first element of $Xs2$, if any, and $append(Xs,Ys,Zs)$ holds, then Zs is ordered.

(4) For *snoc*, query *q*4 states that if *Xs* is ordered, the last element of *Xs*, if any, is less than or equal to *X*, and *snoc*(*Xs*, *X*, *XsX*) holds, then *XsX* is ordered.

$$
\begin{array}{lll}
(q1) & \mathit{false} \leftarrow {\sim}(B1 \Rightarrow B2),\ \mathit{ordered}(Xs,B1),\ \mathit{ordered}(Zs,B2),\ \mathit{ins\_sort}(Xs,Ys,Zs). & \\
(q2) & \mathit{false} \leftarrow {\sim}((B1\ \&\ B2\ \&\ ((B4\ \&\ B5) \Rightarrow L \leq F)\ \&\ (B4 \Rightarrow L \leq Y)) \Rightarrow B3), & \\
 & \quad \mathit{ordered}(Xs2, B2),\ \mathit{ordered}(Zs, B3), & (q2.1) \\
 & \quad \mathit{ordered}(Xs1, B1),\ \mathit{last}(Xs1, B4, L),\ \mathit{first}(Xs2, B5, F), & (q2.2) \\
 & \quad \mathit{ord\_ins}(Y, Xs1, Xs2, Ys, Zs). & \\
(q3) & \mathit{false} \leftarrow {\sim}((B1\ \&\ B2\ \&\ ((B4\ \&\ B5) \Rightarrow L \leq F)) \Rightarrow B3), & \\
 & \quad \mathit{ordered}(Xs, B1),\ \mathit{ordered}(Ys, B2), & \\
 & \quad \mathit{ordered}(Zs, B3),\ \mathit{last}(Xs, B4, L),\ \mathit{first}(Ys, B5, F),\ \mathit{append}(Xs, Ys, Zs). & \\
(q4) & \mathit{false} \leftarrow {\sim}((B1\ \&\ (B2 \Rightarrow L \leq X)) \Rightarrow B3), & \\
 & \quad \mathit{ordered}(Xs, B1),\ \mathit{last}(Xs, B2, L),\ \mathit{ordered}(XsX, B3),\ \mathit{snoc}(Xs, X, XsX). &
\end{array}
$$

Fig. 4. Property clauses for Insertion Sort: the queries.

At this point, we can check whether or not the properties expressed by the queries *q*1–*q*4 do hold by checking the satisfiability of the program clauses together with the property clauses. In order to perform that satisfiability check, we have used the state-of-the-art CHC solver SPACER, based on Z3 [19]. SPACER failed to return an answer within five minutes. The weakness of CHC solvers for examples like the one presented here, motivates our technique based on CHC transformation. This technique produces an equisatisfiable set of CHCs whose satisfiability can hopefully be easier to verify. Indeed, in our example, SPACER succeeds to prove the satisfiability of the new set of CHCs produced by our transformation algorithm. In Sect. 7, we will show that our technique improves the effectiveness of state-of-the-art solvers on a non-trivial benchmark.

4 Catamorphism-Based Queries

As already mentioned, the translation of a program verification problem to CHCs usually generates two disjoint sets of clauses: (i) the set of *program clauses*, and (ii) the set of *property clauses*, with the associated sets of *program predicates*, and *property predicates*, respectively. Without loss of generality, we will assume that property predicates may occur in the property clauses only. Moreover, in order to define a class of CHCs for which our transformation algorithm (see Sect. 6) always terminates, we will consider property predicates that are *catamorphisms*.

In the sequel we need the following definitions. An atom is said to be a *program atom* (or a *catamorphism atom*), if its predicate symbol is a program predicate (or a catamorphism, respectively). When we write a catamorphism atom as *cata*(*X*, *T*, *Y*), we stipulate that *X* is a (tuple of) input basic variable(s), *T* is the input ADT variable, and *Y* is a (tuple of) output basic variable(s).

Definition 2. *A* catamorphism-based query *is a query of the form:*

$$false \leftarrow c,\ cata_1(X_1,T_1,Y_1),\ \ldots,\ cata_n(X_n,T_n,Y_n),\ pred(Z)$$

where: (i) pred is a program predicate and Z is a tuple of distinct variables, (ii) c is a constraint such that $vars(c) \subseteq \{X_1,\ldots,X_n,Y_1,\ldots,Y_n,Z\}$*, (iii)* $cata_1$*, ...,* $cata_n$ *are catamorphism atoms, (iv)* $Y_1,\ldots,Y_n$ *are pairwise disjoint tuples of distinct variables of basic sort not occurring in* $X_1,\ldots,X_n,Z$*, (v)* $T_1,\ldots,T_n$ *are ADT variables occurring in Z.*

The queries of Fig. 4 are examples of catamorphism-based queries. Many interesting program properties can be defined as catamorphism-based queries, although, in general, this might require some ingenuity.

5 Transformation Rules

In this section we present the rules that we use for transforming CHCs. These rules are variants of the usual transformation rules for CHCs (and CLP programs), suitably adapted to our context here, where we use catamorphisms. Then, we prove the soundness and completeness of those rules.

The goal of the transformation rules is to incorporate catamorphisms into program predicates, that is, to derive for each program predicate p, a new predicate *newp* whose definition is given by the conjunction of an atom for p with the catamorphism atoms useful for showing the satisfiability of the query relative to p. Those catamorphisms and their associated constraints may come both from the query relative to p and from the queries relative to predicates on which p depends. In this section we will indicate how this transformation can be done referring to our Insertion Sort example, while in the next section, we will present a transformation algorithm to perform this task in an automatic way.

In what follows, for reasons of simplicity, we assume that for every clause D: $H \leftarrow c, G$, every variable of ADT sort occurring in D also occurs in a program atom in G. This assumption is not restrictive because for every variable X of ADT sort τ, we can add a new program atom $true_\tau(X)$ that holds for every X of sort τ. For instance, if τ is *list*(*int*), we can introduce an atom $true_{list(int)}(X)$ defined by the clauses:

$$true_{list(int)}([\,]). \qquad true_{list(int)}([H|T]) \leftarrow true_{list(int)}(T).$$

A *transformation sequence from* S_0 *to* S_n is a sequence $S_0 \Rrightarrow S_1 \Rrightarrow \ldots \Rrightarrow S_n$ of sets of CHCs such that, for $i=0,\ldots,n-1$, S_{i+1} is derived from S_i, denoted $S_i \Rrightarrow S_{i+1}$, by performing a transformation step consisting in applying one of the following rules R1–R4.

(R1) *Definition Rule.* Let D be a clause of the form $newp(X_1,\ldots,X_k) \leftarrow c, Catas, A$, where: (1) *newp* is a predicate symbol not occurring in the sequence $S_0 \Rrightarrow S_1 \Rrightarrow \ldots \Rrightarrow S_i$ constructed so far, (2) $\{X_1,\ldots,X_k\} = vars(\{Catas, A\})$,

(3) c is a constraint such that $vars(c) \subseteq vars(\{Catas, A\})$, (4) $Catas$ is a conjunction of catamorphism atoms, with $adt\text{-}vars(Catas) \subseteq adt\text{-}vars(A)$, and (5) A is a program atom. By *definition introduction* we may add D to S_i and get the new set $S_{i+1} = S_i \cup \{D\}$.

For $j = 0, \ldots, n$, by $Defs_j$ we denote the set of clauses, called *definitions*, introduced by rule R1 during the construction of the sequence $S_0 \Rrightarrow S_1 \Rrightarrow \ldots \Rrightarrow S_j$. Thus, $Defs_0 = \emptyset$, and for $j = 0, \ldots, n$, $Defs_j \subseteq Defs_{j+1}$.

Example 1. In our Insertion Sort example, the set S_0 consists of all the clauses shown in Figs. 2, 3, and 4, and we start off by introducing the following definition (with constraint *true*), whose body consists of the atoms in the body of query *q1*:

$D1.\ new1(Xs,B1,Zs,B2,Ys) \leftarrow ordered(Xs,B1),\ ordered(Zs,B2),\ ins\_sort(Xs,Ys,Zs).$
Thus, $S_1 = S_0 \cup \{D1\}$. □

The clauses for *newp* are obtained by: (i) first, *unfolding* the definition of *newp*, then (ii) incorporating some catamorphism atoms and constraints provided by the queries into the clauses derived by unfolding, and (iii) finally, folding using suitable new definitions.

Now, we introduce an unfolding rule (see R2 below), which actually is the composition of some unfolding steps followed by the application of the functionality property presented in previous work [11]. Let us first define the notion of the one-step unfolding which is a step of symbolic evaluation performed by applying once the resolution rule.

Definition 3 *(One-step Unfolding). Let D: $H \leftarrow c, L, A, R$ be a clause, where A is an atom, and let P be a set of definite clauses with $vars(D) \cap vars(P) = \emptyset$. Let Cls: $\{K_1 \leftarrow c_1, B_1,\ \ldots,\ K_m \leftarrow c_m, B_m\}$, with $m \geq 0$, be the set of clauses in P, such that: for $j = 1, \ldots, m$, (i) there exists a most general unifier ϑ_j of A and K_j, and (ii) the conjunction of constraints $(c, c_j)\vartheta_j$ is satisfiable. The* one-step unfolding *produces the following set of CHCs:*

$$Unf(D, A, P) = \{(H \leftarrow c, c_j, L, B_j, R)\vartheta_j \mid j = 1, \ldots, m\}.$$

In the following Rule R2 and in the sequel, *Catas* denotes a conjunction of catamorphism atoms.

(R2) *Unfolding Rule.* Let D: $newp(U) \leftarrow c, Catas, A$ be a definition in $S_i \cap Defs_i$, where A is a program atom, and P be the set of definite clauses in S_i. We derive a new set *UnfCls* of clauses by the following three steps.

Step 1. (*One-step unfolding of the program atom*) $UnfCls := Unf(D, A, P)$;
Step 2. (*Unfolding of the catamorphism atoms*)
while there exists a clause E: $H \leftarrow d, L, C, R$ in *UnfCls*, for some conjunctions L and R of atoms, such that C is a catamorphism atom whose argument of ADT sort is not a variable **do**
$UnfCls := (UnfCls \setminus \{E\}) \cup Unf(E, C, P)$;

Step 3. (Applying Functionality)
while there exists a clause E: $H \leftarrow d, L, cata(X,T,Y1), cata(X,T,Y2), R$ in *UnfCls*, for some catamorphism *cata* **do**
$UnfCls := (UnfCls - \{E\}) \cup \{H \leftarrow d, Y1{=}Y2, L, cata(X,T,Y1), R\}$;

Then, by *unfolding D* we derive $S_{i+1} = (S_i \setminus \{D\}) \cup UnfCls$.

Example 2. For instance, in our Insertion Sort example, by unfolding definition $D1$, at Step 1 we replace $D1$ by:
$E1.\ new1(A,B,A,C,[\,]) \leftarrow ordered(A,B),\ ordered(A,C).$
$E2.\ new1(A,B,C,D,[E|F]) \leftarrow ordered(A,B),\ ordered(C,D),$
$empty\_list(G),\ ord\_ins(E,G,A,F,C).$
Step 2 of the unfolding rule is not performed in this example. At Step 3, clause $E1$ is replaced by:
$E3.\ new1(A,B,A,C,[\,]) \leftarrow B{=}C, ordered(A,B).$
Thus, $S_2 = S_0 \cup \{E2, E3\}$. □

By the *query-based strengthening rule* we can use the queries occurring in the set of CHCs whose satisfiability is under verification for strengthening the body of the other clauses with the addition of catamorphism atoms and constraints.

(R3) *Query-based Strengthening Rule.* Let $S_i = P \cup Q$, where P is a set of definite clauses obtained by applying the unfolding rule, and Q is a set of catamorphism-based queries, and let C: $H \leftarrow c, Catas^C, A_1, \ldots, A_m$ be a clause in P, whose program atoms are $A_1, \ldots, A_m$. Let E be the clause derived from C as follows:

for $k = 1, \ldots, m$ **do**

- consider program atom A_k; let $Catas^C_k$ be the conjunction of every catamorphism atom F in $Catas^C$ such that $adt\text{-}vars(A_k) \cap adt\text{-}vars(F) \neq \emptyset$;
- **if** in Q there exists a query (modulo variable renaming) of the form
q_k: $false \leftarrow c_k, cata_1(X_1,T_1,Y_1), \ldots, cata_n(X_n,T_n,Y_n), A_k$
where $Y_1, \ldots, Y_n$ do not occur in C, and the conjunction $cata_1(X_1,T_1,Y_1)$, $\ldots$, $cata_n(X_n,T_n,Y_n)$ can be split into two subconjunctions B_1 and B_2 such that: (i) a variant of B_1 is a subconjunction of $Catas^C_k$, and (ii) for every catamorphism atom $cata_i(X_i,T_i,Y_i)$ in B_2, we have that in $Catas^C_k$ there is no catamorphism atom of the form $cata_i(V,T_i,W)$ (that is, there is no catamorphism atom acting on the same ADT variable T_i)
then add the conjunction $\sim c_k, B_2$ to the body of C.

Then, by *query-based strengthening of C using Q*, we get $S_{i+1} = (S_i \setminus \{C\}) \cup \{E\}$.

Example 3. In our Insertion Sort example, let us consider clause $E2$ and the set Q of catamorphism-based queries listed in Fig. 4. Clause $E2$ has two program atoms: $empty\_list(G)$ and $ord\_ins(E,G,A,F,C)$. Now, in the set Q there exists no query associated with $empty\_list(G)$, while there exists one, namely

q_2, associated with $ord\_ins(E,G,A,F,C)$. By query-based strengthening of $E2$ using Q (in particular, we use query $q2$ only), we get the new clause:

$E4.\ new1(A,B,C,D,[E|F]) \leftarrow$
$\quad \big(L \,\&\, B \,\&\, ((J\&H) \Rightarrow K \leq I) \,\&\, (J \Rightarrow K \leq E)\big) \Rightarrow D,$
$\quad ordered(A,B),\ ordered(C,D),$ $\qquad (B_1)$
$\quad ordered(G,L),\ last(G,J,K),\ first(A,H,I),$ $\qquad (B_2)$
$\quad empty\_list(G),\ ord\_ins(E,G,A,F,C).$

where: (i) subconjunctions B_1 and B_2, mentioned in Rule R3, have been written in lines (B_1) and (B_2), respectively, and (ii) the variables of sort *list*(*int*) are A, C, G, and F. Subconjunctions B_1 and B_2 are suitable variants of the catamorphisms occurring in lines ($q2.1$) and ($q2.2$), respectively, of query $q2$, as shown in Fig. 4. Thus, we get the new set of clauses $S_3 = S_0 \cup \{E3, E4\}$. □

The *folding rule* allows us to replace a conjunction of a program atom and catamorphisms by a single atom, whose predicate has been introduced in a previous application of the Definition Rule.

(R4) *Folding Rule.* Let C: $H \leftarrow c, Catas^C, A_1, \ldots, A_m$ be a clause in S_i, where either H is *false* or C has been obtained by the unfolding rule, possibly followed by query-based strengthening. For $k = 1, \ldots, m$,

- let $Catas^C_k$ be the conjunction of every catamorphism atom F in $Catas^C$ such that $adt\text{-}vars(A_k) \cap adt\text{-}vars(F) \neq \emptyset$;
- let D_k: $H_k \leftarrow d_k, Catas^D_k, A_k$ be a clause in $Defs_i$ (modulo variable renaming) such that: (i) $\mathbb{D} \models \forall(c \rightarrow d_k)$, and (ii) $Catas^C_k$ is a subconjunction of $Catas^D_k$.

Then, by *folding C using $D_1, \ldots, D_m$*, we derive clause E: $H \leftarrow c, H_1, \ldots, H_m$, and we get $S_{i+1} = (S_i \setminus \{C\}) \cup \{E\}$.

Example 4. In order to fold clause $E4$, we introduce two new definitions, one for each program atom occurring in the body of that clause, as follows (the new predicate names are introduced by our tool VeriCaT$_{mq}$ (see Sect. 7)):

$D2.\ new2(A,B,C,D,E,F,G,H,I,J,K,L) \leftarrow ordered(A,B),\ ordered(C,D),$
$\quad ordered(E,F),\ last(A,G,H),\ first(C,I,J),\ ord\_ins(K,A,C,L,E).$

$D3.\ new19(A,B,C,D) \leftarrow ordered(A,B),\ last(A,C,D),\ empty\_list(A).$

Thus, $S_4 = S_0 \cup \{E3, E4, D2, D3\}$. Now, we apply Rule R4, and from clause $E4$ we get:

$E5.\ new1(A,B,C,D,[E|F]) \leftarrow \big(G \,\&\, B \,\&\, ((H\&I) \Rightarrow J \leq K) \,\&\, (H \Rightarrow J \leq E)\big) \Rightarrow D,$
$\quad new2(L,G,A,B,C,D,H,J,I,K,E,F),\ new19(L,G,H,J).$

Then, $S_5 = S_0 \cup \{E3, E5, D2, D3\}$. Also, query $q1$ can be folded using definition $D1$, and we get:

$E6.\ false \leftarrow {\sim}(B1 \Rightarrow B2),\ new1(Xs, B1, Zs, B2, Ys).$

Thus, $S_6 = (S_0 \setminus \{q1\}) \cup \{E3, E5, E6, D2, D3\}$. Then, the transformation will continue by looking for the clauses relative to the newly introduced predicates $new2$ (see Definition $D2$) and $new19$ (see Definition $D3$). □

The key for understanding our transformation technique is to observe that in our Insertion Sort example, the new predicate *new*1, defined by clauses $E3$ and $E5$, incorporates the program predicate *ins_sort* together with the catamorphisms occurring in $q1$. Note also that clause $E5$ has constraints that come from query-based strengthening using query $q2$. Indeed, $E5$ is obtained by folding $E4$, which in turn has been derived by strengthening $E2$.

The advantage of performing this transformation is that, when checking the satisfiability of the new set of clauses where $q1$ has been replaced by $E6$, the solver can look for a model of $new1(Xs,B1,Zs,B2,Ys)$ where the constraint $B1 \Rightarrow B2$ holds, instead of looking in a separate way for models of $ordered(Xs,B1)$, $ordered(Zs,B2)$, and $ins\_sort(Xs,Ys,Zs)$ whose conjunction implies $B1 \Rightarrow B2$.

The following theorem, whose proof is sketched in Appendix 1 of a preliminary version of this paper [10], states the correctness of the transformation rules.

Theorem 1 (Soundness and Completeness of the Rules). *Let* $S_0 \Mapsto S_1 \Mapsto \ldots \Mapsto S_n$ *be a transformation sequence using rules* R1–R4. *Then,* S_0 *is satisfiable if and only if* S_n *is satisfiable.*

Note that the applicability conditions of R3 disallow the application of that rule to a query. Otherwise, we could easily get a satisfiable clause from an unsatisfiable one. Indeed, we could transform $false \leftarrow c(Y), cata(X,Y), p(X)$ into $false \leftarrow {\sim} c(Y) \,\&\, c(Y), cata(X,Y), p(X)$. Note also that folding a clause using itself is not allowed, thus avoiding the transformation of $H \leftarrow c, Catas, A$ into the trivially satisfiable clause $H \leftarrow c, H$.

The applicability conditions of the rules force a sequence of the transformation rules which is fixed, once the new definitions to be introduced are provided. The algorithm that we will present in the next section shows how these definitions can be introduced in an automatic way.

6 Transformation Algorithm

In this section we present an algorithm, called $\mathcal{T}_{mq}$, which given a set P of definite clauses and a set Q of queries, introduces a set of new predicates and transforms $P \cup Q$ into a new set $P' \cup Q'$ such that: (i) $P \cup Q$ is satisfiable if and only if so is $P' \cup Q'$, and (ii) each new predicate defined in $P' \cup Q'$ is equivalent to the conjunction of a program predicate and catamorphisms occurring in $P \cup Q$. As an effect of the application of the query-based strengthening rule, the transformed clauses also exploit the interdependencies among the queries in Q. This transformation is effective, in particular, in the presence of mutually recursive predicates, like in our Insertion Sort example, where we are able to get new clauses for checking the satisfiability of query $q1$ that incorporate the catamorphisms and constraints of query $q2$, and vice versa.

The set of new definitions needed by $\mathcal{T}_{mq}$ is computed as the least fixpoint of an operator $\tau_{P,Q}$ that transforms a set $\varDelta$ of definitions into a new set $\varDelta'$. For introducing that operator, we need some preliminary definitions.

Definition 4. *A* generalization *of a pair* (c_1, c_2) *of constraints is a constraint, denoted* $\alpha(c_1, c_2)$*, such that* $\mathbb{D} \models \forall(c_1 \rightarrow \alpha(c_1, c_2))$ *and* $\mathbb{D} \models \forall(c_2 \rightarrow \alpha(c_1, c_2))$ *[12]. The* projection *of a constraint c onto a tuple V of variables is a constraint* $\pi(c, V)$ *such that: (i)* $vars(\pi(c, V)) \subseteq V$ *and (ii)* $\mathbb{D} \models \forall(c \rightarrow \pi(c, V))$.

A set Δ of definitions is *monovariant* if, for each program predicate p, there is at most one definition in Δ having an occurrence of p in its body.

Definition 5. *Let* D_1*:* $newp1(U_1) \leftarrow c_1, Catas_1, p(Z)$ *and* D_2*:* $newp2(U_2) \leftarrow c_2, Catas_2, p(Z)$ *be two definitions for the same predicate p. We say that* D_2 *is an* extension *of* D_1*, written* $D_1 \sqsubseteq D_2$*, if (i)* $Catas_1$ *is a subconjunction of* $Catas_2$*, and (ii)* $\mathbb{D} \models \forall(c_1 \rightarrow c_2)$*. Let* Δ_1 *and* Δ_2 *be two monovariant sets of definitions. We say that* Δ_2 *is an* extension *of* Δ_1*, written* $\Delta_1 \sqsubseteq \Delta_2$*, if for each* D_1 *in* Δ_1 *there exists* D_2 *in* Δ_2 *such that* $D_1 \sqsubseteq D_2$.

Given a set *Cls* of clauses and a set Δ of definitions, the *Define* function (see Fig. 5) derives a set Δ' of definitions that can be used for folding all clauses in *Cls*. If Δ is monovariant, then also Δ' is monovariant. In particular, due to the (Project) case, Δ' contains a definition for each program predicate occurring in the body of the clauses in *Cls*. Due to the (Extend) case, $\Delta \sqsubseteq \Delta'$.

The *Unfold* and *Strengthen* functions (see Fig. 5) apply the unfolding and query-based strengthening rules, respectively, to sets of clauses.

Now, we define the operator $\tau_{P,Q}$ as follows:

$$\tau_{P,Q}(\Delta) = \begin{cases} \mathit{Define}(Q, \emptyset) & \text{if } \Delta = \emptyset \\ \mathit{Define}(\mathit{Strengthen}(\mathit{Unfold}(\Delta, P), Q), \Delta) & \text{otherwise} \end{cases}$$

In the case where Δ is empty, $\tau_{P,Q}(\Delta)$ introduces by the *Define* function (Project case), a new definition for each program predicate occurring in a query in Q. In the case where Δ is not empty, $\tau_{P,Q}(\Delta)$ is an extension of Δ obtained by first unfolding all clauses in Δ, then applying the query-based strengthening rule to the clauses derived by unfolding, and finally applying the *Define* function.

The *Define* function is parametric with respect to the generalization operator α (see the Extend case). In our implementation we use an operator based on widening [12] that ensures *stabilization*, that is, for any infinite sequence $c_0, c_1, \ldots$ of constraints and any sequence defined as (i) $d_0 = c_0$, and (ii) $d_{k+1} = \alpha(d_k, c_{k+1})$, there exists $m \geq 0$ such that $d_m = d_{m+1}$. The *Strengthen* function ensures that there is a bound on the number of catamorphisms that can be present in a definition. Since, as already mentioned, $\tau_{P,Q}$ is monotonic with respect to $\sqsubseteq$, its least fixpoint $lfp(\tau_{P,Q})$ is equal to $\tau^n_{P,Q}(\emptyset)$, for some finite number n of iterations. Note that, by construction, $lfp(\tau_{P,Q})$ is monovariant.

Once we have computed the set $lfp(\tau_{P,Q})$ of definitions, we can use those definitions for folding all clauses derived by unfolding and strengthening by applying the *Fold* function. Thus, the transformation algorithm is defined as follows:

$$\mathcal{T}_{mq}(P, Q) = \mathit{Fold}\big(\mathit{Strengthen}\big(\mathit{Unfold}(\Delta, P), Q\big), lfp(\tau_{P,Q})\big)$$

Function *Define*(Cls, Δ): a set *Cls* of clauses; a monovariant set Δ of definitions.
Define(Cls, Δ) returns a monovariant set Δ' of new definitions computed as follows.
$\Delta' := \Delta$;
for each clause C: $H \leftarrow c, G$ in *Cls*, where G contains at least one ADT variable **do**
for each program atom A in G **do**
let $Catas_A$ be the conjunction of every catamorphism atom F in G such that $adt\text{-}vars(A) \cap adt\text{-}vars(F) \neq \emptyset$

• **(Skip) if** in Δ' there is a clause $newp(U) \leftarrow d, B, A$, for any conjunction B of catamorphism atoms, such that: (i) $Catas_A$ is a subconjunction of B, and (ii) $\mathbb{D} \models \forall(c \rightarrow d)$, **then** skip;

• **(Extend) else if** the definition for the predicate of A in Δ' is the clause D: $newp(U) \leftarrow d, B, A$, where B is a conjunction of catamorphism atoms, and either (i) $Catas_A$ is *not* a subconjunction of B, or (ii) $\mathbb{D} \not\models \forall(c \rightarrow d)$, **then** introduce definition $ExtD$: $extp(V) \leftarrow \alpha(d, c), A, B'$, where: (i) $extp$ is a new predicate symbol, (ii) $V = vars(\{\alpha(d, c), B', A\})$, and B' is the conjunction of the distinct catamorphism atoms occurring either in B or in $Catas_A$;
$\Delta' := (\Delta' \setminus \{D\}) \cup \{ExtD\}$;

• **(Project) else if** there is no clause in Δ' of the form $K \leftarrow d, B, A$, for any conjunction B of catamorphism atoms,
then introduce definition D: $newp(U) \leftarrow \pi(c, I), A, Catas_A$, where: (i) $newp$ is a new predicate symbol, (ii) I are the input variables of basic sort in $\{A, Catas_A\}$, and (iii) $U = vars(\{\pi(c, I), A, Catas_A\})$;
$\Delta' := \Delta' \cup \{D\}$;

Function *Unfold*(Δ, P): a set $\Delta = \{D_1, \ldots, D_n\}$ of definitions; a set P of definite clauses.
Unfold$(\Delta, P) = \bigcup_{i=1}^{n} C_i$, where C_i is the set of clauses derived by unfolding D_i.

Function *Strengthen*(Cls, Q): a set $Cls = \{C_1, \ldots, C_n\}$ of clauses; a set Q of catamorphism-based queries with at most one query for each program predicate in *Cls*.
Strengthen$(Cls, Q) = \{E_i \mid E_i$ is derived from C_i by query-based strengthening using $Q\}$

Function *Fold*(Cls, Δ): a set $Cls = \{C_1, \ldots, C_n\}$ of clauses; a monovariant set Δ of definitions.
Fold$(Cls, \Delta) = \{E_i \mid E_i$ is derived from C_i by folding C_i using the definitions in $\Delta\}$.

Fig. 5. The *Define*, *Unfold*, *Strengthen*, and *Fold* functions.

The termination of $\mathcal{T}_{mq}$ follows immediately from the fact that $lfp(\tau_{P,Q})$ is computed in a finite number of steps.

Theorem 2 (Termination of Algorithm $\mathcal{T}_{mq}$). *Algorithm $\mathcal{T}_{mq}$ terminates for any set P of definite clauses and set Q of catamorphism-based queries.*

By the soundness and completeness of the transformation rules (see Theorem 1), we also get the following result.

Theorem 3 (Soundness and Completeness of Algorithm $\mathcal{T}_{mq}$). *For any set P of definite clauses and Q of catamorphism-based queries, $P \cup Q$ is satisfiable if and only if $\mathcal{T}_{mq}(P, Q)$ is satisfiable.*

We conclude this section by showing the sequence of definitions for the program predicate *snoc* computed by iterating the applications of $\tau_{P,Q}$ in the Insertion Sort example. The definitions in $lfp(\tau_{P,Q})$ are listed in Appendix 2 of a preliminary version of this paper [10].

(α) $new4(A,B,C,D,E,F,G) \leftarrow ordered(A,B),\ last(A,C,D),$
$ordered(E,F),\ snoc(A,G,E).$

(β) $new5(A,B,C,D,E,F,G,H,I,J,K) \leftarrow ordered(E,J),\ last(E,F,G),$
$ordered(A,D),\ first(E,H,I),\ first(A,B,C),\ snoc(E,K,A).$

(γ) $new13(A,B,C,D,E,F,G,H,I,J,K,L,M) \leftarrow ordered(A,B), last(A,C,D),$
$ordered(E,F), last(E,G,H), first(E,J,K), first(A,L,M), snoc(A,I,E).$

Note that these three definitions are in the $\sqsubseteq$ relation, that is, $\alpha \sqsubseteq \beta \sqsubseteq \gamma$.

7 Experimental Evaluation

We have implemented algorithm $\mathcal{T}_{mq}$ in a tool, called VeriCaT$_{mq}$, which extends VeriCaT [11] by guaranteeing a sound and complete transformation. VeriCaT$_{mq}$ is based on: (i) VeriMAP [8] for transforming CHCs, and (ii) SPACER (with Z3 4.11.2) to check the satisfiability of the transformed CHCs.

We have considered 170 problems, as sets of CHCs, with 470 queries in total, equally divided between the class of satisfiable (sat) problems and unsatisfiable (unsat) problems (for each class we have 85 problems and 235 queries).

The problems considered refer to programs that manipulate: (i) lists of integers by performing concatenation, permutation, reversal, and sorting, and (ii) binary search trees, by inserting and deleting elements. For list manipulating programs, we have considered properties such as: list length, minimum and maximum element, sum of elements, list content as sets or multisets of elements, and list sortedness (in ascending or descending order). For trees, we have considered tree size, tree height, minimum and maximum element, tree content and the binary search tree property.

The problems considered here are derived from those of the benchmark set of previous work [11] with the following important differences. We have considered additional satisfiable problems (for instance, those related to Heapsort). In addition to satisfiable problems, we have also considered unsatisfiable problems that have been obtained from their satisfiable counterparts by introducing bugs in the programs: for instance, by not inserting an element in a list, or adding an extra constraint, or replacing a non-empty tree by an empty one. Note also that the transformed CHCs produced by $\mathcal{T}_{mq}$ contain both basic variables and ADT variables, whereas those produced by the method presented in previous work [11] contain basic variables only.

For comparing the effectiveness of our method with that of a state-of-the-art CHC solver, we have also run SPACER (with Z3 4.11.2) on the original, non-transformed CHCs, in SMT-LIB format.

In Table 1 we summarize the results of our experiments.[1] The first three columns report the name of the program, the total number of problems and queries for each program. The fourth and fifth columns report the number of satisfiable and unsatisfiable problems proved by SPACER before transformation, whereas the last two columns report the number of satisfiable and unsatisfiable problems proved by VeriCaT$_{mq}$.

In summary, VeriCaT$_{mq}$ was able to prove all the 170 considered problems whereas SPACER was able to prove the properties of 84 'unsat' problems out of 85, and none of the 'sat' problems. The total time needed for transforming the CHCs was 275 s (1.62 s per problem, on average), and checking the satisfiability of the transformed CHCs took about 174 s in total (about 1s average time, 0.10 s median time). For comparison, SPACER took 30.27 s for checking the unsatisfiability of 84 problems (0.36 s average time, 0.15 s median time). The benchmark and the tool are available at https://fmlab.unich.it/vericatmq.

Table 1. Programs and problems proved by SPACER and VeriCaT$_{mq}$.

Program	Problems	Queries	SPACER		VeriCaT$_{mq}$	
			sat	unsat	sat	unsat
List Membership	2	6	0	1	1	1
List Permutation	8	24	0	4	4	4
List Concatenation	18	18	0	8	9	9
Reverse	20	40	0	10	10	10
Double Reverse	4	12	0	2	2	2
Reverse w/Accumulator	6	18	0	3	3	3
Bubblesort	12	36	0	6	6	6
Heapsort	8	48	0	4	4	4
Insertionsort	12	24	0	6	6	6
Mergesort	18	84	0	9	9	9
Quicksort (version 1)	12	38	0	6	6	6
Quicksort (version 2)	12	36	0	6	6	6
Selectionsort	14	42	0	7	7	7
Treesort	4	20	0	2	2	2
Binary Search Tree	20	24	0	10	10	10
Total	170	470	0	84	85	85

For instance, for all the list sorting programs we have considered (Bubblesort, Heapsort, Insertionsort, Mergesort, Quicksort, Selectionsort and Treesort), VeriCaT$_{mq}$ was able to prove properties stating that the output list is sorted

[1] Experiments have been performed on an Intel Xeon CPU E5-2640 2.00 GHz with 64GB RAM under CentOS with a time limit of 300 s per problem.

and has the same multiset of elements of the input list. Similarly, VeriCaT_{mq} was able to prove that those properties do *not* hold, if extra elements are added to the output list, or some elements are not copied from the input list to the output list, or a wrong comparison operator is used.

The results obtained by the VeriCaT_{mq} prototype implementation of our method are encouraging and show that, when used in combination with state-of-the-art CHC solvers, it can greatly improve their effectiveness to prove satisfiability of sets of CHCs with multiple queries, while it does not inhibit their remarkable ability to prove unsatisfiability, although some extra time due to the transformation process may be required.

8 Conclusions and Related Work

Many program verification problems can be translated into the satisfiability problem for sets of CHCs that include more than one query. A notable example is the case where we want to verify the correctness of programs made out of several functions, each of which has its pre-/post-conditions [14]. We have proposed an algorithm, called $\mathcal{T}_{mq}$, for transforming a set of CHCs with multiple queries into a new, equisatisfiable set of CHCs that incorporate suitable information about the set of queries contained in the initial set. The advantage gained is that, in order to prove the satisfiability of the transformed CHCs, the CHC solver may exploit the mutual interactions among the satisfiability proofs of the various queries. We have identified a class of queries that specify program properties using catamorphisms on ADTs, such as lists and trees, for which $\mathcal{T}_{mq}$ terminates. We have implemented algorithm $\mathcal{T}_{mq}$ and shown that it improves the effectiveness of the state-of-the-art CHC solver SPACER [19] on a non trivial benchmark. As shown by the experimental results, our technique is particularly effective on satisfiability problems where several recursively defined predicates may have mutual dependencies. In these problems, catamorphisms are defined by structural recursion, but program predicates need not. For instance, the *quicksort* predicate is recursively defined on lists computed by the *partition* predicate, and those lists need not be subterms of the input list.

Algorithm $\mathcal{T}_{mq}$ improves over the transformation algorithm $\mathcal{T}_{cata}$ presented in a previous paper [11], which works by *eliminating* ADTs from sets of CHCs. Instead of the contracts handled by $\mathcal{T}_{cata}$, algorithm $\mathcal{T}_{mq}$ considers queries, and thus, its input is simply a set of CHCs. More importantly, $\mathcal{T}_{mq}$ is sound and complete, in the sense that the set of the initial and transformed CHCs are equisatisfiable sets, whereas $\mathcal{T}_{cata}$ is only sound (indeed, it can be seen as computing an *abstraction* of the initial clauses), and thus if the transformed clauses are unsatisfiable we cannot infer anything about the satisfiability of the initial CHCs. A key point to achieve completeness is that we use a definition rule that introduces new predicates whose arguments are *all* the variables occurring in their defining body. Correspondingly, the folding rule will not remove the ADT variables. Completeness is very important in practice, because proving that a set of clauses is unsatisfiable and finding a counterexample can help identify program bugs. On the theoretical side, a new contribution of this paper with respect previous work, is that the set of new definitions needed for the transformation is computed as the least fixpoint of a suitable operator. This approach is related to techniques that combine in a single theoretical framework both partial evaluation and abstract interpretation of logic programs [21].

The experimental evaluation reported in Sect. 7 shows that our transformation-based verification technique is able to dramatically improve the effectiveness of the SPACER solver for satisfiable sets of CHCs (where the results of SPACER are very poor), while retaining the excellent results of the solver for unsatisfiable sets of CHCs (for which SPACER is, at least in principle, complete).

Decision procedures for suitable classes of first order formulas defined on catamorphisms [23, 24] have been used in program verifiers [25]. However, we do not propose here any specific decision procedure for catamorphisms and, instead, we transform a set of CHCs with catamorphisms into a new set of CHCs where catamorphisms are, in a sense, compiled away.

Type-based norms, which are a special kind of integer-valued catamorphisms, were used for proving termination of logic programs [6] and for *resource analysis* [1] via *abstract interpretation*. Similar abstract interpretation techniques are also implemented in the CiaoPP preprocessor [15] of the Ciao logic programming system. In our approach we do not need to specify *a priori* any abstract domain where to perform the analysis, and instead, by transformation, we generate new CHCs which incorporate the relations defined by the constraints on the catamorphisms. The problem of showing the satisfiability of CHCs defined on ADTs is a very hot topic and various approaches have been proposed in recent work, including: (i) a proof system that combines inductive theorem proving with CHC solving [26], (ii) lemma generation based on syntax-guided synthesis from user-specified templates [27], (iii) invariant discovery based on finite tree automata [20], and (iv) use of suitable abstractions [13]. In particular, Govind *et al.* [13] propose an algorithm based on IC3 [5] for solving CHCs modulo catamorphisms. Their algorithm processes one query at a time, and their definition of catamorphisms is slightly more restrictive than the one presented here.

A limitation of our approach is that the effectiveness of the transformation may depend on the set of properties specified by the queries. For instance, it may happen that programmers provide *partial* program specifications (e.g., for a subset of the program functions), and therefore queries for some program predicates only, such as the main program predicates (e.g., *ins_sort* and *ord_ins* in our example). If this is the case, it is essential to have a mechanism that is able to infer from the queries the unspecified catamorphisms for the remaining program predicates (e.g., *append* and *snoc* on which *ins_sort* and *ord_ins* depend). For future work, we plan to extend $\mathcal{T}_{mq}$ so as to be able to propagate the catamorphisms specified in the queries to those program predicates for which no query has been provided.

Acknowledgement. The authors warmly thank the anonymous reviewers for their helpful comments and suggestions. The authors are members of the INdAM Research Group GNCS.

References

1. Albert, E., Genaim, S., Gutiérrez, R., Martin-Martin, E.: A transformational approach to resource analysis with typed-norms inference. Theory Pract. Log. Program. **20**(3), 310–357 (2020). https://doi.org/10.1017/S1471068419000401
2. Barrett, C.W., Sebastiani, R., Seshia, S.A., Tinelli, C.: Satisfiability modulo theories. In: Handbook of Satisfiability, Frontiers in Artificial Intelligence and Applications, vol. 185, pp. 825–885. IOS Press (2009). https://doi.org/10.3233/978-1-58603-929-5-825
3. Bjørner, N., Gurfinkel, A., McMillan, K., Rybalchenko, A.: Horn clause solvers for program verification. In: Beklemishev, L.D., Blass, A., Dershowitz, N., Finkbeiner, B., Schulte, W. (eds.) Fields of Logic and Computation II. LNCS, vol. 9300, pp. 24–51. Springer, Cham (2015). https://doi.org/10.1007/978-3-319-23534-9_2
4. Blicha, M., Fedyukovich, G., Hyvärinen, A.E.J., Sharygina, N.: Transition power abstractions for deep counterexample detection. In: TACAS 2022. LNCS, vol. 13243, pp. 524–542. Springer, Cham (2022). https://doi.org/10.1007/978-3-030-99524-9_29
5. Bradley, A.R.: SAT-based model checking without unrolling. In: Jhala, R., Schmidt, D. (eds.) VMCAI 2011. LNCS, vol. 6538, pp. 70–87. Springer, Heidelberg (2011). https://doi.org/10.1007/978-3-642-18275-4_7
6. Bruynooghe, M., Codish, M., Gallagher, J.P., Genaim, S., Vanhoof, W.: Termination analysis of logic programs through combination of type-based norms. ACM Trans. Program. Lang. Syst. **29**(2), 10-es (2007). https://doi.org/10.1145/1216374.1216378
7. De Angelis, E., Fioravanti, F., Gallagher, J.P., Hermenegildo, M.V., Pettorossi, A., Proietti, M.: Analysis and transformation of constrained Horn clauses for program verification. Theory Pract. Log. Program. **22**(6), 974–1042 (2022). https://doi.org/10.1017/S1471068421000211
8. De Angelis, E., Fioravanti, F., Pettorossi, A., Proietti, M.: VeriMAP: A tool for verifying programs through transformations. In: Ábrahám, E., Havelund, K. (eds.) TACAS 2014. LNCS, vol. 8413, pp. 568–574. Springer, Heidelberg (2014). https://doi.org/10.1007/978-3-642-54862-8_47
9. De Angelis, E., Govind, V.K.H.: CHC-COMP 2022: Competition report. In: Proceedings 9th Workshop on Horn Clauses for Verification and Synthesis and 10th International Workshop on Verification and Program Transformation. EPTCS, vol. 373, pp. 44–62. Open Publishing Association (2022). https://doi.org/10.4204/EPTCS.373.5
10. De Angelis, E., Proietti, M., Fioravanti, F., Pettorossi, A.: Multiple query satisfiability of constrained Horn clauses. In: arXiv, Computing Research Repository (2022). https://doi.org/10.48550/ARXIV.2211.15207
11. De Angelis, E., Proietti, M., Fioravanti, F., Pettorossi, A.: Verifying catamorphism-based contracts using constrained Horn clauses. Theory Pract. Log. Program. **22**(4), 555–572 (2022). https://doi.org/10.1017/S1471068422000175
12. Fioravanti, F., Pettorossi, A., Proietti, M., Senni, V.: Generalization strategies for the verification of infinite state systems. Theory Pract. Log. Program. **13**(2), 175–199 (2013). https://doi.org/10.1017/S1471068411000627
13. Govind, V.K.H., Shoham, S., Gurfinkel, A.: Solving constrained Horn clauses modulo algebraic data types and recursive functions. In: Proceedings of the ACM on Programming Languages, POPL 2022, vol. 6, pp. 1–29 (2022). https://doi.org/10.1145/3498722

14. Grebenshchikov, S., Lopes, N.P., Popeea, C., Rybalchenko, A.: Synthesizing software verifiers from proof rules. In: 33rd ACM SIGPLAN Conf. Programming Language Design and Implementation, PLDI 2012, pp. 405–416 (2012). https://doi.org/10.1145/2345156.2254112
15. Hermenegildo, M.V., Puebla, G., Bueno, F., López-García, P.: Integrated program debugging, verification, and optimization using abstract interpretation (and the Ciao system preprocessor). Sci. Comput. Program. **58**(1–2), 115–140 (2005). https://doi.org/10.1016/j.scico.2005.02.006
16. Hinze, R., Wu, N., Gibbons, J.: Unifying structured recursion schemes. In: International Conference on Functional Programming, ICFP 2013, pp. 209–220. ACM (2013). https://doi.org/10.1145/2500365.2500578
17. Hojjat, H., Rümmer, P.: The ELDARICA Horn solver. In: Formal Methods in Computer Aided Design, FMCAD 2018, pp. 1–7. IEEE (2018). https://doi.org/10.23919/FMCAD.2018.8603013
18. Jaffar, J., Maher, M.: Constraint logic programming: a survey. J. Log. Program. **19**(20), 503–581 (1994). https://doi.org/10.1016/0743-1066(94)90033-7
19. Komuravelli, A., Gurfinkel, A., Chaki, S.: SMT-based model checking for recursive programs. Formal Methods Syst. Des. **48**(3), 175–205 (2016). https://doi.org/10.1007/s10703-016-0249-4
20. Kostyukov, Y., Mordvinov, D., Fedyukovich, G.: Beyond the elementary representations of program invariants over algebraic data types. In: Conference on Programming Language Design and Implementation, PLDI 2021, pp. 451–465. ACM (2021). https://doi.org/10.1145/3453483.3454055
21. Leuschel, M.: A framework for the integration of partial evaluation and abstract interpretation of logic programs. ACM Trans. Program. Lang. Syst. **26**(3), 413–463 (2004). https://doi.org/10.1145/982158.982159
22. Meijer, E., Fokkinga, M., Paterson, R.: Functional programming with bananas, lenses, envelopes and barbed wire. In: Hughes, J. (ed.) FPCA 1991. LNCS, vol. 523, pp. 124–144. Springer, Heidelberg (1991). https://doi.org/10.1007/3540543961_7
23. Pham, T.-H., Gacek, A., Whalen, M.W.: Reasoning about algebraic data types with abstractions. J. Autom. Reason. **57**(4), 281–318 (2016). https://doi.org/10.1007/s10817-016-9368-2
24. Suter, P., Dotta, M., Kuncak, V.: Decision procedures for algebraic data types with abstractions. In: Symposium on Principles of Programming Languages, POPL 2010, pp. 199–210. ACM (2010). https://doi.org/10.1145/1706299.1706325
25. Suter, P., Köksal, A.S., Kuncak, V.: Satisfiability modulo recursive programs. In: Yahav, E. (ed.) SAS 2011. LNCS, vol. 6887, pp. 298–315. Springer, Heidelberg (2011). https://doi.org/10.1007/978-3-642-23702-7_23
26. Unno, H., Torii, S., Sakamoto, H.: Automating induction for solving Horn clauses. In: Majumdar, R., Kunčak, V. (eds.) CAV 2017. LNCS, vol. 10427, pp. 571–591. Springer, Cham (2017). https://doi.org/10.1007/978-3-319-63390-9_30
27. Yang, W., Fedyukovich, G., Gupta, A.: Lemma synthesis for automating induction over algebraic data types. In: Schiex, T., de Givry, S. (eds.) CP 2019. LNCS, vol. 11802, pp. 600–617. Springer, Cham (2019). https://doi.org/10.1007/978-3-030-30048-7_35

Formalizing and Reasoning About Supply Chain Contracts Between Agents

Dylan Flynn[1], Chasity Nadeau[1], Jeannine Shantz[1], Marcello Balduccini[1], Tran Cao Son[2(✉)], and Edward R. Griffor[3]

[1] Saint Joseph's University, Philadelphia, PA, USA
{df752850,cnadeau,js486075,mbalducc}@sju.edu

[2] Department of Computer Science, New Mexico State University, Las Cruces, NM, USA
stran@nmsu.edu

[3] National Institute of Standards and Technologies, Gaithersburg, MD, USA
edward.griffor@nist.gov

Abstract. Inspired by the recent problems in supply chains, we propose an approach to declarative modeling of contracts between agents that will eventually support reasoning about resilience of and about ways to improve supply chains. Specifically, we present a high-level language for specifying and reasoning about contracts over action domains of agents. We assume that the behavior of the agents can be formally expressed through action theories and view a contract as a collection of constraints. Each constraint specifies the responsibility of an agent to achieve a certain result by a deadline. Each agent also has a mapping between constraints and the agent's *concerns*, i.e. issues that the agent is concerned about, which are modeled in accordance with the CPS Framework proposed by the National Institute of Standards and Technology. We discuss how common questions related to the fulfillment of a contract or the concerns of the agents can be answered and computed via Answer Set Programming.

Keywords: Specifying and reasoning about contracts · Supply chain · Cyber-Physical System Framework · Answer set programming

1 Introduction

Supply chains have historically been optimized with respect to costs and other specific attributes, including the provisioning of materials, manufacturing processes, and distribution logistics. This high degree of optimization makes supply chains inherently brittle, in that their optimized network of exchanges is sensitive to sudden or extreme changes in demand. Alarming demonstrations of this brittleness have been experienced during the COVID-19 pandemic with the supply chain's inability to respond to the surge in demand for masks and ventilators. As these recent events demonstrated, supply chains nowadays constitute a widely distributed and critical infrastructure with legs

Portions of this publication and research effort are made possible through the help and support of NIST via cooperative agreement 70NANB21H167. Son Tran was also partially supported by the NSF grants 1812628 and 1914635.

M. Hanus and D. Inclezan (Eds.): PADL 2023, LNCS 13880, pp. 144–160, 2023.
https://doi.org/10.1007/978-3-031-24841-2_10

into economy and public welfare. Finding effective methods of curbing their brittleness in response to sudden changes and surges in demand is therefore paramount.

In this paper, we report on our progress in an investigation into methodologies that are ultimately aimed at making supply chains more resilient, and specifically for the identification of critical dependencies and for the evaluation, verification and restoration of properties of the supply chain.

As a first step, this paper proposes an approach to declarative modeling of supply chains that views a supply chain as a collection of contracts between agents, where a contract is a set of constraints. Each constraint specifies the responsibility of an agent to achieve a certain result by a deadline. The approach aims at describing, and reasoning about, the evolution of the state of the supply chain over time in response to events. Thus, we assume that the behavior of the agents can be formalized declaratively through action theories.

Typically, a supply chain involves a multitude of stakeholders (e.g., C-suite positions but also people involved in different level of supply, production, inventory, etc.) with substantially different types of expertise and goals. For this reason, a critical challenge with designing and managing resilient supply chains is that one needs not only to clearly identify all the interdependencies among the relevant elements, but also to formalize them in such a way that their relevance and ramifications are understandable by multiple stakeholders regardless of their different views.

To overcome this challenge, we build upon the notion of *concern* from the CPS Framework proposed by the National Institute of Standards and Technology (NIST) [4]. Our approach provides each agent with a mapping between constraints and the agent's concerns, i.e. issues that the agent is concerned about. While resilience may have many faces, we hypothesize that these faces can be captured within the concerns provided by the CPS Framework. An important reason that led us to rely on the CPS Framework is that it is designed to enable meaningful and grounded discussion among stakeholders from different backgrounds and with different objectives – supported in particular by a rich hierarchy of broadly-applicable concerns. This ensures broad applicability of our approach to a variety of types of supply chains, problems, and mix of stakeholders. All in all, our approach consists in viewing a supply chain as a large and complex CPS, and in capturing the supply chain's interdependencies by means of the elements of the NIST CPS Framework.

Because resilience is a broad and multi-faceted topic, we will focus on the three reasoning problems in this paper, which we view as fundamental stepping stones towards reasoning about resilience: *Contract feasibility*: given an initial state of the world, can a given contract be successfully executed? (If that is not the case, then the contract is set up for failure.) *Clause satisfaction check*: assuming that the contract has been in execution for some time, has any agent violated any clause of the contract? If so, then how can the problem be mitigated? *Concern satisfaction check*: assuming that the contract has been in execution for some time, which (and whose) concerns are not satisfied? What can be done to mitigate the problem? Besides the formalization of the problem and of the reasoning mechanisms, we also discuss the implementation of our approach, which is based on the declarative knowledge representation formalism of Answer Set Programming (ASP) [10,12].

2 Preliminaries

2.1 The NIST CPS Framework

An important challenge with supply chains is that stakeholders of varying backgrounds may use different terminology when discussing a supply chain and likely have different, possibly even conflicting, goals, which can lead to challenges under normal circumstances. When the unexpected occurs, disruptions in communication and conflicts among objectives may be magnified, making it more difficult to ensure resilience. Establishing a structure incorporating primitives that promote a common vocabulary and meaningful, grounded discussion among stakeholders may mitigate risks. To create a "common foundation", we propose viewing the supply chain as a large, complex Cyber Physical System (CPS) and leveraging the NIST CPS Framework as a lens through which we can look at a supply chain. With this framework the processes surrounding developing, verifying and delivering products can be formalized and more easily understood by a diverse group of stakeholders [4]. By design, the scope of the CPS Framework is very broad so that it may be adopted by a broad range of applications.

The NIST Framework for Cyber-Physical Systems, referred to as "NIST CPS Framework" or simply "Framework" below, comprises a set of concerns and facets related to the system under design or study. This section briefly clarifies the intent and purpose of the framework. The interested reader is directed to SP 1500-201, SP 1500-202 and SP 1500-203, available on the NIST website.

The CPS Framework provides the taxonomy and methodology for designing, building, and assuring CPS that meet the expectations and concerns of system stakeholders, including engineers, users, and the community that benefits from the system's functions. The concerns of the Framework are represented in a forest, where branching corresponds to the decomposition of concerns. We refer to each tree as a *concern tree* of the CPS Framework. The concerns at the roots of this forest are called *aspects*. For instance, the sub-concerns of the Trustworthiness aspect are Privacy, Reliability, Resilience, Safety, and Security. In turn, the Security concern has sub-concerns Cybersecurity and Physical Security, and the Cybersecurity concern has sub-concerns Confidentiality, Integrity and Availability. The Framework comprises nine aspects. In this paper, we will mainly focus on Business, Functional, and Trustworthiness. A concern about a given system reflects consensus thinking about method or practice, involved in addressing the concern, and in some cases consensus-based standards describing that method or practice. Associated with each concern is a set of *requirements* that address the concern in question. For example, in a CPS that stores personally identifiable information, the system's designers may agree that the requirement to use encrypted memory addresses the Confidentiality concern. Because the Confidentiality concern is a descendant of the Trustworthiness aspect, this requirement, together with other relevant ones, addresses the CPS's Trustworthiness aspect as well. The dependencies among concerns and between requirements and concerns can be formally represented by means of an ontology. Leveraging the ontology, tasks related to reasoning about the satisfaction of concerns can be reduced to: (a) identifying which requirements are satisfied in the current state of the system and which ones are not, and (b) propagating this information up the concern forest, ultimately determining the satisfaction of the aspects. For details on this approach, we refer the interested reader to [11]. For the purpose of this paper, it is sufficient to mention the existence of algorithms for determining whether a requirement or concern γ is satisfied given the ontology $\mathcal{O}$ and a current state s of the CPS. Below, we will write $\mathcal{O} \cup s \models \gamma$ to denote that γ is satisfied under s.

2.2 Action Language $\mathcal{B}$

An action domain in the action language $\mathcal{B}$ [7] is defined over two disjoint sets, a set of actions **A** and a set of fluents **F**. A *fluent literal* is either a fluent $f \in \mathbf{F}$ or its negation $\neg f$. A *fluent formula* is a propositional formula constructed from fluent literals. An action domain is a set of laws of the following form:

$$\textit{Executability condition:} \quad \textbf{executable } a \textbf{ if } \varphi \tag{1}$$

$$\textit{Dynamic law:} \quad a \textbf{ causes } \psi \textbf{ if } \varphi \tag{2}$$

$$\textit{Static Causal Law:} \quad \psi \textbf{ if } \varphi \tag{3}$$

where ψ and φ are fluent formulas and a is an action. Intuitively, an executability condition of the form (1) states that a can only be executed if φ holds. (2), referred to as a *dynamic causal law*, states that ψ is caused to be true after the execution of a in any state of the world where φ is true. (3) represents a *static causal law*, i.e., a relationship between fluents. It conveys that whenever the fluent formula φ holds then so is ψ. For an action domain D, we denote the set of laws of the form (3) by K.

Let D be a domain. A set of fluent literals is said to be *consistent* if it does not contain f and $\neg f$ for some fluent f. An *interpretation* I of the fluents in D is a maximal consistent set of fluent literals of D. A fluent f is said to be true (resp. false) in I iff $f \in I$ (resp. $\neg f \in I$). The truth value of a fluent formula φ in I is defined recursively over the propositional connectives in the usual way. $I \models \varphi$ indicates that φ is true in I.

Let u be a consistent set of fluent literals and K a set of static causal laws. We say that u is closed under K if for every static causal law "ψ **if** φ" in K, if $u \models \varphi$ then $u \models \psi$. By $Cl_K(u)$ we denote the least consistent set of literals from D that contains u and is also closed under K. It is worth noting that $Cl_K(u)$ might be undefined. For instance, if u contains both f and $\neg f$ for some fluent f, then $Cl_K(u)$ cannot contain u and be consistent; another example is that if $u = \{f, g\}$ and K contains both "f **if** h" and "$\neg h$ **if** f, g" then $Cl_K(u)$ does not exist because it has to contain both h and $\neg h$, which means that it is inconsistent. For a formula η, $Cl_K(\eta)$ denotes the set $\{Cl_K(u) \mid u \text{ is a set of fluent literals such that } u \models \eta\}$. Formally, a *state* of D is an interpretation of the fluents in **F** that is closed under the set of static causal laws K of D.

An action a is *executable* in a state s if there exists an executability proposition **executable** a **if** φ in D such that $s \models \varphi$. The *direct effect of an action a* in a state s is the set $e(a, s) = \bigwedge_{a \textbf{ causes } \psi \textbf{ if } \varphi \in D, s \models \varphi} \psi$. For a domain D, $\Phi_D(a, s)$, the set of states that may be reached by executing a in s, is defined as follows: (*i*) if a is executable in s, then $\Phi_D(a, s) = \{s' \mid s' \text{ is a state and } s' \in Cl_K(e(a, s) \wedge \bigwedge_{l \in s \cap s'} l)\}$; and (*ii*) if a is not executable in s, then $\Phi_D(a, s) = \emptyset$. Φ_D is unique for each domain D and is called the *transition function of D*.

Given a domain D, an alternate sequence of states and actions $\alpha = s_0 a_0 s_1 \ldots a_{n-1} s_n$, where s_i's are states and a_i's are actions, is a *trajectory* over the domain D if $s_{i+1} \in \Phi_D(a_i, s_i)$ for every $i = 0, \ldots, n-1$. We say that n is the length of α and s_0 is the starting state of α. Furthermore, α satisfies a fluent formula φ over the set of fluents in D_A, denoted by $\alpha \models \varphi$, if s_n satisfies φ.

3 A Motivating Scenario

Let us consider a scenario involving two agents: XYZ Homes and Lumber Yard A. XYZ Homes aims at building a certain number of homes and contracts with Lumber Yard A to provide suitable lumber.

Example 1 (A Contract Between XYZ Homes and Lumber Yard A). XYZ Homes builds eight to nine 2,000 square foot homes each month for new home buyers. Each new home requires 16,000 board feet of Number 2 Common grade lumber. In order to complete eight to nine homes, XYZ Homes must purchase 144,000 board feet of Number 2 Common grade lumber each month. Lumber Yard A is the preferred supplier of this lumber.

In the first part of our scenario, we look at the agreement that XYZ Homes contracts with Lumber Yard A for the required lumber. The agreement specifies the responsibilities of each agent. It is formalized as a set of constraints on how the work is to be conducted. These constraints can be viewed as requirements (in the sense of the CPS Framework), and each requirement is mapped to one or more of an agent's concerns. A sample of these constraints and concerns includes:

1. Lumber Yard A will produce a total of 144,000 board feet of lumber for XYZ Homes. This constraint addresses the functionality concern of XYZ Homes.
2. Lumber Yard A guarantees to schedule the transport and delivery of 14–16 tractor trailers worth of lumber in one month to XYZ Homes. This constraint addresses the time to market concern.
3. The lumber delivered to XYZ Homes will be at or above Number 2 Common grade. This constraint addresses several concerns including physical, reliability, quality and trustworthiness. For example, if Lumber Yard A were to provide lumber that is of a lesser quality than Number 2 Common grade, then from XYZ Home's perspective, Lumber Yard A would no longer be trustworthy.
4. The agreed upon cost of lumber is at \$122,000 for 144,000 board feet and the transport and delivery cost will be at or below \$500,000 for 144,000 board feet. This constraint addresses the cost concern.

In the context of a contract, agents will normally have to execute actions to fulfill their commitments. For instance, Lumber Yard A has to *produce* 144,000 board feet and *deliver* them to XYZ Homes. On the other hand, XYZ Homes has to pay for the board, etc. We therefore proposed to formalize contacts between agents where each agent is associated with an action domain. Intuitively, the agent's domain describes the actions that the agent can execute, when can an action be executed, and what are the effects of an action. We believe that the action language $\mathcal{B}$ is sufficiently expressive enough for us to represent domains in supply chains.

In using $\mathcal{B}$, we can easily encode *functional fluents* that frequently occurred in action domains of agents involved in supply chain. For example, the number of board feet of lumber that a company possessed is a functional fluent whose domain is between 0 and 5000.

For later use, we encode a simple set of actions for Lumber Yard A and XYZ Homes below. Lumber Yard A has an abstract action for producing lumber and it is represented as follows:

$$produce(X, Z) \ \textbf{causes} \ \exists Y.[Y \leq X : board(Y, Z)]$$

Here, $board(Y, Z)$ is a functional fluent denoting that Y board feet of lumber of quality Z is available for Lumber Yard A.

Lumber Yard A also has the following action:

$$deliver(Y, Z) \ \textbf{causes} \ delivered(Y, Z) \wedge board(X - Y, Z) \ \textbf{if} \ \ board(X, Z), Y \leq X$$

This law states that if Lumber Yard A delivers Y board feet of lumber of quality Z, then they are delivered ($delivered(Y, Z)$) if at least Y board feet of lumber are available ($board(X, Z) \wedge Y \leq X$). As a result of the action, there will be $X - Y$ board feet available after the delivery. Observe that the action domain can be refined to include, for example, the specific detail on shipping such as the need for tractor trailers, the capacity of the trailers, etc.

XYZ Homes needs to receive the board and pay. These actions are represented below

$$receive(X, Z) \ \textbf{causes} \ available\_board(X, Z) \ \textbf{if} \ \ delivered(X, Z)$$

which says that the company would have X board feet of lumber of quality Z if it executes the action $receive(X, Z)$. This action can only be executed if the said amount of board feet of lumber is delivered. In addition, XYZ Homes also has the action

$$\begin{array}{r} pay(X, C) \ \textbf{causes} \ payment(X, C) \wedge available\_funds(Y - X) \\ \textbf{if} \ \ available\_funds(Y), X \geq Y \end{array}$$

where C is either $board$ or $shipping$. The law says that XYZ Homes pays an amount X for the category C and the available fund will be reduced by X. This effect is achieved only if sufficient funds are available to XYZ Hones.

The above representation encodes the actions and their effects under normal circumstances. We will assume that for each action a in our discussion, the executability condition of a is of the form

$$\textbf{executable} \ a \ \textbf{if} \ \neg ab(a), \varphi$$

which, intuitive, says that a can be executed whenever φ is true and $ab(a)$ is false where $ab(a)$ denotes that some abnormal condition under which a cannot be executed. For example, for the action $produce(X, Y)$, we have

$$\textbf{executable} \ produce(X, Z) \ \textbf{if} \ \ \neg ab(produce(X, Z))$$

In the following, we denote by D_L and D_H the action domains of Lumber Yard A and XYZ Homes, respectively.

4 Formalizing a Contract

We now introduce $\mathcal{L}_c$, a high-level language for the specification of contracts between agents with focus on supply chain management. The language is built on a set of agents and the action domains associated with these agents. In formalizing a contract, we assume that each agent is aware of the state of the world and can observe the changes within its environment that it is interested in. To each agent, the contract has two facets, the public part encodes the agreement between the agent and another agent, while the private part details its concerns. For example, the contract between XYZ Homes and Lumber Yard A contains:

- the statement "Lumber Yard A will produce a total of 144,000 board feet of lumber for XYZ Homes". This is a public part of the contract that is known to both parties;
- the statement that the above is a constraint addressing the functionality concern is a private part of the contract (that both sides happen to agree upon);
- the statement "the lumber delivered to XYZ Homes will be at or above Number 2 Common grade" is related to the trustworthiness concern of XYZ Homes; it is not necessarily related to a concern of Lumber Yard A.

Observe that each of the above clauses specifies a goal, the agent responsible for the achievement of the goal, and the deadline. Motivated by this observation, we develop $\mathcal{L}_c$ as follows. From now on until the end of this section, we assume two fixed agents A and B, whose action domains are D_A and D_B, respectively.

4.1 Syntax of $\mathcal{L}_c$

We formalize the public part of a contract between A and B over D_A and D_B using clauses of the form:

$$ref\_id: \quad agent \quad \textbf{responsible\_for} \quad goal \quad \textbf{when} \quad time\_expression \tag{4}$$

Intuitively, a clause of the form (4) is associated with a reference identifier ref_id and says that $agent \in \{A, B\}$ is responsible for achieving $goal$ within the time constraint specified by $time\_expression$, where:

- $goal$ is a fluent formula constructed over fluents appearing in $D_A \cup D_B$; and
- a time constraint is a simple temporal expression of one the following forms:

$$always \mid eventually \mid per\_unit[n \dots m] \mid by\_unit\, n \tag{5}$$

where $unit$ can be any time unit such as day, week, etc., n and m are integers, $n \leq m$, and $[n \dots m]$ denotes the range $[n, n+1, \dots, m]$.

Given the public part of a contract C between A and B, the private part of C for either agent A or B is represented by statements of the form

$$ref\_id: \quad \rho \tag{6}$$

where ref_id is a reference identifier C and ρ is a requirement. We assume that ontology $\mathcal{O}$ associates each requirement with one or more concerns (of the agent) from the concern forest defined in the CPS Framework, or any customized concern forest specific to the agent. Formally, a contract between two agents[1] is defined as follows:

Definition 1. *A contract $\mathcal{C}$ between two agents A and B constructed over two actions domains D_A and D_B is a triple (C, P_A, P_B) where*

- *C is a set of clauses of the form* (4)*; and*
- *P_A (resp. P_B) is a set of statements of the form* (6) *for A (resp. B).*

Intuitively, (C, P_A) or (C, P_B) is the contract under A or B's perspective, respectively, and it will be used by A or B to evaluate the progress of the contract. Observe that A (resp., B) does not necessarily know about P_B (resp., P_A).

Example 2. Let L and H denote Lumber Yard A and XYZ Homes, respectively. The public part of the contract in Example 1 is encoded by the following clauses:

$$C1 : L\ \textbf{responsible\_for}\ board(144K, Q) \wedge 1 \leq Q\ \textbf{when}\ by\_week\ 4 \tag{7}$$

$$C2 : L\ \textbf{responsible\_for}\ delivered(144K, Q) \wedge 2 \leq Q\ \textbf{when}\ by\_week\ 4 \tag{8}$$

$$C3 : H\ \textbf{responsible\_for}\ payment(122K, board)\ \textbf{when}\ by\_week\ 4 \tag{9}$$

$$C4 : H\ \textbf{responsible\_for}\ \exists X \leq 500K.[payment(X, shipping)]\ \textbf{when}\ by\_week\ 4 \tag{10}$$

The first clause $C1$ states that L is responsible for producing 144K board feet of lumber by the end of the project (week 4). $C2$ says that L needs to deliver, i.e., responsible for the shipping of, 144K board feet of lumber of quality 2 or greater (also by week 4). $C3$ and $C4$ indicate that H must pay \$122K for the board feet of lumber and for shipping, respectively, and the cost of shipping must be no greater than \$500K[2].

The links between clauses and each agent's requirements are encoded by the following sets of statements:[3]

- $P_L = \{C2 : \textit{match-customer-expected-grade}, C4 : \textit{receive-due-payment}\}$ where requirement *match-customer-expected-grade* addresses the Quality concern from the CPS Framework's Business aspect, and requirement *receive-due-payment* addresses the Reliability concern under the Trustworthiness aspect and the Cost concern from the Business aspect.

---

[1] Throughout the paper, we only discuss contracts between two agents but the formalization is easily adapted for contracts among multiple agents.

[2] Observe that we have simplified the contract slightly as there is no mention about the tractor trailers. This can be easily encoded if we extend the action domains of H and L to consider the shipping company.

[3] A thorough discussion on requirements is beyond the scope of this paper. Thus, for compactness, we use short requirement names, although in practice a requirement would be spelled out in more details, e.g. the requirement that here we call *match-customer-expected-grade* would likely be expressed by a statement "lumber shall be produced in a grade matching the customer's expectations.".

– $P_H = \{C1$: *sufficient-material-for-building*, $C2$: *material-safe-for-building*, $C2$: *material-sufficiently-durable*, $C4$: *promptly-send-payment*, $C3$: *acceptable-shipping-cost*$\}$ *sufficient-material-for-building* addresses the TimeToMarket concern (Business aspect), *material-safe-for-building* addresses Safety and Reliability (both under the Trustworthiness aspect), *material-sufficiently-durable* addresses Performance (Functional aspect), *promptly-send-payment* addresses Policy (Business aspect), and *acceptable-shipping-cost* addresses Cost (also under the Business aspect).

Observe that concerns are not symmetrical between agents.

4.2 Semantics of $\mathcal{L}_c$

Given a contract $\mathcal{C} = (C, P_A, P_B)$ between two agents A and B. The semantics of $\mathcal{L}_c$ is defined over pairs of trajectories over D_A and D_B of the form (H_A, H_B). For simplicity of the presentation, we assume that the action $wait$, which can always be executed and has no effect, belongs to every domain. Therefore, whenever we refer to two trajectories H_A and H_B over D_A and D_B, respectively, without the loss of generality, we will assume that H_A and H_B have the same length.

Given D_A and D_B, a *joint state s over* D_A *and* D_B (or, *joint state*, for short) is an interpretation over the set of fluents in $D_A \cup D_B$ that is closed with respect to the set of static causal laws in $D_A \cup D_B$. For a joint state s over $D_A \cup D_B$, by s_A (or s_B) we denote the restriction of s over the fluents in D_A (or D_B), respectively. Obviously, s_A (s_B) is a state in D_A (D_B). The truth value of a formula φ over the language of $D_A \cup D_B$ in a joint state s is defined as usual. We will next define the satisfaction of a contract given (H_A, H_B). To do so, we need the following notion.

Definition 2. *Given two agents A and B whose action domains are* D_A *and* D_B*, respectively, and two trajectories* $H_A = s_0^A a_0^A \dots s_{n-1}^A a_{n-1}^A s_n^A$ *over* D_A *and* $H_B = s_0^B a_0^B \dots s_{n-1}^B a_{n-1}^B s_n^B$ *over* D_B*, we say that* H_A *and* H_B *are* compatible *if* $s_i^A \cup s_i^B$ *is a joint state for every* $i = 0, \dots, n$.

The satisfaction of a clause is defined next.

Definition 3. *Given two compatible trajectories* $H_A = s_0^A a_0^A \dots s_{n-1}^A a_{n-1}^A s_n^A$ *in* D_A *and* $H_B = s_0^B a_0^B \dots s_{n-1}^B a_{n-1}^B s_n^B$ *in* D_B*, a clause*

$$ref\_id : x \ \textbf{\textit{responsible\_for}} \ \varphi \ \textbf{\textit{when}} \ time\_exp$$

is satisfied *by* (H_A, H_B)*, denoted by* $(H_A, H_B) \models ref\_id$*, if*

– φ *is true in every* $s_i = s_i^A \cup s_i^B$ *for* $i = 0, \dots, n$ *when* $time\_exp$ *is* $always$*; or*
– φ *is true in* $s_i = s_i^A \cup s_i^B$ *for some* $i = 0, \dots, n$ *when* $time\_exp$ *is* $eventual$*; or*
– φ *is true in* $s_i = s_i^A \cup s_i^B$ *for* $i = u, \dots, l$ *when* $time\_exp$ *is* $per\_unit\ [u \dots l]$*; or*
– φ *is true in* $s_k = s_k^A \cup s_k^B$ *when* $time\_exp$ *is* $by\_unit\ k$.

We say that ref_id *is* violated *by* (H_A, H_B) *if* $(H_A, H_B) \not\models ref\_id$.

Building on the above definition, the satisfaction of a contract is defined as follows.

Definition 4. *Given two compatible trajectories* $H_A = s_0^A a_0^A \ldots s_{n-1}^A a_{n-1}^A s_n^A$ *in* D_A *and* $H_B = s_0^B a_0^B \ldots s_{n-1}^B a_{n-1}^B s_n^B$ *in* D_B *and a contract* (C, P_A, P_B) *between A and B, we say that C is satisfied by* (H_A, H_B) *if every clause in C is satisfied by* (H_A, H_B)*.*

Definition 4 allows for the reasoning about the satisfaction of the public part of a contract. The satisfaction of the private part of a contract with respect an ontology $\mathcal{O}$ is defined next.

Definition 5. *Given two compatible trajectories* $H_A = s_0^A a_0^A \ldots s_{n-1}^A a_{n-1}^A s_n^A$ *in* D_A *and* $H_B = s_0^B a_0^B \ldots s_{n-1}^B a_{n-1}^B s_n^B$ *in* D_B *and a contract* (C, P_A, P_B) *between A and B. Let* $X \in \{A, B\}$ *and* $s(P_X) = \{reg \mid ref\_id : reg \in P_X\}$*. We say that a concern c of agent X is satisfied by* (H_A, H_B) *if* $\mathcal{O} \cup s(P_X) \models c$*.*

4.3 Reasoning About Contracts

Let $\mathcal{C} = (C, D_A, D_B)$ be a contract between A and B. We will next discuss how the semantics of $\mathcal{L}_c$ can be employed in evaluating $\mathcal{C}$ from the perspective of the agents. Naturally, a contract can be evaluated at different time points such as at the time the contract is signed or after some actions have been taken by the agents. Let us briefly discuss the questions that would be of interest to the agents.

- **Q1** *Contract feasibility*: given an initial state of the world, can $\mathcal{C}$ be successfully fulfilled? A different perspective of this question is whether there exists any state of the world in which $\mathcal{C}$ is satisfied. If there is none, it is clear that the contract is set up for failure!
- **Q2** *Clause satisfaction check*: assuming that $\mathcal{C}$ has been in execution for some time, has any agent violated a clause in $\mathcal{C}$? If so, then how can the problem be mitigated?
- **Q3** *Concern satisfaction check*: assuming that $\mathcal{C}$ has been in execution for some time, which/whose concerns are not satisfied? What can be done to mitigate the problem?

The above definitions in the previous section allow us to answer questions **Q1**–**Q3** by

- **Q1** computing two compatible trajectories H_A and H_B that start from the given initial joint state and satisfy the contract (or determining a joint state such that two compatible trajectories, that start from this state and satisfy the contract, can be identified);
- **Q2** identifying the clauses of the contract that are violated; and, to mitigate the problem caused by a violation of a clause (by an agent), new compatible trajectories, which start from the joint state at the end of the given trajectories (e.g., $s_n^A \cup s_n^B$ given the two trajectories in Definition 3) and satisfy the contract, need to be computed.
- **Q3** determining the concerns that are satisfied (or not satisfied). To mitigate the problem, similar approach as described in **Q2** needs to be adopted.

We illustrate the aforementioned idea using the running example.

Example 3 (Illustration). Let us consider the situation where Lumber Yard A (L) has not produced any board feet and XYZ Homes (H) has \$1M on its account. Furthermore, all factories of Lumber Yard A are in good operational order. The initial state for L and H can be represented by the set $s_0^L = \{board(0, Y) \mid Y =$

$1, 2, 3\} \cup \{\neg ab(production)\} \cup \{delivered(0, Y) \mid Y = 1, 2, 3\}$ and $s_0^H = \{available\_board(0, Y) \mid Y = 1, 2, 3\} \cup \{payment(0, C) \mid C = board, shipping\} \cup \{available\_funds(1M)\}$, respectively.

In this case, we can check that the two compatible trajectories satisfying the clauses (C1)–(C3) in the contract specified in Example 2 are:

$$H_L = s_0^L\ produce(144K, 2)\ s_1^L\ deliver(144K, 2)\ s_2^L\ wait\ s_3^L\ wait\ s_4^L$$
$$H_H = s_0^H\ wait\ s_1^H\ wait\ s_2^H\ receive(144K, 2)\ s_3^H\ pay(122K, board)\ s_4^H$$

In the above trajectories, we have that $s_2^L = s_3^L = s_4^L$ and $s_0^H = s_1^H = s_2^H$ and $board(144K, 2)$ is true in s_1^L, $board(0, 2)$ is true in s_2^L, $available\_board(144K, 2)$ is true in s_3^H, $available\_fund(878K)$ is true in s_4^H, etc. Assuming that the unit on the trajectories is week, it is easy to check that these trajectories satisfy the three clauses (C1)–(C3) of the contract in Example 2. One can also see that they can be extended to satisfy clause (C4), provided that the action domain of Lumber Yard A is extended with proper actions for billing the cost for shipping, organizing the delivery, and receiving money. We omit this discussion for brevity.

Consider a situation where XYZ Homes does not have any money in the initial state, i.e., $available\_funds(0)$ belongs to s_0^H, but it can borrow any amount from the bank. In this case, replacing one of the *wait* actions in H_H with *borrow(122K)*, which represents the action of borrowing 122K from the bank by XYZ Homes, will result in H'_H that is compatible with H_L and (H_L, H'_H) satisfies (C1)–(C3).

Let us consider yet another situation in which COVID-19 forces Lumber Yard A to close all of its factories right after the signing of the contract. It means that the initial state for L changes to $u_0^L = s_0^L \setminus \{\neg ab(production)\} \cup \{ab(production)\}$. In this case, H_L is no longer a valid trajectory for L. It is easy to see that there exists no pair of compatible trajectories for H and L that can satisfy clauses (C1)–(C2) of the contract. If Lumber Yard A could, for example, purchase the lumber from some other companies, then alternative trajectories could be identified and the contract can be fulfilled. Observe also that the action domain allows XYZ Homes to pay regardless of whether it receives the board feet, clause (C3) can still be satisfied!

5 Reasoning About Contracts Using Answer Set Programming

Answer Set Programming. [10,12] is a declarative programming paradigm based on logic programming under the answer set semantics. A logic program Π is a set of rules of the form: $c_0 \vee \ldots \vee c_k \leftarrow a_1, \ldots, a_m, not\ b_1, \ldots, not\ b_n$ where c_i's, a_i's, and b_i's are atoms of a propositional language[4] and not represents (default) negation. $c_0 \vee \ldots \vee c_k$ can be absent.

The semantics of a logic program Π is defined via a special class of models called *answer sets* [5]. A program Π can have several answer sets, one answer set, or no answer set. Π is said to be consistent if it has at least one answer set; it is inconsistent otherwise. Several extensions (e.g., *choice atoms*, *aggregates*, etc.) have been introduced to simplify the use of ASP. We will use and explain them when needed.

[4] For convenience, we often use first order logic literals under the assumption that they represent all suitable ground instantiations.

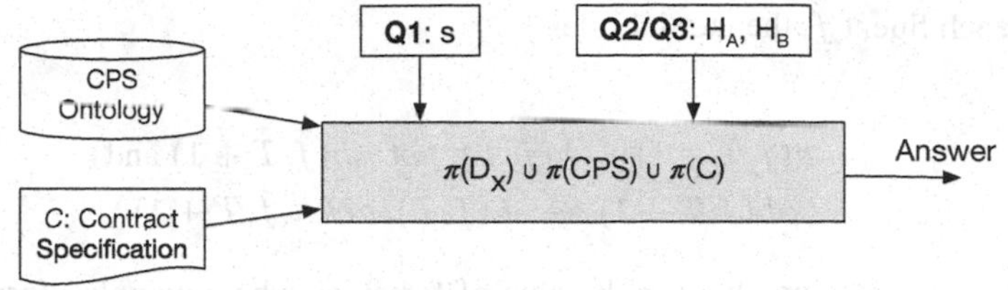

Fig. 1. Overview of Components used in Answering Queries **Q1–Q3** for Agent $X \in \{A, B\}$

5.1 Answer Set Programming for Reasoning About Contracts

This section presents an ASP encoding given a contract between two agents for reasoning about contracts and concerns of the agents, building on the work on planning in ASP and on formalizing CPS (e.g., [1,6,11]). Throughout this section, we assume that $\mathcal{C} = (C, P_A, P_B)$, where A and B are two agents with action domains D_A and D_B, respectively, is given. Figure 1 gives an overview of the components used by agent X in answering Queries **Q1–Q3** (gray box). Depending on the queries, the input is given to them might be an initial joint state (**Q1**) or two trajectories (**Q2–Q3**) that implicitly specify a joint state as well.

We assume that n is a constant in the unit of time used in $\mathcal{C}$ and denotes the maximal length of the trajectories considered by the two agents. Let s_0 be a joint state. For each agent X, we create a programs $\pi(D_X)$, $\pi(CPS)$, and $\pi(\mathcal{C})$ as follows[5].

- $\pi(D_X)$ is the program for reasoning about actions and changes (e.g., [6]) over predicates $h(f,t)$ (fluent f is true at time step t) and $occ(a,t)$ (action a occurs at step t) and consists of the following rules[6]:
  - declaration of steps $step(0..n)$;
  - for each fluent f, the declaration $fluent(f)$;
  - for each action a, the declaration $action(a)$;
  - for each fluent f that is true in s_0 then $h(f,0)$ belongs to $\pi(D_X)$ and the rule

$$\neg h(f,0) \leftarrow not\ h(f,0);$$

  - for each executability law **executable** a **if** φ, the constraint that prevents a to be executed when its precondition is not satisfied: $\leftarrow occ(a,T), not\ h(\varphi,T)$
  - for each dynamic law a **causes** ψ **if** φ, the rule encoding the effect of a

$$h(\psi, T+1) \leftarrow occ(a,T), h(\varphi,T)$$

  - for each static law ψ **if** φ, a rule stating that ψ must be true whenever φ is true

$$h(\psi,T) \leftarrow h(\varphi,T)$$

[5] We note that given an action domain D, the ontology CPS, and a contract $\mathcal{C}$, these three programs can be automatically generated by a Prolog program similar to that provided in https://www.cs.nmsu.edu/~tson/ASPlan/Knowledge/.

[6] In each rule with T as variable, we omit $step(T)$ from the right hand side.

- for each fluent f, the inertial rules

$$h(f, T+1) \leftarrow h(f, T), not\ \neg h(f, T+1) \text{ and}$$
$$\neg h(f, T+1) \leftarrow \neg h(f, T), not\ h(f, T+1).$$

- rules for reasoning about truth value of formulas where complex formulas are encoded using a set of atoms using fresh constants and membership functions, e.g., $f \wedge g$ is encoded by a fresh constant fg and the set of atoms $\{conjunction(fg), member(f, fg), member(g, fg)\}$ and rules such as

$$\begin{aligned} h(conjunction(F), T) \leftarrow\ & conjunction(F), \\ & N = \#count\{1, X : member(X, F)\}, \\ & C = \#count\{1, G : member(G, F), h(G, T)\}, N == C. \end{aligned}$$

Note that this rule uses the $\#count$ aggregate which counts the number of atoms satisfying a condition (e.g., the number of members of a formula). We omit other rules for brevity.

- $\pi(CPS)$ contains atoms encoding concerns, requirements, and relationships among all of them (e.g., $requirement(R)$, $concern(C)$ or $subCo(X, Y)$ indicating that R is a requirement, that C is a concern, and that Y is a subconcern of X) that can be obtained from a translation of the CPS ontology to ASP facts (see, e.g., [11]) and the rules for reasoning about the satisfaction of requirements and concerns. For example, given that a formula F over requirements addresses a concern C (i.e., contributes to its satisfaction), encoded by $addrBy(C, F)$, the following rules determine whether the concern C is satisfied at step T.
$\neg h(sat(C), T) \leftarrow concern(C), addrBy(C, F), not\ h(F, T).$
$\neg h(sat(X), T) \leftarrow subCo(X, Y), not\ h(sat(Y), T), concern(X), concern(Y).$
$\neg h(sat(X), T) \leftarrow subCo(X, Y), \neg h(sat(Y), T), concern(X), concern(Y).$
$h(sat(C), T) \leftarrow not\ \neg(sat(C), T), concern(C).$
- $\pi(\mathcal{C})$ encodes $\mathcal{C}$ and P_X in $\mathcal{C}$ and is constructed as follows.
  - for each clause of the form (4), the program contains the atom $clause(ref\_id)$, the set of atoms encoding v_g, where v_g is the name of the formula $goal$, and the following rules:

$time\_expression$	$\pi(\mathcal{C})$ contains
$always$	$\neg h(sat(ref\_id), n) \leftarrow not\ h(v_g, T).$ $h(sat(ref\_id), n) \leftarrow not\ \neg h(sat(ref\_id), n)$
$eventually$	$h(sat(ref\_id), n) \leftarrow h(v_g, T).$
$per\_unit[u \ldots l]$	$\neg h(sat(ref\_id), n) \leftarrow not\ h(v_g, T), u \leq T, T \leq l.$ $h(sat(ref\_id), n) \leftarrow not\ \neg h(sat(ref\_id), n)$
$by\_unit[k]$	$h(sat(ref\_id), n) \leftarrow h(v_g, k).$

where $h(sat(ref\_id), n)$ says that ref_id is satisfied at step n.

- for each requirement r that occurs in P_X,
  * if there is only one element of the form $ref\_id : \rho$ in P_X that contains r, then $\pi(\mathcal{C})$ contains the atom $addrBy(r, ref\_id)$;
  * if there is more than one element of the form $ref\_id : \rho$ in P_X, then $\pi(\mathcal{C})$ contains the atoms $addrBy(cg, ref\_id)$ and $conjunction(cg)$ and the set $\{member(c, cg) \mid ref\_id : r \in P_X\}$ where cg is a fresh constant.

  This set of atoms helps propagating the satisfaction of clauses of contracts to the satisfaction of concerns of the agent.

Let $H_X = s_0^X a_0^X \ldots s_{n-1}^X a_{n-1}^X s_n^X$ be an alternate sequence of states and actions in D_X. Define $e(H_X) = \{occ(a_i, i) \mid i = 0, \ldots, n-1\} \cup \{h(f, i) \mid f \text{ is true in } s_i\}$. It holds that

Proposition 1. – *$\pi(D_X) \cup e(H_X)$ has a unique answer set if H_X is a trajectory over D_X and a fluent formula φ is true in s_i iff $\pi(D_X) \cup e(H_X) \models h(\varphi, i)$.*
- *if H_X is a trajectory over D_X then H_X satisfies a clause ref_id in $\mathcal{C}$ iff $\pi(D_X) \cup \pi(\mathcal{C}) \cup e(H_X)$ has a unique answer set S that contains $h(sat(ref\_id), n)$.*
- *if H_X is a trajectory over D_X and $ref\_id : c$ in P_X then concern c is satisfied at the end of the trajectory iff $\pi(D_X) \cup \pi(CPS) \cup \pi(\mathcal{C}) \cup e(H_X)$ has a unique answer set S that contains $h(sat(c), n)$.*

Intuitively, the first property ensures that $\pi(D_X)$ correctly encodes the transition function in D_X. The second property shows that checking whether a clause is satisfied by H_X can be reduced to computing an answer set of $\pi(D_X) \cup \pi(\mathcal{C}) \cup e(H_X)$. Similarly, the third property shows that determining whether a concern is satisfied by a trajectory can be reduced to computing an answer set of $\pi(D_X) \cup \pi(CPS) \cup \pi(\mathcal{C}) \cup e(H_X)$. These properties help us answer questions related to the satisfaction of a contract or concerns given the compatible trajectories H_A and H_B. Observe that the only requirement for an agent X to reason about the satisfaction of the contract or a concern is the compatibility of H_A and H_B. X does not need to know the actions that the other agent executes. This is important when finding plans to satisfy a contract of an agent, which we discuss next.

To compute trajectories satisfying a contract, we observe that agents do usually have to plan by themselves. Furthermore, observe that in the context of this paper, the action domain of an agent (e.g., D_H of XYZ Homes in Example 2) might contain fluents which cannot be changed by XYZ Homes, such as $delivered(b, q)$. Therefore, any trajectory created by an agent will need to assume that certain properties that it cannot affect by its actions must be established by the other agent. Formally, we say that fluent f is *exogenous* in D_X if there exists no state s and action a such that f is true/false in s and false/true in some state belonging to $\Phi(a, s)$. Given a formula φ, let $\pi(D_X, \varphi)$ be the program $\pi(D_X)$ extended with the following rules

- $1\{h(f, 0); \neg h(f, 0)\}1$ for each exogenous fluent f;
- $1\{occ(A, T) : action(A)\}1 \leftarrow step(T)$ (exactly one action occurs at one step);
- the goal constraint $\leftarrow not\ h(\varphi, n)$ (φ must be true at step n);
- $\leftarrow not\ h(sat(ref\_id), n)$ for each clause of the form $ref\_id : X$ in C, i.e., X is responsible for the satisfaction of the clause ref_id in C

It can be shown that there is a 1-to-1 correspondence between answer sets of $\pi(D_X, \varphi)$ and trajectories satisfying φ and all the clauses for which X is responsible for. Furthermore, for each answer set S, the set of assumptions made by the agent is indicated by the set $\{h(f, 0) \mid f$ is an exogenous fluent and $h(f, 0) \in S\}$. Additional constructs can be added to minimize the set of assumptions (e.g., if the assumption is made then it must be utilized in at least one state; for example, XYZ Homes can assume that Lumber Yard A will make fluent $delivered(140K, 2)$ true and plan to receive the board and pay; on the other hand, XYZ Homes does not need to assume $board(140K, 2)$ is true since its actions do not refer to this fluent). It is important to point out that when an agent needs to replan, it might not need to take into consideration *all* clauses that belong to its responsibilities; for example, if Lumber Yard A has already produced 140K board feet of lumber Number 2 Common grade (i.e., (7) is already satisfied) but has not been able to deliver the boards then the replanning process should only consider the other clauses. In this sense, the proposed framework supports the resilience of the contract execution. Similar construction can be done so that the obtained trajectory satisfies all the concerns of agent X. We omit it due to the space limitation.

6 Conclusions and Related Work

In this paper, we discussed an approach for the modeling, and reasoning about, a supply chain as a collection of contracts. We view this as a stepping stone towards enabling reasoning about supply chains' resilience and about ways to improve it. We focused on the development of a framework that supports various reasoning tasks in contract realization such as the feasibility of a contract, the satisfaction of clauses, requirements or concerns of agents of a contract, and the possible plans for mitigating an unsatisfied clause or concern. The framework assumes that each agent operates in accordance to its own action domain and has full observability of its environment. It formalizes a contract as a set of public clauses specifying the responsibility of each agent and, for each agent, a set of statements relating the public clauses with the agent's private concerns. By exploiting knowledge representation and reasoning techniques, we show that all of the above reasoning tasks can be reduced to the task of computing answer sets of suitable programs encoding the action domains and the contract. To the best of our knowledge, this is the first attempt at combining the modeling of supply chain contracts via action languages and ASP, and at linking the contracts with the relevant concerns through the CPS Framework. Having said that, there has been a large amount of research on representing and reasoning about contracts and related topics. Due to space constraints, we provide here limited highlights.

Within the ASP community, [13] addressed the problem of traceability in the supply chain. However, their work did not extend to modeling of contracts and stakeholder concerns, nor reasoning about their satisfaction. Others employed ASP to formalize negotiation – e.g., [17] focuses on establishing a contract (an exchange) between agents that can be satisfied by both parties. In other words, the goal of negotiation is different from that of reasoning about contracts. From the implementation perspective, the ASP-based system for multi-agent planning described in [16] could be useful in computing compatible trajectories for different agents. A different logic-based approach is used in

[14], which discusses the representation and reasoning about contracts through deontic-logic based language $\mathcal{CL}$, with a focus on preserving many of the natural properties and concepts relevant to legal contracts.

In a different area of the spectrum of supply chain research, [3] discusses resilience through an equifinality lens to demonstrate that there are different pathways to supply resilience, with a focus on studying combinations of low and high redundancy scenarios. A related direction of research is around service level agreements (e.g., [2]), which are typically focused on models and protocols for managing the negotiations surrounding access to resources in a distributed system and their use. Researchers have also focused on representing agreements via commitments rather than messaging protocols, see e.g. [18]. Finally, another line of research has been focused on standardized representations of contracts, as in [8,9]. Those approaches are focused on a rich representation of the relationships among contracts, but do not address the challenges posed by the multiplicity and diversity of stakeholders, and do not reason about the evolution of the state of the system over time.

In the future, we plan to investigate the relationship between our approach and smart contracts, whose formalization has recently gained attention (see, e.g., [15]).

References

1. Balduccini, M., Griffor, E., Huth, M., Vishik, C., Burns, M., Wollman, D.A.: Ontology-based reasoning about the trustworthiness of cyber-physical systems. ArXiv abs/1803.07438 (2018)
2. Czajkowski, K., Foster, I., Kesselman, C., Sander, V., Tuecke, S.: SNAP: a protocol for negotiating service level agreements and coordinating resource management in distributed systems. In: Feitelson, D.G., Rudolph, L., Schwiegelshohn, U. (eds.) JSSPP 2002. LNCS, vol. 2537, pp. 153–183. Springer, Heidelberg (2002). https://doi.org/10.1007/3-540-36180-4_9
3. Dube, N., Li, Q., Selviaridis, K., Jahre, M.: One crisis, different paths to supply resilience: the case of ventilator procurement for the COVID-19 pandemic. J. Purch. Supply Manag. **28**(5) (2022)
4. Edward, Greer, C., Wollman, D.A., Burns, M.J.: Framework for cyber-physical systems: vol. 1, overview (2017)
5. Gelfond, M., Lifschitz, V.: Logic programs with classical negation. In: Warren, D., Szeredi, P. (eds.) Logic Programming: Proceedings of the Seventh International Conference, pp. 579–597 (1990)
6. Gelfond, M., Lifschitz, V.: Representing actions and change by logic programs. J. Log. Program. **17**(2,3,4), 301–323 (1993)
7. Gelfond, M., Lifschitz, V.: Action languages. Electron. Trans. Artif. Intell. **3**(6), 193–210 (1998)
8. Governatori, G.: Representing business contracts in RuleML. Int. J. Coop. Inf. Syst. **14**(2–3), 181–216 (2005)
9. Governatori, G., Rotolo, A., Sartor, G.: Normative autonomy and normative co-ordination: declarative power, representation, and mandate. AI and Law **12**(1–2), 53–81 (2004)
10. Marek, V., Truszczyński, M.: Stable models and an alternative logic programming paradigm. In: Apt, K.R., Marek, V.W., Truszczynski, M., Warren, D.S. (eds.) The Logic Programming Paradigm, pp. 375–398. Artificial Intelligence. Springer, Berlin, Heidelberg (1999). https://doi.org/10.1007/978-3-642-60085-2_17

11. Nguyen, T.H., Bundas, M., Son, T.C., Balduccini, M., Garwood, K.C., Griffor, E.R.: Specifying and reasoning about CPS through the lens of the NIST CPS framework. TPLP (2022)
12. Niemelä, I.: Logic programming with stable model semantics as a constraint programming paradigm. Ann. Math. Artif. Intell. **25**(3,4), 241–273 (1999)
13. Nogueira, M., Greis, N.P.: Supply chain tracing of multiple products under uncertainty and incomplete information–an application of answer set programming. In: KEOD 2013, pp. 399–406 (2013)
14. Prisacariu, C., Schneider, G.: CL: An action-based logic for reasoning about contracts. In: WoLLIC, pp. 335–349 (2009)
15. Singh, A., Parizi, R.M., Zhang, Q., Choo, K.K.R., Dehghantanha, A.: Blockchain smart contracts formalization: approaches and challenges to address vulnerabilities. Comput. Secur. **88** (2020)
16. Son, T.C., Pontelli, E., Nguyen, N.-H.: Planning for multiagent using ASP-Prolog. In: Dix, J., Fisher, M., Novák, P. (eds.) CLIMA 2009. LNCS (LNAI), vol. 6214, pp. 1–21. Springer, Heidelberg (2010). https://doi.org/10.1007/978-3-642-16867-3_1
17. Son, T.C., Pontelli, E., Nguyen, N., Sakama, C.: Formalizing negotiations using logic programming. ACM Trans. Comput. Log. **15**(2), 12 (2014)
18. Wan, F., Singh, M.P.: Formalizing and achieving multiparty agreements via commitments. In: Proceedings of the Fourth International Joint Conference on Autonomous Agents and Multiagent Systems (AAMAS 2005), pp. 770–777 (2005)

From Starvation Freedom to All-Path Reachability Problems in Constrained Rewriting

Misaki Kojima(✉) and Naoki Nishida

Graduate School of Informatics, Nagoya University, Nagoya 4648601, Japan
k-misaki@trs.css.i.nagoya-u.ac.jp, nishida@i.nagoya-u.ac.jp

Abstract. An all-path reachability problem of a logically constrained term rewrite system is a pair of constrained terms representing state sets, and is demonically valid if every finite execution path from any state in the first set to a terminating state includes a state in the second set. We have proposed a framework to reduce the non-occurrence of specified error states in a transition system represented by a logically constrained term rewrite system to an all-path reachability problem of the system. In this paper, we extend the framework to verification of starvation freedom of asynchronous integer transition systems with shared variables such that some processes enter critical sections.

Keywords: Term rewriting · Program transformation · Program verification · Runtime error · Coinduction

1 Introduction

Recently, approaches to program verification by means of logically constrained term rewrite systems (LCTRSs, for short) [17] are well investigated [5,8,12,13, 21,28]. LCTRSs are useful as computation models of not only functional but also imperative programs. For instance, equivalence checking by means of LCTRSs is useful to ensure the correctness of terminating functions (cf. [8]). The method of transforming sequential programs into LCTRSs has been extended to concurrent programs with semaphore-based exclusive control [16].

It is worth applying verification techniques for LCTRSs to the verification of practical programs such as automotive embedded systems. As for equivalence verification, verification of the non-occurrence of a specified runtime error in a given system is an important task in e.g., developing concurrent systems. In previous work [15], we have proposed a framework to reduce the non-occurrence of a specified runtime error in a concurrent program to an *all-path reachability* problem (APR problem, for short) of the LCTRS obtained from the concurrent

This work was partially supported by JSPS KAKENHI Grant Number 18K11160 and DENSO Corporation.

M. Hanus and D. Inclezan (Eds.): PADL 2023, LNCS 13880, pp. 161–179, 2023.
https://doi.org/10.1007/978-3-031-24841-2_11

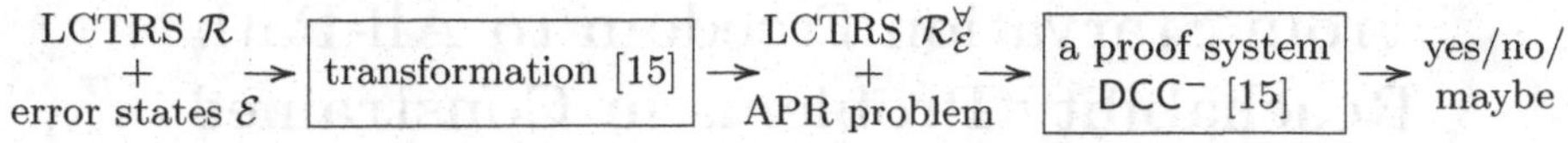

LCTRS $\mathcal{R}$ + error states $\mathcal{E}$	$\rightarrow$	transformation [15]	$\rightarrow$	LCTRS $\mathcal{R}_{\mathcal{E}}^{\forall}$ + APR problem	$\rightarrow$	a proof system DCC^{-} [15]	$\rightarrow$	yes/no/ maybe

Fig. 1. The framework for runtime-error verification [15].

program (Fig. 1), where DCC^{-} is a simplified variant of DCC [5], a proof system for APR problems. An APR problem of a transition system is a pair $P \Rightarrow Q$ of state sets P, Q and is *demonically valid* if every finite execution path—a transition sequence starting with a state in P and ending with a terminating state—includes a state in Q. For a concurrent program with exclusive control, the framework can reduce e.g., the race freedom of mutual exclusion to an APR problem of the corresponding LCTRS.

Unfortunately, the framework is not applicable to the verification of *starvation freedom* which is a typical *liveness* property. Starvation for a process w.r.t. its critical section is a situation where the process wants to enter but never enters the critical section. The framework takes as an input a finite set of constrained terms representing error states. For LCTRSs obtained by the aforementioned transformation [16] from concurrent programs, unlike the race condition, the occurrence of starvation cannot be represented by any finite set of single states that only include information at the moment: A state itself indicates neither that a waiting process may not enter its critical section nor that starvation has already occurred. To characterize the occurrence of starvation at a state, a sequence of transitions from some past state to the state is necessary. For this reason, we cannot represent states with the occurrence of starvation as error states by a finite set of constrained terms without any extra information.

In this paper, to develop a verification method for starvation freedom by means of APR problems, we adapt the framework in [15] (Sect. 4) to the verification of starvation freedom of *asynchronous integer transition systems with shared variables* (AITSs, for short) such that some processes enter their critical sections. The framework in [15] is applicable to LCTRSs obtained from e.g., concurrent programs with semaphore-based exclusive control [16], which are written in a concurrent extension of SIMP [7,12]. To concentrate on starvation freedom, however, this paper deals with AITSs as concurrent programs with mutual exclusion, which are usually transformed into simpler LCTRSs than those obtained from concurrent SIMP programs. In addition, we do not impose any *fairness* constraints, i.e., AITSs in this paper may have *unfair* transitions which are not excluded in solving APR problems.

Given an LCTRS as an AITS (Sect. 3), we first modify the rewrite rules of the LCTRS so as to count the waiting time for a process entering its critical section (Sect. 5.1). Then, we characterize the terminating states (i.e., the ground normal forms of the original LCTRS) by a finite set of constrained terms (Sect. 5.2). Finally, we apply the framework in [15] to the modified LCTRS and the following error states (Sect. 5.3): (1) states with the waiting time exceeding a specified upper limit, and (2) terminating states where the process is waiting. Note that the starvation freedom is reduced to an APR problem of the modified LCTRS. The framework in [15] is extended by combining with the above modification of LCTRSs, which performs as a preprocess of the reduction.

The reduction of starvation freedom to APR problems is not complete because error states are specified by a given upper limit for the waiting time. For this reason, for a starvation-free system, our approach may fail if the limit is too small or there is no fixed upper limit for the system. This may happen under any fairness constraints. On the other hand, the approach with the upper limit is practical in some cases such as real-time systems.

The main contribution of this paper is to show how to apply the APR framework which works for *safety* properties and LCTRSs naively modeling AITSs, to *liveness* properties, i.e., to show an instance of approaches to reduction of verification of liveness properties to that of safety ones. Technical contributions are to introduce to an LCTRS a counter for proving starvation freedom, and to characterize the set of ground normal forms of the LCTRS by finitely many constrained terms.

2 Preliminaries

In this section, we briefly recall LCTRSs [8,17] and all-path reachability [5,24,25]. Familiarity with basic notions on term rewriting [1,22] is assumed.

2.1 Logically Constrained Rewriting

Let $\mathcal{S}$ be a set of *sorts* and $\mathcal{V}$ a countably infinite set of *variables*, each of which is equipped with a sort. A *signature* Σ disjoint from $\mathcal{V}$ is a set of *function symbols* f, each of which is equipped with a *sort declaration* $\iota_1 \times \cdots \times \iota_n \Rightarrow \iota$, written as $f : \iota_1 \times \cdots \times \iota_n \Rightarrow \iota$, where $\iota_1, \ldots, \iota_n, \iota \in \mathcal{S}$. In the rest of this section, we fix $\mathcal{S}$, Σ, and $\mathcal{V}$ and use them without notice. We denote the set of well-sorted *terms* over Σ and $\mathcal{V}$ by $T(\Sigma, \mathcal{V})$. We may write $s : \iota$ if s has sort ι. The set of variables occurring in $s_1, \ldots, s_n$ is denoted by $\mathcal{V}ar(s_1, \ldots, s_n)$. Given a term t and a *position* p (a sequence of positive integers) of t, $t|_p$ denotes the subterm of t at position p, and $s[t]_p$ denotes s with the subterm at position p replaced by t, where the sorts of $s|_p$ and t coincide.

A *substitution* γ is a sort-preserving total mapping from $\mathcal{V}$ to $T(\Sigma, \mathcal{V})$, and naturally extended for a mapping from $T(\Sigma, \mathcal{V})$ to $T(\Sigma, \mathcal{V})$. The *domain* $\mathcal{D}om(\gamma)$ of γ is the set of variables x with $\gamma(x) \neq x$, and the *range* of γ is denoted by $\mathcal{R}an(\gamma)$. The restriction of γ w.r.t. a set X of variables is denoted by $\gamma|_X$: $\gamma|_X(x) = \gamma(x)$ if $x \in X$, and otherwise $\gamma|_X(x) = x$. For two substitutions γ and θ, their composition $\gamma\theta$ is given by $x(\gamma\theta) = \theta(\gamma(x))$ for all variables x. The application of γ to term s is denoted by $s\gamma$.

To define an LCTRS over a signature Σ, we consider the following signatures, mappings, and constants: Two signatures Σ_{terms} and Σ_{theory} such that $\Sigma = \Sigma_{terms} \cup \Sigma_{theory}$, a mapping $\mathcal{I}$ that assigns to each sort ι occurring in Σ_{theory} a set $\mathcal{I}_\iota$, a mapping $\mathcal{J}$ that assigns to each $f : \iota_1 \times \cdots \times \iota_n \Rightarrow \iota \in \Sigma_{theory}$ a function $f^{\mathcal{J}}$ in $\mathcal{I}_{\iota_1} \times \cdots \times \mathcal{I}_{\iota_n} \Rightarrow \mathcal{I}_\iota$, and a set $\mathcal{V}al_\iota \subseteq \Sigma_{theory}$ of *value-constants* $a : \iota$ for each sort ι occurring in Σ_{theory} such that $\mathcal{J}$ gives a bijective mapping from $\mathcal{V}al_\iota$ to $\mathcal{I}_\iota$. Note that for each sort occurring in Σ_{theory}, $\mathcal{I}$ specifies the universe, and for each symbol in Σ_{theory}, $\mathcal{J}$ specifies the interpretation. We denote $\bigcup_{\iota \in \mathcal{S}} \mathcal{V}al_\iota$ by $\mathcal{V}al$. We require that $\Sigma_{terms} \cap \Sigma_{theory} \subseteq \mathcal{V}al$. The sorts occurring in Σ_{theory} are called *theory sorts*, and the symbols *theory symbols*. Symbols in $\Sigma_{theory} \setminus \mathcal{V}al$ are *calculation symbols*, for which we may use infix notation. A term in $T(\Sigma_{theory}, \mathcal{V})$ is called a *theory term*. For ground theory terms, we define the *interpretation* $[\![\cdot]\!]$ as $[\![f(s_1, \ldots, s_n)]\!] = \mathcal{J}(f)([\![s_1]\!], \ldots, [\![s_n]\!])$. Note that for every ground theory term s, there is a unique value-constant c such that $[\![s]\!] = [\![c]\!]$.

We typically choose a theory signature with $\Sigma_{theory} \supseteq \Sigma^{core}_{theory}$, where Σ^{core}_{theory} includes $\mathtt{bool}$, a sort of *Booleans*, such that $\mathcal{V}al_{\mathtt{bool}} = \{\mathsf{true}, \mathsf{false}\}$ and $\mathcal{I}(\mathtt{bool}) = \{\top, \bot\}$, $\Sigma^{core}_{theory} = \mathcal{V}al_{\mathtt{bool}} \cup \{\wedge, \vee, \implies : \mathtt{bool} \times \mathtt{bool} \Rightarrow \mathtt{bool}, \neg : \mathtt{bool} \Rightarrow \mathtt{bool}\} \cup \{=_\iota, \neq_\iota : \iota \times \iota \Rightarrow \mathtt{bool} \mid \iota \text{ is a theory sort}\}$, and $\mathcal{J}$ interprets these symbols as expected: $\mathcal{J}(\mathsf{true}) = \top$ and $\mathcal{J}(\mathsf{false}) = \bot$. We omit the sort subscripts from $=$ and $\neq$ when they are clear from context. The standard integer signature $\Sigma^{\mathtt{int}}_{theory}$ is $\Sigma^{core}_{theory} \cup \{+, -, \times, \mathsf{exp}, \mathsf{div}, \mathsf{mod} : \mathtt{int} \times \mathtt{int} \Rightarrow \mathtt{int}\} \cup \{\geq, > : \mathtt{int} \times \mathtt{int} \Rightarrow \mathtt{bool}\} \cup \mathcal{V}al_{\mathtt{int}}$ where $\mathcal{S} \supseteq \{\mathtt{int}, \mathtt{bool}\}$, $\mathcal{V}al_{\mathtt{int}} = \{\mathsf{n} \mid n \in \mathbb{Z}\}$, $\mathcal{I}(\mathtt{int}) = \mathbb{Z}$, and $\mathcal{J}(\mathsf{n}) = n$. Note that we use n (in sans-serif font) as the function symbol for $n \in \mathbb{Z}$ (in *math* font). We define $\mathcal{J}$ in the natural way. As a syntactic sugar for readability, we add the *ite* operator $\mathsf{ite} : \mathtt{bool} \times \mathtt{int} \times \mathtt{int} \Rightarrow \mathtt{int}$ to $\Sigma^{\mathtt{int}}_{theory}$ such that $\mathcal{J}(\mathsf{ite})(\top, n_1, n_2) = n_1$ and $\mathcal{J}(\mathsf{ite})(\bot, n_1, n_2) = n_2$. Note that formula $\psi[\mathsf{ite}(\phi, e_1, e_2)]$ is equivalent to, e.g., $(\phi \implies \psi[e_1]) \wedge (\neg\phi \implies \psi[e_2])$.

A *constrained rewrite rule* is a triple $\ell \to r\ [\phi]$ such that ℓ and r are terms of the same sort, ϕ is a constraint, and ℓ has the form $f(\ell_1, \ldots, \ell_n)$ that is not a theory term. If $\phi = \mathsf{true}$, then we may write $\ell \to r$. We define $\mathcal{LV}ar(\ell \to r\ [\phi])$ as $\mathcal{V}ar(\phi) \cup (\mathcal{V}ar(r) \setminus \mathcal{V}ar(\ell))$. We say that a substitution γ *respects* $\ell \to r\ [\phi]$ if $\mathcal{R}an(\gamma|_{\mathcal{LV}ar(\ell \to r\ [\phi])}) \subseteq \mathcal{V}al$ and $[\![\phi\gamma]\!] = \top$. Note that it is allowed to have $\mathcal{V}ar(r) \not\subseteq \mathcal{V}ar(\ell)$, but fresh variables in the right-hand side may only be instantiated with *value-constants* (see the definition of $\to_{\mathcal{R}}$ below). Note that we do not deal with *calculation rules* [8] because for any rewrite rule in this paper, no calculation symbol appears in the left- or right- hand sides. The *rewrite relation* $\to_{\mathcal{R}}$ is a binary relation over terms, defined as follows: For a term s, $s[\ell\gamma]_p \to_{\mathcal{R}} s[r\gamma]_p$ if and only if $\ell \to r\ [\phi] \in \mathcal{R}$ and γ respects $\ell \to r\ [\phi]$.

Now we define a *logically constrained term rewrite system* (LCTRS, for short) as an abstract reduction system $(T(\Sigma, \mathcal{V}), \to_{\mathcal{R}})$, simply denoted by $\mathcal{R}$, where $\mathcal{R}$ is a set of constrained rewrite rules. An LCTRS is usually given by supplying Σ, $\mathcal{R}$, and an informal description of $\mathcal{I}$ and $\mathcal{J}$ if these are not clear from context. The set of *normal forms* of $\mathcal{R}$ is denoted by $NF_{\mathcal{R}}$.

Example 2.1. Let $\mathcal{S} = \{\mathtt{int}, \mathtt{bool}\}$, $\Sigma = \Sigma_{terms} \cup \Sigma^{\mathtt{int}}_{theory}$ and $\Sigma_{terms} = \{\mathsf{fact} : \mathtt{int} \Rightarrow \mathtt{int},\ \mathsf{subfact} : \mathtt{int} \times \mathtt{int} \Rightarrow \mathtt{int}\} \cup \{\mathsf{n} : \mathtt{int} \mid n \in \mathbb{Z}\}$. To implement an LCTRS calculating the *factorial* function over $\mathbb{Z}$, we use the signature Σ above and the following LCTRS: $\mathcal{R}_{fact} = \{\ \mathsf{fact}(x) \to \mathsf{subfact}(x, \mathsf{1}),\ \ \mathsf{subfact}(x, y) \to y\ [x \leq \mathsf{0}],\ \ \mathsf{subfact}(x, y) \to \mathsf{subfact}(x', y')\ [\neg(x \leq \mathsf{0}) \wedge x' = x - \mathsf{1} \wedge y' = x \times y]\ \}$.

The term $\mathsf{fact}(3)$ is reduced by $\mathcal{R}_{fact}$ to 6: $\mathsf{fact}(3) \to_{\mathcal{R}_{fact}} \mathsf{subfact}(3,1) \to_{\mathcal{R}_{fact}} \mathsf{subfact}(2,3) \to_{\mathcal{R}_{fact}} \mathsf{subfact}(1,6) \to_{\mathcal{R}_{fact}} \mathsf{subfact}(0,6) \to_{\mathcal{R}_{fact}} 6$.

A *constrained term* is a pair $\langle t \mid \phi \rangle$ of a term t and a constraint ϕ. Note that the sort of $\langle t \mid \phi \rangle$ is the same as that of t. The set of all ground instances of $\langle t \mid \phi \rangle$ is denoted by $[\![\langle t \mid \phi \rangle]\!]$: $[\![\langle t \mid \phi \rangle]\!] = \{t\gamma \mid \mathcal{D}om(\gamma) \supseteq \mathcal{V}ar(t),\ \mathcal{R}an(\gamma) \subseteq T(\Sigma),\ \gamma \text{respects} \phi\}$. We consider constrained terms as sets of ground terms. The set of *derivatives* of a constrained term is defined as follows: $\Delta_{\mathcal{R}}(\langle t \mid \phi \rangle) = \bigcup_{\ell \to r\ [\psi] \in \mathcal{R}} \Delta_{\ell,r,\psi}(\langle t \mid \phi \rangle)$ where $\ell \to r\ [\psi]$ has no shared variable with $\langle t \mid \phi \rangle$[1] and $\Delta_{\ell,r,\psi}(\langle t \mid \phi \rangle) = \{\langle (t[r]_p)\gamma \mid (\phi \wedge \psi)\gamma \rangle \mid p \text{ is a position of } t,\ t|_p \text{ and } \ell \text{ are unifiable, } \gamma \text{ is an mgu of } t|_p \text{ and } \ell,\ \mathcal{R}an(\gamma|_{\mathcal{V}ar(\phi,\psi)}) \subseteq \mathcal{V}al \cup \mathcal{V},\ (\phi \wedge \psi)\gamma \text{ is satisfiable}\}$. A constrained term $\langle t \mid \phi \rangle$ is called $\mathcal{R}$-*derivable* if $\Delta_{\mathcal{R}}(\langle t \mid \phi \rangle) \neq \emptyset$.

2.2 All-Path Reachability

Let $(M, \rightharpoonup)$ be a transition system with $\rightharpoonup \subseteq M \times M$. An *execution path* (of $(M, \rightharpoonup)$) is either an infinite transition sequence or a finite transition sequence ending with an irreducible state. A *state predicate* is a state set P ($\subseteq M$). The predicate P is said to be *runnable* (w.r.t. $\rightharpoonup$) if $P \neq \emptyset$ and for each $t \in P$ there exists $t' \in M$ such that $t \rightharpoonup t'$. Note that a runnable state predicate does not include any irreducible state. A *reachability predicate* (w.r.t. $(M, \rightharpoonup)$) is a pair $P \Rightarrow Q$ of state predicates P, Q.

An execution path τ is said to *satisfy* a reachability predicate $P \Rightarrow Q$, written $\tau \models^{\forall} P \Rightarrow Q$, if τ starts with a state in P and τ includes a state in Q, whenever τ is finite. A reachability predicate $P \Rightarrow Q$ is said to be *demonically valid* (w.r.t. $(M, \rightharpoonup)$) if $\tau \models^{\forall} P \Rightarrow Q$ for any execution path τ starting from a state in P. The demonical validity of $P \Rightarrow Q$ means that every finite execution path starting from a state in P eventually reaches a state in Q.

Let $\mathcal{R}$ be an LCTRS and $\langle t_\ell \mid \phi_\ell \rangle, \langle t_r \mid \phi_r \rangle$ constrained terms. We call the pair $\langle t_\ell \mid \phi_\ell \rangle \Rightarrow \langle t_r \mid \phi_r \rangle$ an *all-path reachability problem* (APR problem, for short) of $\mathcal{R}$. Note that $\langle t_\ell \mid \phi_\ell \rangle$ and $\langle t_r \mid \phi_r \rangle$ may have shared variables. We say that $\mathcal{R}$ *demonically satisfies* $\langle t_\ell \mid \phi_\ell \rangle \Rightarrow \langle t_r \mid \phi_r \rangle$,[2] written $\mathcal{R} \models^{\forall} \langle t_\ell \mid \phi_\ell \rangle \Rightarrow \langle t_r \mid \phi_r \rangle$, if $[\![\langle t_\ell\gamma \mid \phi_\ell\gamma \rangle]\!] \Rightarrow [\![\langle t_r\gamma \mid \phi_r\gamma \rangle]\!]$ is demonically valid w.r.t. $(T(\Sigma), \to_{\mathcal{R}})$ for any substitution γ such that $\mathcal{D}om(\gamma) = \mathcal{V}ar(t_\ell, \phi_\ell) \cap \mathcal{V}ar(t_r, \phi_r)$, and $\mathcal{R}an(\gamma) \subseteq T(\Sigma)$, and $\mathcal{R}an(\gamma|_{\mathcal{V}ar(\phi_\ell) \cap \mathcal{V}ar(\phi_r)}) \subseteq \mathcal{V}al$.

Next, we recall a simplified variant of the proof system DCC [5] for a certain class of APR problems [15]. An APR problem $\langle t_\ell \mid \phi_\ell \rangle \Rightarrow \langle t_r \mid \phi_r \rangle$ is called *constant-directed* if t_r is a constant normal form and ϕ_r is satisfiable. In the following, w.l.o.g., we assume that $\phi_r = \mathsf{true}$.[3] Given an LCTRS $\mathcal{R}$ and a finite

[1] When there exists a shared variable, we rename the variables in $\ell \to r\ [\psi]$.

[2] We also say that $\langle t_\ell \mid \phi_\ell \rangle \Rightarrow \langle t_r \mid \phi_r \rangle$ is *demonically valid* w.r.t. $\mathcal{R}$.

[3] A constant-directed problem $\langle t_\ell \mid \phi_\ell \rangle \Rightarrow \langle c \mid \phi_r \rangle$ is equivalent to the constant-directed problem $\langle t_\ell \mid \phi_\ell \wedge (\exists \vec{x}.\ \phi_r) \rangle \Rightarrow \langle c \mid \mathsf{true} \rangle$, where $\{\vec{x}\} = \mathcal{V}ar(\phi_r) \setminus \mathcal{V}ar(t_\ell, \phi_\ell)$.

$$\frac{\phi_\ell \text{ is unsatisfiable}}{\langle t_\ell \mid \phi_\ell\rangle \Rightarrow \langle c \mid \mathsf{true}\rangle}\ (\mathsf{axiom}) \qquad \frac{t_\ell = c}{\langle t_\ell \mid \phi_\ell\rangle \Rightarrow \langle c \mid \mathsf{true}\rangle}\ (\mathsf{subs})$$

$$\frac{\begin{array}{l}\langle t_1 \mid \phi_1\rangle \Rightarrow \langle c \mid \mathsf{true}\rangle \\ \quad\vdots \\ \quad\langle t_n \mid \phi_n\rangle \Rightarrow \langle c \mid \mathsf{true}\rangle \quad \langle t_\ell \mid \phi_\ell\rangle \text{ is } \mathcal{R}\text{-derivable} \quad [\![\langle t_\ell \mid \phi_\ell\rangle]\!] \cap NF_{\mathcal{R}} = \emptyset\end{array}}{\langle t_\ell \mid \phi_\ell\rangle \Rightarrow \langle c \mid \mathsf{true}\rangle}\ (\mathsf{der})$$

where $\Delta_{\mathcal{R}}(\langle t_\ell \mid \phi_\ell\rangle) = \{\langle t_i \mid \phi_i\rangle \mid 1 \leq i \leq n\}$ for some $n > 0$.

$$\frac{\exists (\langle t'_\ell \mid \phi'_\ell\rangle \Rightarrow \langle c \mid \mathsf{true}\rangle) \in G.\ [\![\langle t_\ell \mid \phi_\ell\rangle]\!] \subseteq [\![\langle t'_\ell \mid \phi'_\ell\rangle]\!]^4}{\langle t_\ell \mid \phi_\ell\rangle \Rightarrow \langle c \mid \mathsf{true}\rangle}\ (\mathsf{weak\ circ})$$

Fig. 2. The proof rules of DCC^-.

set G of constant-directed APR problems, the proof system $\mathsf{DCC}^-(\mathcal{R}, G)$ [15] consists of the proof rules in Fig. 2[4].

Theorem 2.2 (soundness of DCC^- [15]). *Let $\mathcal{R}$ be an LCTRS, and G a finite set of constant-directed APR problems. Suppose that for each problem $(\langle t_\ell \mid \phi_\ell\rangle \Rightarrow \langle c \mid \mathsf{true}\rangle) \in G$, there exists a proof tree $\mathcal{T}$ under $\mathsf{DCC}^-(\mathcal{R}, G)$, each circ node of which has a der node as an ancestor. Then, $\mathcal{R} \models^\forall \langle t_\ell \mid \phi_\ell\rangle \Rightarrow \langle c \mid \mathsf{true}\rangle$ for all problems $(\langle t_\ell \mid \phi_\ell\rangle \Rightarrow \langle c \mid \mathsf{true}\rangle) \in G$.*

In proving an APR problem, as for *cyclic proofs* [3], it is enough to construct a single proof tree under $\mathsf{DCC}^-(\mathcal{R}, G)$ such that the root node is the APR problem, G includes the APR problem and all der nodes in the tree, and for each weak circ node, the tree includes a der node, the conclusion of which is the same as the weak circ node.

3 LCTRSs for Asynchronous ITSs with Shared Variables

In this section, we briefly explain how to represent *asynchronous integer transition systems with shared variables* [2] (AITS, for short) by LCTRSs (cf. [4,20]). In the following, we consider assignments for variables in transition systems and program graphs as substitutions. The *update* of an assignment η w.r.t. a variable x and its new value v can be represented by the composition $\{x \mapsto v\}\eta$ of substitutions $\{x \mapsto v\}$ and η.

We consider an AITS consisting of n processes $P_1, \ldots, P_n$ with m shared integer or Boolean variables $x_1, \ldots, x_m$. We denote the variable sequences $x_1, \ldots, x_m$ by $\vec{x}$. We use finitely many constants for representing locations of processes. For simplicity, we do not deal with any local variable of processes, and use shared

[4] If there exists a substitution γ such that $\mathcal{R}an(\gamma|_{\mathcal{V}ar(\phi'_\ell)}) \subseteq T(\Sigma_{theory}, \mathcal{V}ar(\phi_\ell))$, $t_\ell = t'_\ell\gamma$, and $\phi_\ell \iff \phi'_\ell\gamma$ is valid, then $[\![\langle t_\ell \mid \phi_\ell\rangle]\!] \subseteq [\![\langle t'_\ell \mid \phi'_\ell\rangle]\!]$ [15, Proposition 5.8].

variables instead of local ones.[5] The transition of a process can be represented by a *program graph* $(Loc, \hookrightarrow, Loc_0, \phi_0)$ such that *Loc* is a finite set of *locations*, $\hookrightarrow \subseteq Loc \times Fol \times Act \times Loc$ is a conditional transition relation, Loc_0 ($\subseteq Loc$) is a set of initial locations, and ϕ_0 ($\in Fol$) is the initial condition, where *Act* is a set of multiple assignments $\langle x_{i_1}, \ldots, x_{i_j} \rangle := \langle e_{i_1}, \ldots, e_j \rangle$ for pairwise distinct variables $x_{i_1}, \ldots, x_{i_j} \in \{\vec{x}\}$ and integer or Boolean expressions $e_1, \ldots, e_j$ over $x_1, \ldots, x_m$, and *Fol* is the set of formulas whose free variables are in $\{x_1, \ldots, x_m\}$. Note that a single assignment $\langle x_{i_1} \rangle := \langle e_{i_1} \rangle$ may be written as $x_{i_1} := e_{i_1}$. The notation $\ell \overset{\phi:\alpha}{\hookrightarrow} \ell'$ is used as shorthand for $(\ell, \phi, \alpha, \ell') \in \hookrightarrow$; if ϕ is a tautology, we omit ϕ, writing $\ell \overset{\alpha}{\hookrightarrow} \ell'$; if α is the empty assignment $\langle\rangle := \langle\rangle$, then we may omit it, writing $\ell \overset{\phi}{\hookrightarrow} \ell'$ or $\ell \hookrightarrow \ell'$.

In the following, we use $p_1, \ldots, p_n$ as variables for locations of $P_1, \ldots, P_n$, respectively. We denote the sequence $p_i, p_{i+1}, \ldots, p_j$ ($i \leq j$) by $\overrightarrow{p_{i..j}}$, and the full sequence $p_1, \ldots, p_n$ by $\vec{p}$.

Let $PG_i = (Loc_i, \hookrightarrow_i, Loc_{0,i}, \phi_{0,i})$ be a program graph for process P_i ($i \in \{1, \ldots, n\}$). The AITS which is an interleaving of $PG_1, \ldots, PG_n$ can be represented by the following LCTRS: $\bigcup_{i=1}^{n} \{ \mathsf{state}(\overrightarrow{p_{1..i-1}}, \ell, \overrightarrow{p_{i+1..n}}, \vec{x}) \to \mathsf{state}(\overrightarrow{p_{1..i-1}}, \ell', \overrightarrow{p_{i+1..n}}, (\vec{x})\{x_{k_j} \mapsto x'_{k_j} \mid 1 \leq j \leq d\}) \; [\phi \wedge \bigwedge_{j=1}^{d} x'_{k_j} = e_j] \mid \ell \overset{\phi:\alpha}{\hookrightarrow}_i \ell',\ \alpha = \langle x_{k_1}, \ldots, x_{k_d} \rangle := \langle e_1, \ldots, e_d \rangle \}$, where

- the sorts are $\{\mathtt{int}, \mathtt{state}, \mathtt{loc}\}$,
- $\mathsf{state} : \overbrace{\mathtt{loc} \times \cdots \times \mathtt{loc}}^{n} \times typ_1 \times \cdots \times typ_m \to \mathtt{state}$, where for each $i \in \{1, \ldots, m\}$, if x_i is an integer variable, then $typ_i = \mathtt{int}$, and otherwise, $typ_i = \mathtt{bool}$,
- $\ell : \mathtt{loc}$ for $\ell \in \bigcup_{i=1}^{n} Loc_i$, and
- $p_1, \ldots, p_n, x_1, \ldots, x_m, x'$ are pairwise distinct variables.

The initial states are $\bigcup_{\ell_1 \in Loc_1, \ldots, \ell_n \in Loc_n} [\![\langle \mathsf{state}(\ell_1, \ldots, \ell_n, \vec{x}) \mid \bigwedge_{i=1}^{n} \phi_{i,0} \rangle]\!]$. Note that each rewrite rule of the above LCTRS represents a transition of exactly one process. To simplify discussion in later sections, we assume w.l.o.g. that $Loc_1, \ldots, Loc_n$ are pairwise disjoint sets, and all $Loc_{0,1}, \ldots, Loc_{0,n}$ are singleton.

Example 3.1. Consider the program graph in Fig. 3 for *Peterson's mutual exclusion algorithm* [2, Example 2.25]. Two processes P_0, P_1[6] share Boolean variables b_0, b_1 and an integer variable x; b_i indicates that P_i wants to enter the critical section crit_i; x stores the identifier of the process that has priority for the critical section at that time. The LCTRS $\mathcal{R}_1$ for the AITS consisting of PG_0 and PG_1 is illustrated in Fig. 4. The initial state is the term $\mathsf{state}(\mathsf{noncrit}_0, \mathsf{noncrit}_1, \mathsf{false}, \mathsf{false}, 0)$. The AITS has 72 states, some of which may not be reachable from any initial state; each process has two local states; there

[5] To represent local variables $y_1, \ldots, y_k$ of process P_i with locations $\ell_1, \ldots$, we use a ground term $\ell_j(v_1, \ldots, v_k)$ for the local state with location ℓ_j and integers $v_1, \ldots, v_k$ that are assigned to $y_1, \ldots, y_k$, respectively.

[6] For simplicity, in some examples, we use P_0 instead of P_n, i.e., use $P_0, \ldots, P_{n-1}$.

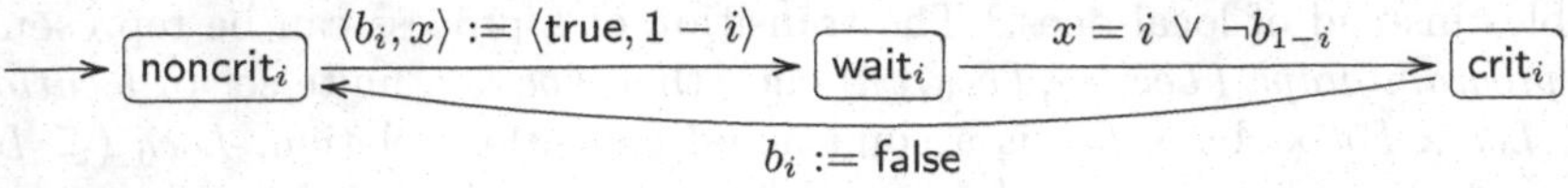

$$\longrightarrow \boxed{\mathsf{noncrit}_i} \xrightarrow{\langle b_i, x\rangle := \langle \mathsf{true}, 1-i\rangle} \boxed{\mathsf{wait}_i} \xrightarrow{x = i \lor \neg b_{1-i}} \boxed{\mathsf{crit}_i}$$

$b_i := \mathsf{false}$

Fig. 3. Program graph PG_i $(i = 0, 1)$ for Peterson's mutual exclusion algorithm.

$$\mathcal{R}_1 = \left\{ \begin{array}{llll} \mathsf{state}(\mathsf{noncrit}_0, p_1, & b_0, b_1, x) \to \mathsf{state}(\mathsf{wait}_0, & p_1, & b_0', b_1, x')\ [b_0' = \mathsf{true} \land x' = 1] \\ \mathsf{state}(\mathsf{wait}_0, \ \ p_1, & b_0, b_1, x) \to \mathsf{state}(\mathsf{crit}_0, & p_1, & b_0, b_1, x\ )\ [x = 0 \lor \neg b_1] \\ \mathsf{state}(\mathsf{crit}_0, \ \ p_1, & b_0, b_1, x) \to \mathsf{state}(\mathsf{noncrit}_0, p_1, & & b_0', b_1, x\ )\ [b_0' = \mathsf{false}] \\ \mathsf{state}(p_0, \ \ \mathsf{noncrit}_1, & b_0, b_1, x) \to \mathsf{state}(p_0, & \mathsf{wait}_1, & b_0, b_1', x')\ [b_1' = \mathsf{true} \land x' = 0] \\ \mathsf{state}(p_0, \ \ \mathsf{wait}_1, & b_0, b_1, x) \to \mathsf{state}(p_0, & \mathsf{crit}_1, & b_0, b_1, x\ )\ [x = 1 \lor \neg b_0] \\ \mathsf{state}(p_0, \ \ \mathsf{crit}_1, & b_0, b_1, x) \to \mathsf{state}(p_0, & \mathsf{noncrit}_1, & b_0, b_1', x\ )\ [b_0' = \mathsf{false}] \end{array} \right\}$$

Fig. 4. LCTRS $\mathcal{R}_1$ for the AITS consisting of PG_0 and PG_1 in Fig. 3.

are three shared variables, each of which may take three kinds of integers (cf. [2, Section 2.3]). If we add a shared variable z that is incremented in moving from crit_i to $\mathsf{noncrit}_i$, then the AITS has infinitely many states.

4 Reducing Non-occurrence of Error States to APR Problems

In this section, we recall the framework in [15] to reduce the non-occurrence of error states to an APR problem, where the error states are specified by finitely many constrained terms. In the rest of the paper, we use $\mathcal{R}$ and $\langle s_0 \mid \phi_0 \rangle$[7] as the LCTRS and its initial states defined in Sect. 3 for an AITS consisting of n processes $P_1, \ldots, P_n$ with m shared integer or Boolean variables $x_1, \ldots, x_m$ without notice. In addition, we let $s_0 = \mathsf{state}(\ell_{0,1}, \ldots, \ell_{0,n}, \vec{x})$ and use $p_1, \ldots, p_n, x_1, \ldots, x_m$ as pairwise disjoint variables without notice.

Definition 4.1 ([15]). *Let $\mathcal{E}$ be a finite set of constrained terms which represent error states and have sort* `state`. *Introducing fresh constants* success *and* error *with sort* `state` *into the signature Σ, we define the LCTRS $\mathcal{R}_{\mathcal{E}}^{\forall}$ as follows:*

$$\mathcal{R}_{\mathcal{E}}^{\forall} = \mathcal{R} \cup \{ \mathsf{state}(\vec{p}, \vec{x}) \to \mathsf{success} \} \cup \{ u \to \mathsf{error}\ [\phi] \mid \langle u \mid \phi \rangle \in \mathcal{E} \}$$

Note that u is neither success *nor* error. *Note also that u may be (a renamed variant of) $\mathsf{state}(\vec{p}, \vec{x})$ if all states are considered error ones. We reduce the non-occurrence of any instance of constrained terms in $\mathcal{E}$ to the APR problem $\langle s_0 \mid \phi_0 \rangle \Rightarrow \langle \mathsf{success} \mid \mathsf{true} \rangle$ of $\mathcal{R}_{\mathcal{E}}^{\forall}$.*

[7] By definition, the sort of $\langle s_0 \mid \phi_0 \rangle$ is `state`.

Rule $\mathsf{state}(\vec{p}, \vec{x}) \to \mathsf{success}$ rewrites all states to success. Each rule $u \to \mathsf{error}\ [\phi]$ rewrites all states represented by $\langle u \mid \phi \rangle \in \mathcal{E}$ to error. Note that every finite execution path of $\mathcal{R}_{\mathcal{E}}^{\forall}$ starting from a ground term in $\langle s_0 \mid \phi_0 \rangle$ ends with either success or error. We may omit $\mathcal{E}$ from $\mathcal{R}_{\mathcal{E}}^{\forall}$ because we can know $\mathcal{E}$ from $\mathcal{R}_{\mathcal{E}}^{\forall} \setminus \mathcal{R}$. By proving $\langle s_0 \mid \phi_0 \rangle \Rightarrow \langle \mathsf{success} \mid \mathsf{true} \rangle$ to be demonically valid, we can ensure the non-existence of a finite rewrite sequence of $\mathcal{R}$ that starts with a ground term in $\langle s_0 \mid \phi_0 \rangle$ and includes a ground term in some $\langle u \mid \phi \rangle \in \mathcal{E}$.

Theorem 4.2 ([15]). *For a finite set $\mathcal{E}$ of constrained terms with sort* `state`*, $\mathcal{R}_{\mathcal{E}}^{\forall} \models^{\forall} \langle s_0 \mid \phi_0 \rangle \Rightarrow \langle \mathsf{success} \mid \mathsf{true} \rangle$ if and only if $s \not\to_{\mathcal{R}}^{*} u'$ for any ground term s in $\langle s_0 \mid \phi_0 \rangle$ and any ground term u' in $\bigcup_{\langle u \mid \phi \rangle \in \mathcal{E}} \langle u \mid \phi \rangle$.*

Example 4.3. Let us consider the *race freedom* of mutual exclusion—the two processes do not enter their critical sections simultaneously—for the LCTRS $\mathcal{R}_1$ in Example 3.1. The error states are represented by $\langle \mathsf{state}(\mathsf{crit}_0, \mathsf{crit}_1, b_0, b_1, x) \mid \mathsf{true} \rangle$, and we generate the following LCTRS:

$$\mathcal{R}_1^{\forall} = \mathcal{R}_1 \cup \{\ \mathsf{state}(p_0, p_1, b_0, b_1, x) \to \mathsf{success},\ \mathsf{state}(\mathsf{crit}_0, \mathsf{crit}_1, b_0, b_1, x) \to \mathsf{error}\ \}$$

The race freedom is reduced to the following APR problem of $\mathcal{R}_1^{\forall}$:

$$\langle \mathsf{state}(\mathsf{noncrit}_0, \mathsf{noncrit}_1, \mathsf{false}, \mathsf{false}, 0) \mid \mathsf{true} \rangle \Rightarrow \langle \mathsf{success} \mid \mathsf{true} \rangle$$

The proof system DCC^- succeeds in proving that $\mathcal{R}_1^{\forall}$ demonically satisfies the above APR problem.

The number of states of the AITS represented by $\mathcal{R}_1^{\forall}$ is finite, and thus, the race freedom can easily be proved by computing all reachable states from the initial ones. On the other hand, the approach in this section succeeds in proving the non-occurrence of error states in AITSs such that the number of states is huge or infinite (see [15]).

5 Reducing Starvation Freedom to APR Problems

The framework in Sect. 4 cannot reduce starvation freedom to an APR problem without any extra information introduced to states. In this section, we extend the framework to starvation freedom for a process w.r.t. a critical section of the process. To this end, we add to a state extra information to notice that starvation may occur, and propose a preprocess for a given LCTRS so as to make rewrite rules in the LCTRS update the added extra information. Note that we can verify starvation freedom of the entire system by verifying starvation freedom for every process and every critical section.

In the rest of this section, we consider starvation freedom for process P_i w.r.t. a critical section in $\mathcal{R}$. Let crit be a location which is the entry point of the critical section, and $\mathsf{wait}_1, \ldots, \mathsf{wait}_k$ the locations that transition to crit by one reduction step of $\mathcal{R}$, i.e., for each $j \in \{1, \ldots, k\}$, there is a rewrite rule $\mathsf{state}(\overrightarrow{p_{1..i-1}}, \mathsf{wait}_j, \overrightarrow{p_{i+1..n}}, \vec{x}) \to \mathsf{state}(\overrightarrow{p_{1..i-1}}, \mathsf{crit}, \overrightarrow{p_{i+1..n}}, \vec{x'})\ [\,\phi_j\,]$ in $\mathcal{R}$.

5.1 Modification of Rewrite Rules

Starvation freedom guarantees that P_i can enter the critical section in finite time, whenever P_i wants to enter it. To characterize starvation (freedom) in $\mathcal{R}$, we have to represent a waiting time of P_i at some wait_j until entering the critical section (i.e., transitioning to crit). To this end, we introduce a counter for the waiting time as a new argument of state. We modify rewrite rules in $\mathcal{R}$ so that

- the counting starts when P_i enters a waiting state (some wait_j), and resets when the process enters the critical section (i.e., transitions to crit), and
- the value of the counter is incremented at any transition step of other processes whenever P_i is waiting.

Starvation freedom holds if values of the counter at any reachable state is less than some upper limit. Therefore, states with the value of the counter exceeding the limit are considered as error states, and we reduce starvation freedom to an APR problem of the modified LCTRS and such error states. In our approach, an predicted upper limit c_{max} is specified by a user in advance: If the reduced APR problem with c_{max} is proved to be demonically satisfied, then the existence of the expected upper limit is ensured, i.e., the starvation freedom is proved.

As a new argument of state, we first introduce a counter that represents the time—the number of reduction steps—for P_i waiting for the critical section, where the new sort of state is $\overbrace{\mathtt{loc} \times \cdots \times \mathtt{loc}}^{n} \times \mathit{typ}_1 \times \cdots \times \mathit{typ}_m \times \mathtt{int} \to \mathtt{state}$. Note that the "$n+m+1$"-th argument of state is used for the counter. In the following, we use c, c' as variables for the counter.

Next, we modify the rewrite rules in $\mathcal{R}$ as follows.

- When P_i enters a wait state wait_j, the value of the counter is set to 1: We replace $\mathsf{state}(\overrightarrow{p_{1..i-1}}, \ell, \overrightarrow{p_{i+1..n}}, \vec{x}) \to \mathsf{state}(\overrightarrow{p_{1..i-1}}, \mathsf{wait}_j, \overrightarrow{p_{i+1..n}}, \vec{x'})\ [\phi]$ by
$$\mathsf{state}(\overrightarrow{p_{1..i-1}}, \ell, \overrightarrow{p_{i+1..n}}, \vec{x}, c) \to \mathsf{state}(\overrightarrow{p_{1..i-1}}, \mathsf{wait}_j, \overrightarrow{p_{i+1..n}}, \vec{x'}, c')\ [\phi \wedge c' = 1]$$
- When P_i enters the critical section (i.e., P_i transitions from wait_j to crit), we reset the counter: We replace $\mathsf{state}(\overrightarrow{p_{1..i-1}}, \mathsf{wait}_j, \overrightarrow{p_{i+1..n}}, \vec{x}) \to \mathsf{state}(\overrightarrow{p_{1..i-1}}, \mathsf{crit}, \overrightarrow{p_{i+1..n}}, \vec{x'})\ [\phi_j]$ by
$$\mathsf{state}(\overrightarrow{p_{1..i-1}}, \mathsf{wait}_j, \overrightarrow{p_{i+1..n}}, \vec{x}, c) \to \mathsf{state}(\overrightarrow{p_{1..i-1}}, \mathsf{crit}, \overrightarrow{p_{i+1..n}}, \vec{x'}, c')\ [\phi_j \wedge c' = 0]$$
- The remaining rules of P_i preserve the value of the counter: We replace $\mathsf{state}(\overrightarrow{p_{1..i-1}}, \ell, \overrightarrow{p_{i+1..n}}, \vec{x}) \to \mathsf{state}(\overrightarrow{p_{1..i-1}}, \ell', \overrightarrow{p_{i+1..n}}, \vec{x'})\ [\phi]$ by
$$\mathsf{state}(\overrightarrow{p_{1..i-1}}, \ell, \overrightarrow{p_{i+1..n}}, \vec{x}, c) \to \mathsf{state}(\overrightarrow{p_{1..i-1}}, \ell', \overrightarrow{p_{i+1..n}}, \vec{x'}, c)\ [\phi]$$
where $\ell \notin \{\mathsf{wait}_1, \ldots, \mathsf{wait}_k\}$.
- When the value of the counter is more than $\mathsf{0}$, the counter is incremented each time other processes $P_{j'}$ ($j' \neq i$) make transitions: We replace $\mathsf{state}(\overrightarrow{p_{1..j'-1}}, \ell, \overrightarrow{p_{j'+1..n}}, \vec{x}) \to \mathsf{state}(\overrightarrow{p_{1..j'-1}}, \ell', \overrightarrow{p_{j'+1..n}}, \vec{x'})\ [\phi]$ by
$$\mathsf{state}(\overrightarrow{p_{1..j'-1}}, \ell, \overrightarrow{p_{j'+1..n}}, \vec{x}, c) \to \mathsf{state}(\overrightarrow{p_{1..j'-1}}, \ell', \overrightarrow{p_{j'+1..n}}, \vec{x'}, c')\ [\phi \wedge c' = \mathsf{inc}_{\geq 1}(c)]$$

$$\widetilde{\mathcal{R}_1} = \left\{ \begin{array}{r} \mathsf{state}(\mathsf{noncrit}_0, p_1, b_0, b_1, x, c) \to \mathsf{state}(\mathsf{wait}_0, p_1, b_0', b_1, x', c') \\ [\, b_0' = \mathsf{true} \wedge x' = 1 \wedge c' = 1 \,] \\ \mathsf{state}(\mathsf{wait}_0, p_1, b_0, b_1, x, c) \to \mathsf{state}(\mathsf{crit}_0, p_1, b_0, b_1, x, c') \\ [\, x = 0 \vee \neg b_1 \wedge c' = 0 \,] \\ \mathsf{state}(\mathsf{crit}_0, p_1, b_0, b_1, x, c) \to \mathsf{state}(\mathsf{noncrit}_0, p_1, b_0', b_1, x, c) \;\; [\, b_0' = \mathsf{false} \,] \\ \mathsf{state}(p_0, \mathsf{noncrit}_1, b_0, b_1, x, c) \to \mathsf{state}(p_0, \mathsf{wait}_1, b_0, b_1', x', c') \\ [\, b_1' = \mathsf{true} \wedge x' = 0 \wedge c' = \mathsf{inc}_{\geq 1}(c) \,] \\ \mathsf{state}(p_0, \mathsf{wait}_1, b_0, b_1, x, c) \to \mathsf{state}(p_0, \mathsf{crit}_1, b_0, b_1, x, c') \\ [\, x = 1 \vee \neg b_0 \wedge c' = \mathsf{inc}_{\geq 1}(c) \,] \\ \mathsf{state}(p_0, \mathsf{crit}_1, b_0, b_1, x, c) \to \mathsf{state}(p_0, \mathsf{noncrit}_1, b_0, b_1', x, c') \\ [\, b_1' = \mathsf{false} \wedge c' = \mathsf{inc}_{\geq 1}(c) \,] \end{array} \right\}$$

Fig. 5. LCTRS $\widetilde{\mathcal{R}_1}$ generated by the method in Sect. 5.

where $\mathsf{inc}_{\geq 1}(c)$ denotes $\mathsf{ite}(c \geq 1, c+1, c)$.

We denote the modified LCTRS in the above way by $\widetilde{\mathcal{R}}_{\mathsf{crit}}$. We omit crit from $\widetilde{\mathcal{R}}_{\mathsf{crit}}$, writing $\widetilde{\mathcal{R}}$, because crit is clear from the difference between $\mathcal{R}$ and $\widetilde{\mathcal{R}}_{\mathsf{crit}}$.

Example 5.1. Consider the LCTRS $\mathcal{R}_1$ in Example 3.1 again. Given a positive integer c_{max}, the method in this section generates the LCTRS $\widetilde{\mathcal{R}_1}$ in Fig. 5.

Given an integer c_{max} (> 0), error states along the idea above are represented by $\langle \mathsf{state}(\vec{p}, \vec{x}, c) \mid c > \mathsf{c}_{max} \rangle$. The states in $\langle \mathsf{state}(\vec{p}, \vec{x}, c) \mid c > \mathsf{c}_{max} \rangle$ may not be actual error states because e.g., c_{max} is too small to ensure starvation freedom.

5.2 Characterization of Terminating States by Constrained Terms

The modified LCTRS $\widetilde{\mathcal{R}}$ is correct from viewpoint of starvation (Theorem 5.9 shown later). On the other hand, the error state $\langle \mathsf{state}(\vec{p}, \vec{x}, c) \mid c > \mathsf{c}_{max} \rangle$ in the previous section is not enough. After reaching a *terminating* state, the counter cannot be incremented and the waiting time cannot be measured. In this case, starvation occurs because P_i cannot enter the critical section. Terminating states where P_i is waiting should be error states. The condition that P_i is waiting can be represented by the counter: The counter is greater than zero. Therefore, we characterize *terminating states* where every process is waiting or already halts.

Terminating states are ground normal forms of $\mathcal{R}$ and they are transformed into error states by adding the non-zero value for the counter. To use the framework in Sect. 4, we have to represent error states by a finite set of constrained terms. Since the set of ground normal forms is infinite in general, we cannot use the set to represent error states. For this reason, we construct a finite set of constrained terms for the ground normal forms of $\mathcal{R}$.

To characterize terminating states by a finite set of constrained terms, for each location, we consider the condition under which the transition from the location is possible. In the following, for a finite set $Y = \{y_1, \dots, y_k\}$ of variables, we denote an arbitrary but fixed sequence $y_1, \dots, y_k$ by $\overrightarrow{Y}$.

Definition 5.2. *Let ℓ be a location of P_i. We define $\mathit{DefCnst}_{\mathcal{R}}(\ell)$ as follows:*

$$\mathit{DefCnst}_{\mathcal{R}}(\ell) = \bigvee_{\mathsf{state}(\overrightarrow{p_{1..i-1}}, \ell, \overrightarrow{p_{i+1..n}}, \vec{x}) \to r\ [\phi] \in \mathcal{R}} \exists \overrightarrow{\mathcal{V}ar(\phi) \setminus \{\vec{x}\}}.\ \phi$$

Note that ℓ in Definition 5.2 is not a location of any other process. The applicability of a rewrite rule $\mathsf{state}(\overrightarrow{p_{1..i-1}}, \ell, \overrightarrow{p_{i+1..n}}, \vec{x}) \to r\ [\phi]$ to a term depends on whether the matching substitution for the left-hand side and the term satisfies ϕ, but ϕ may include a variable not appearing in the left-hand side. To decide the applicability only, it is sufficient to existentially quantify the variables in $\mathcal{V}ar(\phi) \setminus \mathcal{V}ar(\vec{x})$.

Example 5.3. Consider $\mathcal{R}_1$ in Example 3.1. For wait_0 and $\mathsf{noncrit}_1$, we have that

- $\mathit{DefCnst}_{\mathcal{R}_1}(\mathsf{wait}_0) \;=\; x = 0 \vee \neg b_1$, and
- $\mathit{DefCnst}_{\mathcal{R}_1}(\mathsf{noncrit}_1) \;=\; \exists b_1', x'.\ (b_1' = \mathsf{true} \vee x' = 0)$.

Proposition 5.4. *Let $\ell_1, \dots, \ell_n$ be locations of $P_1, \dots, P_n$, respectively, and $\gamma = \{x_1 \mapsto v_1, \dots, x_m \mapsto v_m\}$ with integers $v_1, \dots, v_m$. Then, $\mathsf{state}(\ell_1, \dots, \ell_n, \vec{x})\gamma$ is reducible w.r.t. $\mathcal{R}$, if and only if $[\![(\mathit{DefCnst}_{\mathcal{R}}(\ell_i))\,\gamma]\!] = \top$ for some $i \in \{1, \dots, n\}$.*

Proof. Trivial by definition. □

As a consequence of Proposition 5.4, terminating states are characterized as follows.

Corollary 5.5. *A ground term t with sort* `state` *is a normal form of $\mathcal{R}$ if and only if $t \in \bigcup_{\ell_1 \in \mathit{Loc}_1, \dots, \ell_n \in \mathit{Loc}_n} \langle \mathsf{state}(\ell_1, \dots, \ell_n, \vec{x}) \mid \bigwedge_{i=1}^{n} \neg \mathit{DefCnst}_{\mathcal{R}}(\ell_i) \rangle$.*

The set of constrained term in the above corollary may include redundant ones which have unsatisfiable constraints. We do not have to consider such redundant ones, removing from the set.

Definition 5.6. *We define the set of terminating states as follows:*

$$\mathit{TerStates}(R) = \{ \langle \mathsf{state}(\ell_1, \dots, \ell_n, \vec{x}) \mid \phi \rangle \mid \ell_i \in \mathit{Loc}_i,\ \phi = \textstyle\bigwedge_{i=1}^{n} \neg \mathit{DefCnst}_{\mathcal{R}}(\ell_i),\ \phi \text{ is satisfiable} \}$$

By definition, it is clear that $\mathit{TerStates}(R)$ is finite up to variable renaming, and $\bigcup_{\langle t \mid \phi \rangle \in \mathit{TerStates}(\mathcal{R})} \langle t \mid \phi \rangle$ is the set of ground terms with sort `state`.

Example 5.7. Consider the LCTRS $\mathcal{R}_1$ in Example 3.1 again. We have that

$$\mathit{TerStates}(\mathcal{R}_1) = \{ \langle \mathsf{state}(\mathsf{wait}_0, \mathsf{wait}_1, b_0, b_1, x) \mid \neg(x{=}0 \vee \neg b_1) \wedge \neg(x{=}1 \vee \neg b_0) \rangle \}$$

5.3 From Starvation Freedom to APR Problems

In this section, we reduce starvation freedom to an APR problem by means of the framework in Sect. 4.

For a positive integer c_{max} and the LCTRS $\widetilde{\mathcal{R}}$ in Sect. 5.1, we let $\mathcal{E}_{>\mathsf{c}_{max}}$ be the following set:

$$\{\langle \mathsf{state}(\vec{p},\vec{x},c) \mid c > \mathsf{c}_{max}\rangle\} \\ \cup \{\langle \mathsf{state}(\ell_1,\ldots,\ell_n,\vec{x},c) \mid c > \mathsf{0} \wedge \phi\rangle \mid \langle \mathsf{state}(\ell_1,\ldots,\ell_n,\vec{x}) \mid \phi\rangle \in \mathit{TerStates}(\mathcal{R})\}$$

Then, we apply the framework in Sect. 4 to $\widetilde{\mathcal{R}}$ and $\mathcal{E}_{>\mathsf{c}_{max}}$, obtaining the following LCTRS $(\widetilde{\mathcal{R}})^{\forall}_{\mathcal{E}_{>\mathsf{c}_{max}}}$ and APR problem of $(\widetilde{\mathcal{R}})^{\forall}_{\mathcal{E}_{>\mathsf{c}_{max}}}$:

$$(\widetilde{\mathcal{R}})^{\forall}_{\mathcal{E}_{>\mathsf{c}_{max}}} = \widetilde{\mathcal{R}} \cup \{\, \mathsf{state}(\vec{p},\vec{x},c) \to \mathsf{success}\,\} \cup \{\, u \to \mathsf{error}\ [\phi] \mid \langle u \mid \phi\rangle \in \mathcal{E}_{>\mathsf{c}_{max}}\,\}$$

and

$$\langle \mathsf{state}(\ell_{0,1},\ldots,\ell_{0,n},\vec{x},\mathsf{0}) \mid \phi_0\rangle \Rightarrow \langle \mathsf{success} \mid \mathsf{true}\rangle$$

Recall that $\ell_{0,1},\ldots,\ell_{0,n}$ are the initial locations of $P_1,\ldots,P_n$, respectively. We omit $\mathcal{E}$ from $(\widetilde{\mathcal{R}})^{\forall}_{\mathcal{E}_{>\mathsf{c}_{max}}}$, writing $(\widetilde{\mathcal{R}})^{\forall}_{>\mathsf{c}_{max}}$.

Example 5.8. Consider the modified LCTRS $\widetilde{\mathcal{R}_1}$ in Example 5.1. Given $\mathcal{E}_{>\mathsf{c}_{max}} = \{\ \langle \mathsf{state}(p_0,p_1,b_0,b_1,x,c) \mid c > \mathsf{c}_{max}\rangle,\ \langle \mathsf{state}(\mathsf{wait}_0,\mathsf{wait}_1,b_0,b_1,x,c) \mid \neg(x = \mathsf{0} \vee \neg b_1) \wedge \neg(x = \mathsf{1} \vee \neg b_0)\rangle\ \}$, the framework in Sect. 4 generates the following LCTRS and APR problem to verify the starvation freedom for P_0 w.r.t. wait_0:

$$(\widetilde{\mathcal{R}_1})^{\forall}_{>\mathsf{c}_{max}} = \widetilde{\mathcal{R}_1} \cup \left\{ \begin{array}{lr} \mathsf{state}(p_0,p_1,b_0,b_1,x,c) \to \mathsf{success} & \\ \mathsf{state}(p_0,p_1,b_0,b_1,x,c) \to \mathsf{error} & [c > \mathsf{c}_{max},] \\ \mathsf{state}(\mathsf{wait}_0,\mathsf{wait}_1,b_0,b_1,x,c) \to \mathsf{error} & \\ \qquad [\neg(x = \mathsf{0} \vee \neg b_1) \wedge \neg(x = \mathsf{1} \vee \neg b_0) \wedge c > \mathsf{0}] & \end{array} \right\}$$

and $\langle \mathsf{state}(\mathsf{noncrit}_0,\mathsf{noncrit}_1,\mathsf{false},\mathsf{false},\mathsf{0},\mathsf{0}) \mid \mathsf{true}\rangle \Rightarrow \langle \mathsf{success} \mid \mathsf{true}\rangle$. The proof system DCC^- succeeds in proving that $(\widetilde{\mathcal{R}_1})^{\forall}_{>\mathsf{4}}$ demonically satisfies the above APR problem.

Using the method in this section, we cannot conclude that starvation occurs even when the counter value exceeds c_{max}, i.e., when $(\widetilde{\mathcal{R}})^{\forall}_{>\mathsf{c}_{max}}$ does not demonically satisfy the APR problem obtained by the method. This is because the value of c_{max} may not be sufficiently large. For example, if we set c_{max} to 1, then P_0 can wait for only one transition step of P_1, and the initial state reaches error. Therefore, if the proof fails, then we try to prove another upper limit that is larger than the selected value for c_{max}.

Finally, we show the correctness of the method in this section. In the rest of this section, c_{max} denotes a positive integer, $\mathcal{E}_{>\mathsf{c}_{max}}$ denotes $\{\langle \mathsf{state}(\vec{p},\vec{x},c) \mid c > \mathsf{c}_{max}\rangle\} \cup \{\langle \mathsf{state}(\ell_1,\ldots,\ell_n,\vec{x},c) \mid c > \mathsf{0} \wedge \phi\rangle \mid \langle \mathsf{state}(\ell_1,\ldots,\ell_n,\vec{x}) \mid \phi\rangle \in \mathit{TerStates}(\mathcal{R})\}$, and s'_0 denotes $\mathsf{state}(\ell_{0,1},\ldots,\ell_{0,n},\vec{x},\mathsf{0})$, where $\ell_{0,1},\ldots,\ell_{0,n}$ are the initial locations of $P_1,\ldots,P_n$, respectively. Recall that $\langle s_0 \mid \phi_0\rangle$ is the initial states of $\mathcal{R}$. Note that for a ground term t with sort $\mathtt{state}$, $t|_{n+m+1}$ is the value of the counter introduced in constructing $\widetilde{\mathcal{R}}$.

Theorem 5.9. *Suppose that for any ground term $t'_0 \in \langle s'_0 \mid \phi_0 \rangle$, if $t'_0 \to^*_{\widetilde{\mathcal{R}}}$ $\mathsf{state}(\ell_1,\ldots,\ell_n,\vec{v},v_c)$, then both of the following hold:*

- $[\![v_c]\!] \leq [\![\mathsf{c}_{max}]\!]$, *and*
- *if* $\mathsf{state}(\ell_1,\ldots,\ell_n,\vec{v})$ *is a normal form of* $\mathcal{R}$, *then* $v_c = \mathsf{0}$.

Then, for any ground term $t_0 \in \langle s_0 \mid \phi_0 \rangle$ and any ground term t such that $t_0 \to^*_{\mathcal{R}} t$ and $t|_i = \mathsf{wait}_j$, every (possibly infinite) execution path starting from t includes a ground term u such that $u|_i = \mathsf{crit}$.

Proof. We proceed by contradiction. Suppose that there exist ground terms $t_0 = \mathsf{state}(\ell_{0,1},\ldots,\ell_{0,n},\vec{v}) \in \langle s_0 \mid \phi_0 \rangle$ and $t = \mathsf{state}(\ell_1,\ldots,\ell_n,\vec{v'})$ such that $t_0 \to^*_{\mathcal{R}} t$, $t|_i = \mathsf{wait}_j$, and there exists an execution path starting from t and not including a term u such that $u|_i = \mathsf{crit}$. By definition, we have the corresponding execution path of $\widetilde{\mathcal{R}}$ that is obtained by adding the values of the introduced counter. Since the value v_c of the counter in $\mathsf{state}(\ell_1,\ldots,\ell_n,\vec{v'},v_c)$ is more than $\mathsf{0}$ and the execution path does not include a term u such that $u|_i = \mathsf{crit}$, the counter is strictly increasing along the execution path of $\widetilde{\mathcal{R}}$. We make a case analysis depending on whether the execution path is finite or not.

- Consider the case where the execution path of $\mathcal{R}$ is finite. Suppose that the execution path ends with a term $u' = \mathsf{state}(\ell_1,\ldots,\ell_n,\vec{v})$. Then, the corresponding execution path of $\widetilde{\mathcal{R}}$ ends with $\mathsf{state}(\ell_1,\ldots,\ell_n,\vec{v},v'_c)$ such that $v'_c \neq \mathsf{0}$. This contradicts the assumption.
- Consider the remaining case where the execution path of $\mathcal{R}$ is infinite. Since the counter is strictly increasing along the execution path of $\widetilde{\mathcal{R}}$, the path includes a term having the value of the counter that exceeds c_{max}. This contradicts the assumption. □

The following corollary is a direct consequence of Theorems 4.2 and 5.9.

Corollary 5.10. *Suppose that $(\widetilde{\mathcal{R}})^{\forall}_{>\mathsf{c}_{max}} \models^{\forall} \langle s'_0 \mid \phi_0 \rangle \Rightarrow \langle \mathsf{success} \mid \mathsf{true} \rangle$. Then, for any ground term $t_0 \in \langle s_0 \mid \phi_0 \rangle$ and any ground term t such that $t_0 \to^*_{\mathcal{R}} t$ and $t|_i = \mathsf{wait}_j$, every (possibly infinite) execution path starting from t includes a ground term u such that $u|_i = \mathsf{crit}$.*

6 Experiments

We have implemented the proof system DCC^- in a prototype of Crisys2, an interactive proof system based on constrained rewriting induction [8,18].[8] The implemented tool Crisys2apr succeeds in proving the APR problems in Examples 5.8 and other small examples (See footnote 8). In this section, to evaluate efficiency of constructing inference trees based on DCC^-, we briefly report the result of experiments by means of a tool to solve APR problems in our setting.

Table 1 shows the result of our experiments which were conducted on a machine running 64 bit FreeBSD 12.3 on Intel(R) Xeon(R) E-2134 CPU @

[8] https://www.trs.css.i.nagoya-u.ac.jp/~nishida/padl2023/.

Table 1. The result of experiments by means of our implementation.

No.	Examples	Proc.	Rules	States	c_{max}	Strv.-free	Result	Time	Ht.
1	Example 5.8	2	9	72	4	✓	Yes	7.39 s	7
2	+ variable z	2	9	∞	4	✓	Yes	8.11 s	7
3	[2, Example 2.24]	2	9	18	10		Maybe	12.32 s	12
					20		Maybe	28.89 s	22
					30		Maybe	53.65 s	32
4	[2, Example 2.28]	2	8	8	10		Maybe	2.70 s	11
					20		Maybe	5.54 s	21
					30		Maybe	9.14 s	31
5	[2, Example 2.30]	3	15	24	2	✓	Yes	1.00 s	6
6	[2, Exercise 2.7]	2	9	72	4	✓	Yes	4.69 s	7
7		3	23	729	10		Maybe	16511.11 s	13
8	[2, Exercise 2.9]	2	9	∞	4	✓	Yes	8.66 s	6
9	[2, Exercise 2.10]	2	14	128	10		Maybe	2168.20 s	12

3.5 GHz with 64 GB memory; Z3 (ver. 4.8.16) [19] was used as an external SMT solver. The second to fourth columns show the numbers of processes, constrained rewrite rules, and states which may not be reachable from any initial state, respectively. The fifth column shows whether examples are starvation-free or not. The sixth and seventh columns show results and running times, respectively. The eighth column shows the number of initial goals—if it is more than one, then one is the main and the others are auxiliary goals that perform as lemmas. The ninth column shows the value of "c_{max}" that is minimum for success. The tenth and eleventh columns show execution times and heights of inference trees. Note that all the non-starvation-free examples include unfair execution paths, and "Maybe" is the expected result for them.

The infinite numbers of states of the second and eighth examples are not essential: The range of variable z is infinite and thus the number of states of the second example is infinite. In general, the execution time must be exponential on the number of states.

7 Related Work

Reachability is one of the well investigated properties in term rewriting, and there are several works that analyze reachability problems [9–11,26]. A powerful approach to reachability in term rewriting is *automata completion*: Given a tree automaton and a rewrite system, the completion generates a tree automaton that (over-approximately) recognizes reachable ground terms from a ground term recognized by the given tee automaton. This technique is applicable to APR problems of TRSs. To apply the technique to constrained systems such as LCTRSs, we need a further approximation. The simplest way is to ignore constraints of rewrite rules, but this is no longer useful, e.g., the TRS obtained from $\mathcal{R}_1$ by removing constraints is not starvation-free.

The approach in [15] and this paper—the use of APR problems of LCTRSs—is based on [4,5]. A leading verification example in [5] is to prove *partial correctness* represented by APR problems, i.e., to prove that every finite execution path starting with an initial state under the pre-condition ends with a successful terminating state that satisfies the post-condition. The goal of [4] is to reduce *total correctness* to partial ones by introducing a counter to states as in this paper. Unlike our approach, the counter is used to prove termination, and is decreased at every step. To succeed in the proof for a given system, we need to find an appropriate initial value of the counter represented by an arithmetic expression over variables appearing in the system. Since we deal with a different property such as starvation freedom, this paper and [15] show another application of [4,5] to program verification.

Reduction of verification of liveness properties to that of safety ones has been well studied in the context of termination verification. More general counters than ours are used in [6] for termination/non-termination verification, and using a first-order fixpoint logic, the idea has been generalized in [14] for verification of arbitrary temporal properties expressed in the modal μ-calculus.

$\mathbb{K}$ framework [23] is a more general setting of rewriting than LCTRSs, and the proof system for all-path reachability has been implemented in $\mathbb{K}$ [24,25]. The race-freedom of Peterson's mutual exclusion algorithm has been proved in [25] by means of the APR approach, while starvation freedom is not considered. A comparison of all-path reachability logic with CTL* can be seen in [25].

Software model checking has been characterized by *cyclic-proof search* [27]. Given a transition system, the reachability predicate R is defined for reachable states from the initial ones, and the safety verification problem for a property specified by a predicate α to be satisfied by all reachable states is represented by the sequent $R(x) \vdash \alpha(x)$. The cyclic-proof search has similar proof rules to the proof system DCC (and also DCC^-), but has several rules for sequent calculus, e.g., CUT. It is shown in [27] that the cyclic-proof search using only the rule SE based on *symbolic execution* covers bounded model checking: Given a bound k, the proof tree with height k corresponds to the k-step symbolic execution. Since SE is the same as rule der of DCC, the proof tree under DCC or DCC^- also covers k-bounded model checking.

8 Conclusion

In this paper, we adapted the framework in [15] to the verification of starvation freedom of an LCTRS representing an AITS with shared variables. We first modified the rewrite rules of the LCTRS so as to count the waiting time for a process entering its critical section, and then characterized the ground normal forms of the LCTRS by a finite set of constrained terms. Finally, we applied the framework to the modified LCTRS and the set of error states specified by the counter and the characterized set, reducing the starvation freedom to an

APR problem of the modified LCTRS. We have implemented the framework in Sect. 4 and a prototype of the transformation in Sect. 5: (See footnote 8) Given an LCTRS $\mathcal{R}_{\mathcal{E}}^{\forall}$ and an APR problem $\langle s_0 \mid \phi_0 \rangle \Rightarrow \langle \mathsf{success} \mid \mathsf{true} \rangle$, the prototype drops the rules for $\mathcal{E}$, modifies the remaining rules, and generates the rules of error.

As described in Sect. 5, depending on how to choose a positive integer c_{max}, our method may fail even for starvation-free systems. We will study the existence of an appropriate value as c_{max}, and how to compute it, if exists.

In failing to solve APR problems, we often construct inference trees that include leafs with goals, e.g., $\langle \mathsf{error} \mid \mathsf{true} \rangle \Rightarrow \langle \mathsf{success} \mid \mathsf{true} \rangle$, that are not demonically valid. For example, for all the examples in Table 1, our tool halts with such trees. The trees may be useful to refute starvation freedom and/or to extract unfair execution paths. Development of such techniques is one of future directions of this research.

By considering e.g., state as a predicate, LCTRSs representing AITSs must be seen as constrained Horn clauses (CHC, for short), while LCTRSs model functional or imperative programs with e.g., function calls and global variables (see e.g., [12]). In future work, the approach in this paper should be compared with the application of CHC-based approaches in, e.g., [14] from theoretical and/or empirical viewpoint.

References

1. Baader, F., Nipkow, T.: Term Rewriting and All That. Cambridge University Press, Cambridge (1998). https://doi.org/10.1145/505863.505888
2. Baier, C., Katoen, J.: Principles of Model Checking. MIT Press, Cambridge (2008)
3. Brotherston, J.: Cyclic proofs for first-order logic with inductive definitions. In: Beckert, B. (ed.) TABLEAUX 2005. LNCS (LNAI), vol. 3702, pp. 78–92. Springer, Heidelberg (2005). https://doi.org/10.1007/11554554_8
4. Buruiană, A.S., Ciobâcă, Ş.: Reducing total correctness to partial correctness by a transformation of the language semantics. In: Niehren, J., Sabel, D. (eds.) Proceedings of the 5th International Workshop on Rewriting Techniques for Program Transformations and Evaluation. Electronic Proceedings in Theoretical Computer Science, vol. 289, pp. 1–16. Open Publishing Association (2018). https://doi.org/10.4204/EPTCS.289.1
5. Ciobâcă, Ş, Lucanu, D.: A coinductive approach to proving reachability properties in logically constrained term rewriting systems. In: Galmiche, D., Schulz, S., Sebastiani, R. (eds.) IJCAR 2018. LNCS (LNAI), vol. 10900, pp. 295–311. Springer, Cham (2018). https://doi.org/10.1007/978-3-319-94205-6_20
6. Fedyukovich, G., Zhang, Y., Gupta, A.: Syntax-guided termination analysis. In: Chockler, H., Weissenbacher, G. (eds.) CAV 2018. LNCS, vol. 10981, pp. 124–143. Springer, Cham (2018). https://doi.org/10.1007/978-3-319-96145-3_7
7. Fernández, M.: Programming Languages and Operational Semantics – A Concise Overview. Undergraduate Topics in Computer Science. Springer, London (2014). https://doi.org/10.1007/978-1-4471-6368-8
8. Fuhs, C., Kop, C., Nishida, N.: Verifying procedural programs via constrained rewriting induction. ACM Trans. Computat. Log. **18**(2), 14:1–14:50 (2017). https://doi.org/10.1145/3060143

9. Genet, T., Rusu, V.: Equational approximations for tree automata completion. J. Symb. Comput. **45**(5), 574–597 (2010). https://doi.org/10.1016/j.jsc.2010.01.009
10. Genet, T., Tong, V.V.T.: Reachability analysis of term rewriting systems with Timbuk. In: Nieuwenhuis, R., Voronkov, A. (eds.) LPAR 2001. LNCS (LNAI), vol. 2250, pp. 695–706. Springer, Heidelberg (2001). https://doi.org/10.1007/3-540-45653-8_48
11. Jacquemard, F.: Decidable approximations of term rewriting systems. In: Ganzinger, H. (ed.) RTA 1996. LNCS, vol. 1103, pp. 362–376. Springer, Heidelberg (1996). https://doi.org/10.1007/3-540-61464-8_65
12. Kanazawa, Y., Nishida, N.: On transforming functions accessing global variables into logically constrained term rewriting systems. In: Niehren, J., Sabel, D. (eds.) Proceedings of the 5th International Workshop on Rewriting Techniques for Program Transformations and Evaluation. Electronic Proceedings in Theoretical Computer Science, vol. 289, pp. 34–52. Open Publishing Association (2019)
13. Kanazawa, Y., Nishida, N., Sakai, M.: On representation of structures and unions in logically constrained rewriting. In: IEICE Technical Report SS2018-38, IEICE 2019, vol. 118, no. 385, pp. 67–72 (2019). In Japanese
14. Kobayashi, N., Nishikawa, T., Igarashi, A., Unno, H.: Temporal verification of programs via first-order fixpoint logic. In: Chang, B.-Y.E. (ed.) SAS 2019. LNCS, vol. 11822, pp. 413–436. Springer, Cham (2019). https://doi.org/10.1007/978-3-030-32304-2_20
15. Kojima, M., Nishida, N.: On reducing non-occurrence of specified runtime errors to all-path reachability problems. In: Informal Proceedings of the 9th International Workshop on Rewriting Techniques for Program Transformations and Evaluation, pp. 1–16 (2022)
16. Kojima, M., Nishida, N., Matsubara, Y.: Transforming concurrent programs with semaphores into logically constrained term rewrite systems. In: Informal Proceedings of the 7th International Workshop on Rewriting Techniques for Program Transformations and Evaluation. pp. 1–12 (2020)
17. Kop, C., Nishida, N.: Term rewriting with logical constraints. In: Fontaine, P., Ringeissen, C., Schmidt, R.A. (eds.) FroCoS 2013. LNCS (LNAI), vol. 8152, pp. 343–358. Springer, Heidelberg (2013). https://doi.org/10.1007/978-3-642-40885-4_24
18. Kop, C., Nishida, N.: Automatic constrained rewriting induction towards verifying procedural programs. In: Garrigue, J. (ed.) APLAS 2014. LNCS, vol. 8858, pp. 334–353. Springer, Cham (2014). https://doi.org/10.1007/978-3-319-12736-1_18
19. de Moura, L., Bjørner, N.: Z3: an efficient SMT solver. In: Ramakrishnan, C.R., Rehof, J. (eds.) TACAS 2008. LNCS, vol. 4963, pp. 337–340. Springer, Heidelberg (2008). https://doi.org/10.1007/978-3-540-78800-3_24
20. Naaf, M., Frohn, F., Brockschmidt, M., Fuhs, C., Giesl, J.: Complexity analysis for term rewriting by integer transition systems. In: Dixon, C., Finger, M. (eds.) FroCoS 2017. LNCS (LNAI), vol. 10483, pp. 132–150. Springer, Cham (2017). https://doi.org/10.1007/978-3-319-66167-4_8
21. Nishida, N., Winkler, S.: Loop detection by logically constrained term rewriting. In: Piskac, R., Rümmer, P. (eds.) VSTTE 2018. LNCS, vol. 11294, pp. 309–321. Springer, Cham (2018). https://doi.org/10.1007/978-3-030-03592-1_18
22. Ohlebusch, E.: Advanced Topics in Term Rewriting. Springer, New York (2002). https://doi.org/10.1007/978-1-4757-3661-8
23. Rosu, G., Serbanuta, T.: An overview of the K semantic framework. J. Log. Algebraic Program. **79**(6), 397–434 (2010). https://doi.org/10.1016/j.jlap.2010.03.012

24. Ştefănescu, A., Ciobâcă, Ş, Mereuta, R., Moore, B.M., Şerbănută, T.F., Roşu, G.: All-path reachability logic. In: Dowek, G. (ed.) RTA 2014. LNCS, vol. 8560, pp. 425–440. Springer, Cham (2014). https://doi.org/10.1007/978-3-319-08918-8_29
25. Stefanescu, A., Ciobâcă, Ş., Mereuta, R., Moore, B.M., Serbanuta, T., Rosu, G.: All-path reachability logic. Log. Methods Comput. Sci. **15**(2) (2019). https://doi.org/10.23638/LMCS-15(2:5)2019
26. Takai, T., Kaji, Y., Seki, H.: Right-linear finite path overlapping term rewriting systems effectively preserve recognizability. In: Bachmair, L. (ed.) RTA 2000. LNCS, vol. 1833, pp. 246–260. Springer, Heidelberg (2000). https://doi.org/10.1007/10721975_17
27. Tsukada, T., Unno, H.: Software model-checking as cyclic-proof search. Proc. ACM Program. Lang. **6**(POPL), 1–29 (2022). https://doi.org/10.1145/3498725
28. Winkler, S., Middeldorp, A.: Completion for logically constrained rewriting. In: Kirchner, H. (ed.) Proceedings of the 3rd International Conference on Formal Structures for Computation and Deduction. LIPIcs, vol. 108, pp. 30:1–30:18. Schloss Dagstuhl-Leibniz-Zentrum für Informatik (2018). https://doi.org/10.4230/LIPIcs.FSCD.2018.30

SwitchLog: A Logic Programming Language for Network Switches

Vaibhav Mehta(✉), Devon Loehr, John Sonchack, and David Walker

Princeton University, Princeton, NJ 08544, USA
vaibhavm@princeton.edu

Abstract. The development of programmable switches such as the Intel Tofino has allowed network designers to implement a wide range of new in-network applications and network control logic. However, current switch programming languages, like P4, operate at a very low level of abstraction. This paper introduces SwitchLog, a new experimental logic programming language designed to lift the level of abstraction at which network programmers operate, while remaining amenable to efficient implementation on programmable switches. SwitchLog is inspired by previous distributed logic programming languages such as NDLog, in which programmers declare a series of facts, each located at a particular switch in the network. Logic programming rules that operate on facts at different locations implicitly generate network communication, and are updated incrementally, as packets pass through a switch. In order to ensure these updates can be implemented efficiently on switch hardware, SwitchLog imposes several restrictions on the way programmers can craft their rules. We demonstrate that SwitchLog can be used to express a variety of networking applications in a mere handful of lines of code.

Keywords: Programmable networks · Data plane programming · P4 · Logic programming · Datalog

1 Introduction

Programmable switches allow network operators to customize the packet processing of their network, adding powerful new capabilities such as for monitoring [6] and performance-aware routing [8]. The core of a programmable switch is a reconfigurable ASIC – a processor specialized for packet operations at high and guaranteed throughputs.

The *de facto* standard programming language for reconfigurable ASICs is P4 [2]. However, while P4 allows us to program these devices, it does not make it *easy*. P4 can be viewed as an "assembly language" for line-rate switch programming: it provides fine-grained, low-level control over device operations, but there are few abstractions and, to paraphrase Robin Milner, much can "go wrong."

As a result, researchers have begun to investigate the definition of higher-level languages for switch programming such as Domino [15], Sonata [6],

M. Hanus and D. Inclezan (Eds.): PADL 2023, LNCS 13880, pp. 180–196, 2023.
https://doi.org/10.1007/978-3-031-24841-2_12

MARPLE [14], Path Queries [13], MAFIA [9], Chipmunk [5], μP4 [19], Lyra [4], Lucid [10,17], and Π4 [3]. Each of these languages provides useful new ideas to the design space—the ultimate future switch programming language may well include components from each of them, which is why exploration of a diverse range of ideas is so valuable now. For instance, Domino, Chipmunk and Lyra improve the switch programming experience by providing higher-level abstractions and using program synthesis to generate efficient low-level code. Sonata and Path Queries, among others, provide new abstractions for monitoring network traffic. μP4 develops a framework that makes switch programs more modular and compositional. Lucid adds abstractions for events and event-handling. Π4 and Lucid both develop new type systems for detecting user errors in switch programs.

In this paper, we continue to explore the design space of switch programming languages. In particular, we were inspired by Loo's past work on NDLog [11,12] and VMWare's Differential Datalog [1], logic programming languages designed for software network controllers. Loo observed that many key networking algorithms, such as routing, are essentially table-driven algorithms: based on network traffic or control messages, the algorithm constructs one or more tables. Those tables are pushed to switches in the network data plane, which use them to guide packet-level actions such as routing, forwarding, load balancing, or access control. Such table-driven algorithms are easily and compactly expressed as logic programs. Moreover, the VMWare team observed that many networking control software algorithms are naturally incremental, so developing an incremental Datalog system would deliver both high performance and economy of notation. Finally, when programs are developed in this form, they are also amenable to formal verification [20].

With these ideas in mind, we developed SwitchLog, a new, experimental, incremental logic programming language designed to run on real-world switches, such as the Intel Tofino. By running in the data plane instead of the control plane, SwitchLog implementations of network algorithms can benefit from finer-grained visibility into traffic conditions [15] and orders of magnitude better performance [17]. However, Because SwitchLog is designed for switch hardware, its design must differ from NDLog or Differential Datalog, which both run on general-purpose software platforms. In particular, arbitrary joins (which are common in traditional Datalog languages) are simply too expensive to implement in the network. Evaluating an arbitrary join requires iteration over all tuples in a table. On a switch, iteration over a table requires processing *one packet for every tuple*, because the switch's architecture can only access a single memory address (i.e., tuple) per packet.

To better support switch hardware, the incremental updates in a SwitchLog program are designed to only require a bounded, and in many cases constant, amount of work. To minimize the need for joins, SwitchLog extends conventional Datalog with constructs that let programmers take advantage of the hardware-accelerated lookup tables in the switch. In a SwitchLog program, each relation specifies some fields to be *keys*, and the others *values*. The rest of the program is then structured so that: 1) most facts can be looked up by their key (a constant-time operation in a programmable switch), rather than by iterating over all known facts; 2) facts with identical keys can be merged with user provided value-aggregation functions.

Although SwitchLog is certainly more restrictive than conventional Datalogs, these restrictions make it practical to execute logic programs on a completely new class of hardware: line-rate switches. To aid programmers in writing efficient programs, we provide several simple syntactic guidelines for writing programs that both compile to switch hardware, and do not generate excessive work during execution. We believe that these guidelines strike a good balance between efficiency and expressivity – in practice, we have found that they still allow us to efficiently implement a range of interesting network algorithms.

To demonstrate how SwitchLog may integrate with other network programming languages, we have implemented SwitchLog as a sublanguage within Lucid [17], an imperative switch programming language that compiles to P4. Within a Lucid program, programmers use SwitchLog to write queries that generate tables of information for use in a larger network application. Once such tables are *materialized*, the rest of the Lucid components may act on that information (for instance, by routing packets, performing load balancing, or implementing access control). Conversely, execution of SwitchLog components may be directly triggered by events in the Lucid program, e.g., a Lucid program can inform the SwitchLog sub-program of new facts. This back-and-forth makes it possible to use logic programming abstractions where convenient, and fall back on lower-level imperative constructs otherwise. While the current prototype is integrated with Lucid, we note that the same design could be used to integrate SwitchLog with any other imperative switch programming language, including P4.

In the follow sections, we illustrate the design of SwitchLog by example, comparing and contrasting it with previous declarative networking languages like NDLog in Sect. 2, explaining how to compile SwitchLog to raw Lucid in Sect. 3, evaluating our system in Sect. 4, and concluding in Sect. 6.

2 SwitchLog Design

SwitchLog is inspired by other distributed logic programming languages such as Network Datalog (NDLog) [12]. To illustrate the key similarities and differences, we begin our discussion of SwitchLog's design by presenting an implementation of shortest paths routing in NDLog, and then show how the application changes when we move to SwitchLog.

2.1 NDLog

A typical Datalog program consists of a set of *facts*, as well as several *rules* for deriving more facts. Each fact is an element $r(x_1, \ldots, x_n)$ of some predefined relations over data values x_i. Each rule has the form $p_1 :- p_2, p_3 ... p_n$, where the p_i are facts, and may be read logically as "the facts p_2 through p_n together imply

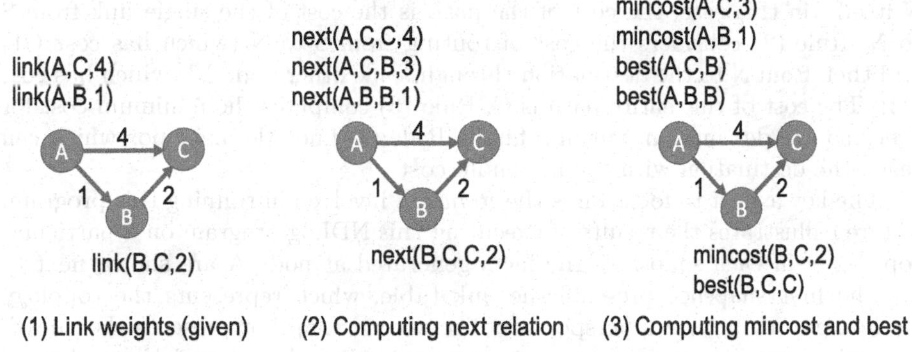

Fig. 1. Sample execution of NDLog program

p_1". Operationally, if facts p_2 through p_n have been derived then fact p_1 will be as well. If a relation never appears on the left-hand side of any rule, it is called a *base* relation; elements of base relations are never derived, and must be supplied externally.

In a *distributed* logic programming language like NDLog, each fact is located in a particular place—by convention, the first argument of a fact indicates where it is stored. Hence, the fact $f(@L, x, y)$ is stored at location L—the @ symbol acts as a mnemonic. We call this argument the *location specifier*. Other parameters of the relation may also be locations, but they have no special meaning.

When a rule has the form $f(@L_2, x) :- g(@L_1, y)$, communication occurs: A message is transmitted from location L_1 to location L_2. Hence, such logic programming languages facilitate description of distributed communication protocols (the essence of networking) very concisely.

As an example, consider implementing shortest paths routing in NDLog. We might use the following relations.

```
link(@S, N, C)       // Cost from S to immediate neighbor N is C
next(@S, D, N, C)    // Cost from S to D through N is C
mincost(@S, D, C)    // Cost of min path from S to D is C
best(@S, D, N)       // Best path from S to D goes through N next
```

We will assume the `link` relation has already been defined—for each node S, the `link` relation determines the cost of sending traffic to each neighbor N. To compute `next`, `mincost`, and `best`, we use the following rules.

```
(1) next(@SELF, N, N, C) :- link(@SELF, N, C).
(2) next(@SELF, D, N, C) :- link(@SELF, N, C1),
                            next(@N, D, N', C2), C=C1+C2.
(3) mincost(@SELF, D, min<C>)  :- next(@SELF, D, N, C)
(4) best(@SELF, D, N) :- mincost(@SELF, D, C), next(@SELF, D, N, C)
```

Rules (1) and (2) consider the cost of routing from S to D through every neighbor N. Rule (1) considers the possibility that the destination is the neighbor

N itself—in this case, the cost of the path is the cost of the single link from S to N. Rule (2) considers the cost of routing from S to N (which has cost C1) and then from N to the destination through some other node N' (which has cost C2). The cost of this entire path is C. Rule (3) computes the minimum cost for every source-destination pair and finally Rule (4) finds the neighbor which can reach the destination with the minimum cost.

The key aspect to focus on is the *iteration* involved in running this program. Figure 1 illustrates the results of executing this NDLog program on a particular topology—in each snapshot, the facts generated at node A are shown next to it. The first snapshot presents the link table, which represents the topology of the network from A's perspective. The second snapshot presents the "next" facts that are computed. Snapshot 3 presents the mincost and best relations computed.

To compute `mincost(A, C, 3)` (via rule 3), the logic program considers *all* tuples with the form `next(A, C,` N`,` X`)` and finds the minimum integer X that occurs. In this case, there were just two such tuples (`next(A,C,C,4)` and `next(A, C, B, 3)`), but in more complex topologies or examples, there could be enormous numbers of such tuples. Iteration over large sets of tuples, while theoretically feasible on a switch via recirculation, is too expensive to support in practice.

Likewise, rule (4) implements a *join* between `mincost` and `next` relations. In principle, any pair of tuples `mincost(`S`,`D`,`C`)` and `next(`S`,`D`,_,`C`)` where values S, D, and C might coincide must be considered. Once again, when implementing such a language on switch, we must be careful to control the kinds of *joins* like this that are admitted—many joins demand iteration over all tuples in a relation and/or require complex data structures for efficient implementation that cannot be realized on switch at line rate.

2.2 SwitchLog

As illustrated in the prior section, NDLog programs are very powerful: they can implement arbitrary aggregation operations (like min over a relation) and complex, computationally-expensive joins. Our new language, SwitchLog, may be viewed as a restricted form of NDLog, limited to ensure efficient implementation on a programmable ASIC switch like the Intel Tofino. An essential goal of our design is to ensure that computations never need to do too much work at once; in particular, we must be very careful about which joins we admit, to avoid excessive amounts of iteration.

We achieve this by imagining that our derived facts are stored in a lookup table (a fundamental construct in computer networks), using some of their fields as a key. Then, if the program's rules have the right form (described in Sect. 2.3), we are able to look up facts by key, rather than iterating over all facts in the database. This allows us to execute such rules in constant time.

Keys and Values. Relations in SwitchLog are declared using the `table` keyword, with the following syntax:

```
// The cost from S to neighbor N is C
table link(@S, loc N : key, int C)
// The best path from S to D is via neighbor N, with cost C
table next(@S, loc D : key, loc N, int C)
```

The type `loc` refers to a location in the network (e.g. a switch identifier). The `key` annotation means that field is a key, so `N` is a key of `link`, and `C` is a value. Similarly, `D` is a key of `next`, and `N` and `C` are values. The location specifier is always implicitly a key. The system maintains the invariant that, for each relation, *there is at most one known fact with a given combination of keys.*

Merging Facts. To enforce this invariant, every SwitchLog rule contains a *merge* operator, which describes what to do if we derive a fact with the same keys as an existing fact. To illustrate this, consider the following variant of the shortest paths problem in SwitchLog.[1]

```
(S1) next(@SELF, N, N, C) with merge min<C> :- link(@SELF, N, C).
(S2) next(@SELF, D, N, C) with merge min<C> :-
            link(@SELF, N, C1), next(@N, D, C2, hop), C=C1+C2.
```

This program implements the same routing protocol as the earlier NDLog program, but does so in a slightly different way.

Rule (S1) says "if there is a link from `SELF` to `N` with cost `C`, then there is a path from `SELF` to `N` with cost `C`, via `N`." Rule (S2) says "if there is a link from `SELF` to `N` with cost `C1`, and there is a path from `N` to `D` with cost `C2`, then there is a path from `SELF` to `D` with cost `C1+C2`, via `N`." The `min<C>` operator tells us that if we discover another path from `SELF` to `D`, we should retain the one with the lower value of C.

Currently, SwitchLog supports the following merge operators:

1. `min<A>` and retains the fact with the lower value of `A`, where `A` is one of the arguments of the fact on the LHS.
2. `Count<A,n>` increments the argument `A` of the LHS by `n` each time a given fact is rederived. `A` must be an integer, and it must be a value field of the relation.
3. `recent` retains the most recent instance of a fact to be derived (e.g. to simulate changing network conditions).

In the rest of this paper, we may omit the merge operator on rules; doing so indicates that the most recently derived fact is preferred (i.e. we default to the `recent` operator).

---

[1] A similar program could be implemented in NDLog. It is not that SwitchLog is more efficient than NDLog necessarily, rather that SwitchLog is restricted so that *only* the efficient NDLog programs may be implemented.

The Link Relation. Switches in a network are rarely connected directly to every other switch. Knowledge of how the switches are connected is important for efficient implementation, since sending messages to a distant destination requires more work than communicating with an immediate neighbor.

SwitchLog incorporates this information through use of a special base relation called `link`. The user may determine the exact arguments of the `link` relation, but the first argument (after the location specifier) must be a location. The fact `link(S, N, ...)` means that switch `S` is connected to switch `N`. When this is true, we refer to `S` and `N` as *neighbors*.

2.3 Guidelines for SwitchLog Rules

In addition to providing merge operators, SwitchLog rules must obey several more restrictions to ensure that they can be executed efficiently on a switch. These constraints are detailed below, along with examples. The examples refer to the following relation declarations:

```
table P1(@S, int Y : key, int Z)
table P2(@S, int Y : key, int Z)
table P3(@S, loc Y : key, loc Q : key, int Z, int W)
```

SwitchLog rules must obey the following constraints:

1. No more than two unique location specifiers may appear in the rule.
   |`P1(@SELF,`$Y, Z) : -P2(@A, Y, Z), P2(@SELF, Y, Z)$| ✓
   |`P1(@SELF,`$Y, Z) : -P2(@A, Y, Z), P2(@B, Y, Z)$| ×

2. If two different location specifiers appear in the rule, then they must be neighbors. (Equivalently, the rule must contain a `link` predicate containing those locations.)

3. If the same relation appears multiple times in the rule, each instance must have a different location specifier.
   $|P1(@SELF, X, Y) : -P1(@A, X, Y), P2(@SELF, X, Y)|$ ✓
   $|P1(@SELF, X, Y) : -P1(@SELF, X, Y), P2(@SELF, X, Y)|$ ×

4. All fields of the LHS must appear on the RHS

5. All predicates on the RHS must have the same set of keys, except that one predicate may have a single additional key, which must be a location.
   |`P1(@SELF,`$Y, Z) : -P2(@SELF, Y, Z), P3(@SELF, O, Y, Z, W)$| ✓
   |`P1(@SELF,`$Y, Z) : -P2(@SELF, Y, Z), P3(@SELF, O, Q, Z, W)$| ×

All of these restrictions are syntactic and can be checked automatically. Although in principle, a SwitchLog program that violates these restrictions can be compiled, the compiler we have implemented throws an error for restrictions 1, 3, 4, and 5. A program violating restriction 2 only generates a warning, but still compiles, even though the behavior is not well-defined.

Choosing Good Guidelines. None of the restrictions above are truly fundamental; in principle, one could compile a program that broke all of them, albeit at the cost of significant extra work. We chose these particular constraints because, in our judgement, they strike a good balance between expressivity and efficiency. We show in our evaluation (Sect. 4) that we have been able to implement a range of networking algorithms, most of which can execute each rule with a constant amount of work. At the same time, we were still able to implement a routing protocol in which additional work is unavoidable (due to the need to synchronize routing information when a switch is added to the network).

Each of our guidelines reflects some constraint of programmable switches, as well as the fact that switch computation is inherently incremental, dealing with only one packet at a time. These will be described in more detail in the next section, but briefly, the reasons for each rule are as follows:

1. Each different location requires a separate query to look up the fact, and queries from different locations are difficult to merge.
2. If queries are sent between switches which are not neighbors, they must be routed like traffic packets, generating additional work.
3. Switch hardware only allows us to access a single fact of each relation per packet, so we cannot lookup the same fact multiple times in a single location without doing extra work.
4. The predicate on the LHS must be fully defined by the facts on the right; i.e. all of its entries must be "bound" on the RHS.
5. If all the predicates on the RHS have the same set of keys, then given any one fact we can use its keys to look up the values of the others. If a predicate has an additional key, then we need to try all possible values of that key. However, since that key has location type, we need only iterate over the number of switches in the network, rather than a table.

3 Compiling SwitchLog

SwitchLog is compiled to Lucid, a high-level language for stateful, distributed data-plane programming that is itself compiled to P4_16 for the Intel Tofino. Lucid provides an event-driven abstraction for structuring applications and coordinating control that makes it easy to express the high-level ideas of SwitchLog.

A Lucid program consists of definitions for one or more *events*, each of which carries user-specified data. An event might represent an incoming traffic packet to process, a request to install a firewall entry, or a probe from a neighboring switch, for example. Each event has an associated *handler* that defines an atomic stateful computation to perform when that event occurs. Lucid programs store persistent state using the type `Array.t`, which represents an integer-indexed array.

To demonstrate the compilation process, consider the following simple SwitchLog program:

```
// Values for the foo and bar relation (1024 entries each)
global Array.t foo_0 = Array.create(1024); // Stores FOO.v1
global Array.t bar_0 = Array.create(1024); // Stores BAR.v2

// Values for the ABC relation
global Array.t abc_0 = Array.create(1024); // Stores ABC.v1
global Array.t abc_1 = Array.create(1024); // Stores ABC.v2

// Mapping from neighbor to port
global Array.t<<16>> nid_port = Array.create(COUNT);
```

Fig. 2. Arrays representing the SwitchLog relations

```
table FOO(@loc, int k1 : key, int v1)
table BAR(@loc, int k1 : key, int v2)
table ABC(@loc, int k1 : key, int v1, int v2)

(R1) ABC(@SELF, k1, v1, v2) with merge min<v2> :-
     FOO(@SELF, k1, v1), BAR(@loc, k1, v2), link(@SELF, loc)
```

Note that (`R1`) obeys the all restrictions on SwitchLog Programs described in Sect. 2.3; in particular, `FOO` and `BAR` have the same set of keys, except for the location specifier of `BAR`.

3.1 Compiling Relations

The first step to compiling a SwitchLog program is allocating space to store our derived facts. We represent each relation as a hash table, with the keys and values of the relation used directly as the keys and values of the table. Each switch has a copy of each hash table, containing those facts which are located at that switch.

Since modern switches do not allow wide, many-bit aggregates to be stored in a single array, we store each value separately; hence each hash table is represented by a series of arrays, one for each value in the relation. In addition to storing the set of derived facts, each switch maintains a `nid_port` array which maps neighbors to the port they're connected to. Figure 2 shows the arrays in Lucid.

Each SwitchLog relation is also associated with a Lucid event of the same name, carrying the data which defines an element of that relation. The event's handler uses the keys of the relation to update the Arrays defined in Fig. 2. Then, since other rules may depend on the new information, we trigger an event to evaluate any dependent rules.

3.2 Evaluating Rules

Evaluating a SwitchLog rule requires us to do two things. The evaluation of the rule is always triggered by the derivation of a new fact which matches a predicate

```
memop store_if_smaller(int x, int y) {
    if (x < y)   { return x; }
    else         { return y; }
}
```

Fig. 3. Memops for the min aggregate

on the RHS of the rule. Hence, the first thing we must do is look up facts which match the remaining RHS predicates. If the starting fact contains all the keys of the predicate (e.g. looking up FOO in R1 given BAR; note that we always know SELF), this is simple; we need only use them to look up the corresponding values, then begin evaluating the body of the rule. This is the behavior of event_bar in Fig. 4.

However, we may not know all the keys in advance. If we start executing R1 with a new FOO fact, we do not know which value of loc will result in the minimum v2. To account for this, we must send queries to *each* neighbor of SELF to get potential values of v2. Once those queries return, we can use them to execute the body of the rule. In Fig. 4, this behavior is split over three events. event_foo begins the process, event_loop_neighbors sends requests to each neighbor, and event_lookup_bar actually performs the lookup at each neighbor and begins executing R1 with the result.

Once we have looked up all the relevant information, the second step is to create the new fact, store it in memory, and trigger any rules that depend on it. This is done in rule_R1.

This example illustrates our general compilation strategy. For each relation, we create (1) arrays to represent it as a hash table, and (2) an event to update those arrays. After each update, we trigger any rules that depend on that relation, either by evaluating them directly (e.g., event_bar), or by first gathering any necessary information, such as unknown keys (e.g., event_foo). The evaluation of rules may trigger further updates, which may trigger further rules, and so on.

The behavior of our compiled code is summarized in Algorithm 1. It is similar to semi-naive evaluation in Datalog in that when a new fact comes in, we only incrementally compute new facts for rules. However, there are two key differences in how this computation is done. The first is that we have to account for inter-switch communication, so we have to query for certain predicates that may be missing, and the continue the evaluation when the query returns. The second is that our key-value semantics ensure that one new fact can generate at most one instance of a predicate p_i on a switch.

3.3 Optimizations

Accessing Memory. Switch hardware places heavy restrictions on memory accesses: only a single index of each array may be accessed by each packet, and then only once per packet. This means that we cannot, for example, read a value

```
// Update bar's value and execute rule (R1) at all neighbors
handle event_bar(int k1, int v2) {
    int idx = hash<<16>>(SEED, k1);
    Array.set(bar_0, idx, v2);
    generate_ports(all_neighbors, rule_R1(k1, v2));
}
// Update foo's value and (eventually) execute rule (R1)
handle event_foo(int k1, int v1) {
    int<<16>> idx = hash<<16>>(SEED, k1, k2);
    Array.set(foo_0, idx, v1);
    generate event_loop_neighbors(0, k1);
}

// Request the value of bar from each neighbor
handle event_loop_neighbors(int i, int k1) {
    int<<16>> port = Array.get(nid_port, i);
    generate_port(port, event_lookup_bar(SELF, k1));
    if(i < neighbor_ct) {
        generate event_loop_neighbors(i+1, k1);
    }
}

// Look up the value of bar, then execute rule (R1) at the
// requester's location
handle event_lookup_bar(int requester, int k1) {
    int<<16>> idx = hash<<16>>(SEED, k1);
    int v2 = Array.get(bar_0, idx);
    int<<16>> port = Array.get(nid_port, requester);
    generate_port(port, rule_R1(k1, v2));
}

// We don't need v1 because we can look it up
handle rule_R1(int k1, int v2) {
    int<<16>> idx = hash<<16>>(SEED, SELF, k1);
    int v1 = Array.get(foo_0, idx);
    Array.set(abc_0, idx, v1);
    // Store v2 only if it's smaller than the current entry
    Array.setm(abc_1, idx, store_if_smaller, v2);
}
```

Fig. 4. Lucid Code for evaluating rules and updating tables. SEED is a seed for the hash function

from memory, perform some computation, and write back to the same memory using a single packet. The general solution is *packet recirculation* – generating a new packet (or event, in Lucid) that does the write in a second pass through the switch. While general, this approach is inefficient because it doubles the number of packets we must process to evaluate the rule.

Fortunately, the hardware provides a better solution if the amount of computation between the read and the write is very small (say, choosing the minimum

Algorithm 1. Evaluation Algorithm

Require: A fact q_n
  for each predicate P_i dependent on q_n **do**
    if P_i can be evaluated at the current switch **then**
      ΔP_i = Evaluate using $\Delta P_0 ... \Delta P_{i-1}$ and rule R_i
      $P_i = P_i \cup \Delta P_i$
      if P_i involves communication **then**
        Broadcast to all switches that need P_i
      end if
    else
      if P_i needs a predicate stored elsewhere **then**
        Store P_i
        Query required switches
      end if
    end if
  end for

of two values). The `memop` construct in Lucid represents the allowed forms of computation as a syntactically-restricted function. The argument `store_if_smaller` to `Array.setm` is a memop (defined in Fig. 3) which compares `v2` to the current value in memory, and stores the smaller one. By using memops, we are able to avoid costly recirulation.

Inlining Events. Often, the compilation process described above produces "dummy" events, which only serve to generate another event, perhaps after performing a small computation of their own. Each generated event results in a new packet we need to process, so we inline the generated events wherever possible. This ensures we are doing the maximum amount of work per packet, and thus minimizes the amount of overhead our rules require.

3.4 Integration with Lucid

While SwitchLog programs can be used independently to generate sets of facts (which could be read from switch memory by network operators or monitoring systems), they are most powerful when used to guide packet forwarding within the switch itself. We enable this by embedding SwitchLog into Lucid. Users write general-purpose Lucid code that can introduce new facts to the SwitchLog sub-program, then read derived facts to guide packet processing decisions. For example, using the `fwd_neighbour` array to lookup values and make decisions based on them. This powerful technique allows users to compute amenable data using SwitchLog's declarative syntax, while also operating on that data with the full expressiveness of Lucid. Figure 5 shows how the next table from the routing example can be used to forward packets.

```
table link(@SELF, int dest : key, int cost)
table fwd(@SELF, int dest : key, int cost, int neighbor)

rule fwd(@SELF, dest, cost, dest) :-
                link(@SELF, dest, cost)

rule fwd(@SELF, dest, cost, next) with merge min<cost> :-
                link(@SELF, neighbor, cost1),
                fwd(@neighbor, dest, cost2, next),
                int cost = cost1 + cost2;

handle packetin(int dst) {
    // check if a path exists.
    int<<16>> idx = hash<<16>>(SEED, SELF, dst);
    // Lookup the `neighbor` value of the stored fwd table
    int next_hop = Array.get(fwd_neighbor, idx);
    if (next_hop != 255) {
        // Find the matching port and forward the packet
        int outport = Array.get(nid_port, next_hop);
        generate packetout(outport);
    }

}
```

Fig. 5. A SwitchLog program integrated with Lucid. The first few lines express a routing protocol in SwitchLog, and the packetin event uses the fwd table

	LoC			**Resources**			
Application	SwitchLog	Lucid	P4	Stages	Tables	sALUs	Recirculation
Mac Learner	7	55	630	8	11	3	$\propto$ new host
Router	4	74	801	9	19	6	$\propto$ link up/down
Netflow Cache [16]	3	21	320	3	8	2	-
Stateful Firewall	3	28	410	5	10	1	-
Distr Hvy Hitter [7]	5	44	548	3	6	1	-
Flow Size Query [14]	3	58	1276	4	15	7	-

Fig. 6. Lines of code and resource utilization of SwitchLog applications compiled to Tofino.

4 Evaluation

We evaluated SwitchLog by using it to implement four representative data-plane applications. The applications have diverse objectives, illustrating the flexibility of SwitchLog.

- **Path computation.** The router and mac learner implement routing algorithms at the core of most modern networks.

- **Monitoring.** The netflow cache implements the data structure at the core of many telemetry systems [6,16,18], a per-flow metric cache, while the host usage query implements a per-host bandwidth measurement query from Marple [14].
- **Security.** Finally, the stateful firewall implements a common security protocol – only allowing packets to enter a local network from the Internet if they belong to connections previously established by internal hosts. The distributed heavy hitter detector, based on [7], identifies heavy hitter flows that split their load across multiple network entry points (for example, DDoS attackers seeking to avoid detection).

Our primary evaluation metrics for SwitchLog were conciseness and efficiency, which we gauged by measuring the lines of code and resource requirements of our programs when compiled to the Intel Tofino. Figure 6 reports the results.

Conciseness. As Fig. 6 shows, SwitchLog applications are around 10X shorter than the Lucid programs that they compile to, and over 100X shorter than the resulting P4. A SwitchLog program is much shorter than its Lucid equivalent because each rule in SwitchLog (a single line) translates into many lines of imperative Lucid code that defines the events necessary to propagate new facts and the handlers/memops necessary to perform the respective updates. The resulting Lucid code is itself 10X smaller than the final P4, simply because Lucid's syntax is itself much more concise than P4.

Resource Utilization. As Fig. 6 shows, all the SwitchLog programs that we implemented fit within the 12-stage processing pipeline of the Tofino. Each stage of the Tofino's pipeline contains multiple kinds of compute and memory resources, for example ternary match-action tables that select instructions to execute in parallel based on packet header values, ALUs that execute instructions over packet headers, and stateful ALUs (sALUs) that update local memory banks that persist across packets. All SwitchLog programs in used under 10% of all stage-local Tofino resources. Our evaluation programs were most resource intensive with respect to the ternary tables and stateful ALUs – Fig. 6 lists the total number of each resource required by the programs.

To get an idea of how well-optimized SwitchLog-generated code is, we hand-optimized the Lucid program generated by SwitchLog for the router application with the goal of reducing the number of stages. We found 2 simple optimizations that reduced the program from 9 stages to 7 stages. First, we saved a stage by deleting event handlers that were no longer needed because they had been inlined. Second, we saved another stage by rewriting a memop on the program's critical path to perform an add operation on a value loaded from memory, which previously took place in a subsequent stage. We were unable to find any other ways to optimize the router at the Lucid level, though it is likely that a lower-level P4 implementation could be further optimized, as the Lucid compiler itself is not optimal.

Another important resource to consider is pipeline processing bandwidth. SwitchLog programs recirculate packets when they generate events, and these packets compete with end-to-end traffic for processing bandwidth. Some amount of recirculation is generally okay – for example the Tofino has a dedicated 100 Gb/s recirculation port and enough processing bandwidth to service that port and all other ports at line rate. As the final column of Fig. 6 summarizes, we found that in our SwitchLog programs, packet recirculation only occurred when the network topology changes. Such events are rare, thus in practice we expect that all of these applications would fit within the recirculation budget of switches in real networks. We also note that our event inlining optimization was particularly important for the netflow cache. Without the optimization, the cache required a recirculation for *every packet*, whereas with the optimization, it required no packet recirculation at all.

5 Limitations and Future Work

The key-value semantics limit the kinds of programs that can be expressed in SwitchLog. For instance, consider the following NDLog program to compute all neighbors and costs.

```
(1) paths(@SELF, N, N, C) :- link(@SELF, N, C).
(2) paths(@SELF, D, N, C) :- link(@SELF, N, C1),
                             paths(@N, D, N', C2), C=C1+C2.
```

The *paths* table can contain multiple neighbors, which means that one set of @SELF, D values can have multiple values of N and C. However, SwitchLog's semantics only allow one set of values to be stored for each set of keys. Therefore, it is not possible to keep an up-to date table that requires storing multiple values for a given set of keys.

There are also several data plane applications that cannot be expressed in SwitchLog either. A Bloom Filter is a probabilistic data structure that is used commonly on network switches. It consists of an Array, and typically uses a number of different hash functions that map each element to an array index. The current syntax and semantics of SwitchLog do not support the use of multiple hash functions. However, if we allowed multiple hash functions, a bloom filter might look as follows:

```
(1) arr1(@SELF, H, B) :- data(@SELF, K1), H = hash1(K1); B = 1;
(2) arr2(@SELF, H, B) :- data(@SELF, K1), H = hash2(K1); B = 1;
(3) arr3(@SELF, H, B) :- data(@SELF, K1), H = hash3(K2); B = 1;
(4) bloom_filter(@SELF, H, B) :- data(@SELF, K1),
                                 arr1(@SELF, hash1(K1), 1),
                                 arr2(@SELF, hash2(K1), 1),
                                 arr3(@SELF, hash3(K1), 1).
```

where *data* is some SwitchLog predicate. Since SwitchLog is can be integrated with Lucid, Rule (4) could also be implemented as a Lucid function, that computes the hashes and looks up the three arrays. A future target for SwitchLog is to have probabilistic data structures like a Bloom Filter or Count-Min Sketch built-in. In particular, the count aggregate might be implemented as a count-min sketch internally rather than an actual counter.

Another limitation of the declarative syntax is the inability to modify packet headers. In Lucid, this is achieved through the *exit* event. However, allowing programmers to specify exit events in SwitchLog programs would lead to undefined program behavior. In particular, in case of communication, the exit event needs to be a request generated by the SwitchLog compiler, and in case there are 2 possible exit events, it is unclear which one to execute.

6 Conclusion

SwitchLog brings a new kind of logic programming abstraction to the network data plane, allowing programmers to construct distributed, table-driven programs at a high-level of abstraction. While SwitchLog was inspired by past declarative network programming languages such as NDLog [12], the computational limitations of current switch hardware necessitate a modified design; in particular, SwitchLog's explicit key/value distinction enables constant-time lookup of most facts. By restricting the form of rules, SwitchLog can ensure that only efficient joins are permitted. With these restrictions, SwitchLog allows programmers to express a variety of useful networking applications as concise logic programs, and execute those programs inside a real network.

References

1. Differential datalog. VMWare (2019). https://github.com/vmware/differential-datalog
2. Bosshart, P., et al.: P4: Programming protocol-independent packet processors. ACM SIGCOMM Comput. Commun. Rev. **44**(3), 87–95 (2014)
3. Eichholz, M., Campbell, E.H., Krebs, M., Foster, N., Mezini, M.: Dependently-typed data plane programming. In: Proceedings of the ACM Programming Languages **6**(POPL) (2022). https://doi.org/10.1145/3498701
4. Gao, J., et al.: Lyra: a cross-platform language and compiler for data plane programming on heterogeneous asics. In: Proceedings of the Annual Conference of the ACM Special Interest Group on Data Communication on the Applications, Technologies, Architectures, and Protocols for Computer Communication, p. 435–450. SIGCOMM'20, Association for Computing Machinery, New York, NY, USA (2020). https://doi.org/10.1145/3387514.3405879
5. Gao, X., et al.: Switch code generation using program synthesis. In: Proceedings of the Annual Conference of the ACM Special Interest Group on Data Communication on the Applications, Technologies, Architectures, and Protocols for Computer Communication, pp. 44–61. SIGCOMM'20, Association for Computing Machinery, New York, NY, USA (2020). https://doi.org/10.1145/3387514.3405852
6. Gupta, A., Harrison, R., Canini, M., Feamster, N., Rexford, J., Willinger, W.: Sonata: query-driven streaming network telemetry. In: Proceedings of the 2018 Conference of the ACM Special Interest Group on Data Communication, pp. 357–371. SIGCOMM'18, Association for Computing Machinery, New York, NY, USA (2018). https://doi.org/10.1145/3230543.3230555

7. Harrison, R., Cai, Q., Gupta, A., Rexford, J.: Network-wide heavy hitter detection with commodity switches. In: Proceedings of the Symposium on SDN Research, pp. 1–7 (2018)
8. Hsu, K.F., Beckett, R., Chen, A., Rexford, J., Walker, D.: Contra: a programmable system for performance-aware routing. In: 17th USENIX Symposium on Networked Systems Design and Implementation (NSDI 20), pp. 701–721 (2020)
9. Laffranchini, P., Rodrigues, L.E.T., Canini, M., Krishnamurthy, B.: Measurements as first-class artifacts. In: 2019 IEEE Conference on Computer Communications, INFOCOM 2019, Paris, France, 29 April - 2 May 2019, pp. 415–423. IEEE (2019). https://doi.org/10.1109/INFOCOM.2019.8737383
10. Loehr, D., Walker, D.: Safe, modular packet pipeline programming. In: Proceedings ACM Programming Languages **6**(POPL) (2022). https://doi.org/10.1145/3498699
11. Loo, B.T.: The design and implementation of declarative networks, p. 210 (2006). http://digicoll.lib.berkeley.edu/record/139082
12. Loo, B.T., Hellerstein, J.M., Stoica, I., Ramakrishnan, R.: Declarative routing: extensible routing with declarative queries. ACM SIGCOMM Comput. Commun. Rev. **35**(4), 289–300 (2005)
13. Narayana, S., Arashloo, M.T., Rexford, J., Walker, D.: Compiling path queries. In: Proceedings of the 13th Usenix Conference on Networked Systems Design and Implementation, pp. 207–222. NSDI'16, USENIX Association, USA (2016)
14. Narayana, S., et al.: Language-directed hardware design for network performance monitoring. In: Proceedings of the Conference of the ACM Special Interest Group on Data Communication, pp. 85–98. SIGCOMM'17, Association for Computing Machinery, New York, NY, USA (2017). https://doi.org/10.1145/3098822.3098829
15. Sivaraman, A., et al.: Packet transactions: high-level programming for line-rate switches. In: Proceedings of the 2016 ACM SIGCOMM Conference, pp. 15–28. SIGCOMM'16, Association for Computing Machinery, New York, NY, USA (2016). https://doi.org/10.1145/2934872.2934900
16. Sonchack, J., Aviv, A.J., Keller, E., Smith, J.M.: Turboflow: information rich flow record generation on commodity switches. In: Proceedings of the 13th EuroSys Conference, pp. 1–16 (2018)
17. Sonchack, J., Loehr, D., Rexford, J., Walker, D.: Lucid: a language for control in the data plane. In: Proceedings of the 2021 ACM SIGCOMM 2021 Conference, pp. 731–747 (2021)
18. Sonchack, J., Michel, O., Aviv, A.J., Keller, E., Smith, J.M.: Scaling hardware accelerated network monitoring to concurrent and dynamic queries with {* Flow}. In: 2018 USENIX Annual Technical Conference (USENIX ATC 18), pp. 823–835 (2018)
19. Soni, H., Rifai, M., Kumar, P., Doenges, R., Foster, N.: Composing dataplane programs with μp4. In: Proceedings of the Annual Conference of the ACM Special Interest Group on Data Communication on the Applications, Technologies, Architectures, and Protocols for Computer Communication, pp. 329–343. SIGCOMM'20, Association for Computing Machinery, New York, NY, USA (2020). https://doi.org/10.1145/3387514.3405872
20. Wang, A., Basu, P., Loo, B.T., Sokolsky, O.: Declarative network verification. In: Gill, A., Swift, T. (eds.) PADL 2009. LNCS, vol. 5418, pp. 61–75. Springer, Heidelberg (2008). https://doi.org/10.1007/978-3-540-92995-6_5

Linear Algebraic Abduction with Partial Evaluation

Tuan Nguyen[1(✉)], Katsumi Inoue[1(✉)], and Chiaki Sakama[2(✉)]

[1] National Institute of Informatics, Tokyo, Japan
{tuannq,inoue}@nii.ac.jp
[2] Wakayama University, Wakayama, Japan
sakama@wakayama-u.ac.jp

Abstract. Linear algebra is an ideal tool to redefine symbolic methods with the goal to achieve better scalability. In solving the abductive Horn propositional problem, the transpose of a program matrix has been exploited to develop an efficient exhaustive method. While it is competitive with other symbolic methods, there is much room for improvement in practice. In this paper, we propose to optimize the linear algebraic method for abduction using partial evaluation. This improvement considerably reduces the number of iterations in the main loop of the previous algorithm. Therefore, it improves practical performance especially with sparse representation in case there are multiple subgraphs of conjunctive conditions that can be computed in advance. The positive effect of partial evaluation has been confirmed using artificial benchmarks and real Failure Modes and Effects Analysis (FMEA)-based datasets.

Keywords: Abduction · Linear algebra · Partial evaluation

1 Introduction

Abduction is a form of explanatory reasoning that has been used for Artificial Intelligence (AI) in diagnosis and perception [15] as well as belief revision [4] and automated planning [8]. Logic-based abduction is formulated as the search for a set of abducible propositions that together with a background theory entails the observations while preserving consistency [7]. Recently, abductive reasoning has gained interests in connecting neural and symbolic reasoning [6] together with explainable AI [14,34].

Recently, several studies have been done to recognize the ability to use efficient parallel algorithms in linear algebra for computing logic programming (LP). For example, high-order tensors have been employed to support both deductive and inductive inferences for a limited class of logic programs [24]. In [29], Sato presented the use of first-order logic in vector spaces for Tarskian semantics, which demonstrates how tensorization realizes efficient computation of Datalog. Using a linear algebraic method, Sakama et al. explore relations between LP and tensor then propose algorithms for computation of LP models [27,28]. In [23], Nguyen et al. have analyzed the sparsity level of program matrices and proposed to employ sparse representation for scalable computation. Following this direction, Nguyen et al. have also exploited the sparse matrix

M. Hanus and D. Inclezan (Eds.): PADL 2023, LNCS 13880, pp. 197–215, 2023.
https://doi.org/10.1007/978-3-031-24841-2_13

representation to propose an efficient linear algebraic approach to abduction that incorporates solving Minimal Hitting Sets (MHS) problems [22].

Partial evaluation was introduced to generate a compiler from an interpreter based on the relationship between a formal description of the semantics of a programming language and an actual compiler [9]. The idea was also studied intensively in [3]. Then, Tamaki and Sato incorporated unfold and fold transformations in LP [33] as partial evaluation techniques, and Lloyd and Shepherdson have developed theoretical foundations for partial evaluation in LP [19]. According to Lloyd and Shepherdson, partial evaluation can be described as producing an equivalent logic program such that it should run more efficiently than the original one for reasoning steps. Following this direction, the idea of partial evaluation has been successfully employed to compute the least models of definite programs using linear algebraic computation [21]. Nguyen et al. have reported a significant improvement in terms of reducing runtime on both artificial data and real data (based on transitive closures of large network datasets) [21].

This paper aims at exploring the use of partial evaluation in abductive reasoning with linear algebraic approaches. We first propose an improvement to the linear algebraic algorithm for solving Propositional Horn Clause Abduction Problem (PHCAP). Then we present the efficiency of the method for solving PHCAP using the benchmarks based on FMEA. The rest of this paper is organized as follows: Sect. 2 reviews the background and some basic notions used in this paper; Sect. 3 presents the idea of partial evaluation using the linear algebraic approach with a theoretical foundation for correctness; Sect. 4 demonstrates experimental results using FMEA-based benchmarks; Sect. 5 discusses related work; Sect. 6 gives final remarks and discusses potential future works.

2 Preliminaries

We consider the language of propositional logic $\mathcal{L}$ that contains a finite set of propositional variables.

A *Horn logic program* is a finite set of *rules* of the form:

$$h \leftarrow b_1 \wedge \cdots \wedge b_m \quad (m \geq 0) \tag{1}$$

where h and b_i are propositional variables in $\mathcal{L}$. Given a program P, the set of all propositional variables appearing in P is the *Herbrand base* of P (written $\mathcal{B}_P$). In (1) the left-hand side of $\leftarrow$ is called the *head* and the right-hand side is called the *body*. A Horn logic program P is called *singly defined* (*SD program*, for short) if $h_1 \neq h_2$ for any two different rules $h_1 \leftarrow B_1$ and $h_2 \leftarrow B_2$ in P where B_1 and B_2 are conjunctions of atoms. That is, no two rules have the same head in an SD program. When P contains more than one rule with the same head $(h \leftarrow B_1), \ldots, (h \leftarrow B_n)$ $(n > 1)$, replace them with a set of new rules:

$$\begin{aligned} & h \leftarrow b_1 \vee \cdots \vee b_n \quad (n > 1) \\ & b_1 \leftarrow B_1, \quad \cdots, \quad b_n \leftarrow B_n \end{aligned} \tag{2}$$

where $b_1, \ldots, b_n$ are new atoms such that $b_i \notin \mathcal{B}_P$ $(1 \leq i \leq n)$ and $b_i \neq b_j$ if $i \neq j$. For convenience, we refer to (1) as an *And*-rule and (2) as an *Or*-rule.

Every Horn logic program P is transformed to $\Pi = Q \cup D$ such that Q is an SD program and D is a set of *Or*-rules. The resulting Π is called a *standardized program*. Therefore, a *standardized program* is a definite program such that there is no duplicate head atom in it and every rule is in the form of either *And*-rule or *Or*-rule. Note that the rule (2) is shorthand of n rules: $h \leftarrow b_1$, ..., $h \leftarrow b_n$, so a standardized program is considered a Horn logic program. Throughout the paper, a program means a standardized program unless stated otherwise. For each rule r of the form (1) or (2), define $head(r) = h$ and $body(r) = \{b_1, \ldots, b_m\}$ (or $body(r) = \{b_1, \ldots, b_n\}$). A rule is called a *fact* if $body(r) = \emptyset$. A rule is called a *constraint* if $head(r)$ is empty. A constraint $\leftarrow b_1 \wedge \cdots \wedge b_m$ is replaced with

$$\bot \leftarrow b_1 \wedge \cdots \wedge b_m$$

where $\bot$ is a symbol representing **False**. When there are multiple constraints, say $(\bot \leftarrow B_1), \ldots, (\bot \leftarrow B_n)$, they are transformed to

$$\bot \leftarrow \bot_1 \vee \cdots \vee \bot_n \quad \text{and} \quad \bot_i \leftarrow B_i \quad (i = 1, \ldots, n)$$

where $\bot_i \notin \mathscr{B}_P$ is a new symbol. An *interpretation* $I \,(\subseteq \mathscr{B}_P)$ is a *model* of a program P if $\{b_1, \ldots, b_m\} \subseteq I$ implies $h \in I$ for every rule (1) in P, and $\{b_1, \ldots, b_n\} \cap I \neq \emptyset$ implies $h \in I$ for every rule (2) in P. A model I is the *least model* of P (written LM_P) if $I \subseteq J$ for any model J of P. We write $P \models a$ when $a \in LM_P$. For a set $S = \{a_1, \ldots, a_n\}$ of ground atoms, we write $P \models S$ if $P \models a_1 \wedge \cdots \wedge a_n$. A program P is *consistent* if $P \not\models \bot$.

Definition 1 Horn clause abduction: A *Propositional Horn Clause Abduction Problem (PHCAP)* consists of a tuple $\langle \mathscr{L}, \mathbb{H}, \mathbb{O}, P \rangle$, where $\mathbb{H} \subseteq \mathscr{L}$ (called *hypotheses* or *abducibles*), $\mathbb{O} \subseteq \mathscr{L}$ (called *observations*), and P is a consistent Horn logic program.

A logic program P is associated with a *dependency graph* (V, E), where the nodes V are the atoms of P and, for each rule from P, there are edges in E from the atoms appearing in the body to the atom in the head. We refer to the node of an *And*-rule and the node of an *Or*-rule as *And*-node and *Or*-node respectively. In this paper, we assume a program P is *acyclic* [1] and in its standardized form. Without loss of generality, we assume that any abducible atom $h \in \mathbb{H}$ does not appear in any head of the rule in P. If there exist $h \in \mathbb{H}$ and a rule $r : h \leftarrow body(r) \in P$, we can replace r with $r' : h \leftarrow body(r) \vee h'$ in P and then replace h by h' in $\mathbb{H}$. If r is in the form (2), then r' is an *Or*-rule, and no need to further update r'. On the other hand, if r is in the form (1), then we can update r' to become an *Or*-rule by introducing an *And*-rule $b_r \leftarrow body(r)$ in P and then replace $body(r)$ by b_r in r'.

Definition 2 Explanation of PHCAP: A set $E \subseteq \mathbb{H}$ is called a *solution* of a PHCAP $\langle \mathscr{L}, \mathbb{H}, \mathbb{O}, P \rangle$ if $P \cup E \models \mathbb{O}$ and $P \cup E$ is consistent. E is also called an *explanation* of $\mathbb{O}$. An explanation E of $\mathbb{O}$ is *minimal* if there is no explanation E' of $\mathbb{O}$ such that $E' \subset E$.

In this paper, the goal is to propose an algorithm finding the set $\mathbb{E}$ of all minimal explanations E for a PHCAP $\langle \mathscr{L}, \mathbb{H}, \mathbb{O}, P \rangle$. Deciding if there is a solution of a PHCAP (or $\mathbb{E} \neq \emptyset$) is *NP*-complete [7,32]. That is proved by a transformation from a satisfiability problem [10].

In PHCAP, P is partitioned into P_{And} - a set of *And*-rules of the form (1), and P_{Or} - a set of *Or*-rules of the form (2). Given P, define $head(P) = \{head(r) \mid r \in P\}$, $head(P_{And}) = \{head(r) \mid r \in P_{And}\}$, and $head(P_{Or}) = \{head(r) \mid r \in P_{Or}\}$.

3 Linear Algebraic Abduction with Partial Evaluation

3.1 Linear Algebraic Computation of Abduction

We first review the method of encoding and computing explanation in vector spaces that has been proposed in [22].

Definition 3. Matrix representation of standardized programs in PHCAP[22]**:** Let $\langle \mathcal{L}, \mathbb{H}, \mathbb{O}, P \rangle$ be a PHCAP such that P is a standardized program with $\mathcal{L} = \{p_1, \ldots, p_n\}$. Then P is represented by a *program matrix* $M_P \in \mathbb{R}^{n \times n}$ $(n = |\mathcal{L}|)$ such that for each element a_{ij} $(1 \leq i, j \leq n)$ in M_P:

1. $a_{ij_k} = \frac{1}{m}$ $(1 \leq k \leq m; 1 \leq i, j_k \leq n)$ if $p_i \leftarrow p_{j_1} \wedge \cdots \wedge p_{j_m}$ is in P_{And} and $m > 0$;
2. $a_{ij_k} = 1$ $(1 \leq k \leq l; 1 \leq i, j_k \leq n)$ if $p_i \leftarrow p_{j_1} \vee \cdots \vee p_{j_l}$ is in P_{Or};
3. $a_{ii} = 1$ if $p_i \leftarrow$ is in P_{And} or $p_i \in \mathbb{H}$;
4. $a_{ij} = 0$, otherwise.

Compared with the program matrix definition in [28], Definition 3 has an update in the condition 3 that we set 1 for all abducible atoms $p_i \in \mathbb{H}$. The program matrix is used to compute deduction, while in abductive reasoning, we do it in reverse. We then exploit the matrix for deduction to define a matrix that we can use for abductive reasoning.

Definition 4. Abductive matrix of PHCAP [22]**:** Suppose that a PHCAP has P with its *program matrix* M_P. The *abductive matrix* of P is the transpose of M_P represented as $M_P{}^T$.

In our method, we distinguish *And*-rules and *Or*-rules and handle them separately. Thus, it is crucial to have a simpler version of the abductive matrix for efficient computation.

Definition 5. Reduct abductive matrix of PHCAP: We can obtain a *reduct abductive matrix* $M_P(P^r_{And})^T$ from the abductive matrix $M_P{}^T$ by:

1. Removing all columns *w.r.t.* *Or*-rules in P_{Or}.
2. Setting 1 at the diagonal corresponding to all atoms which are heads of *Or*-rules.

We should note that this is a proper version of the previous definition in [22] that we will explain in detail later in this section. The reduct abductive matrix is the key component to define the partial evaluation method.

The goal of PHCAP is to find the set of minimal explanations $\mathbb{E}$ according to Definition 2. Therefore, we need to define a representation of explanations in vector spaces.

Definition 6. Vector representation of subsets in PHCAP[22]**:** Any subset $s \subseteq \mathcal{L}$ can be represented by a vector v of the length $|\mathcal{L}|$ such that the i-th value $v[i] = 1$ $(1 \leq i \leq |\mathcal{L}|)$ iff the i-th atom p_i of $\mathcal{L}$ is in s; otherwise $v[i] = 0$.

Using Definition 6, we can represent any $E \in \mathbb{E}$ by a column vector $E \in \mathbb{R}^{|\mathscr{L}| \times 1}$. To compute E, we define an *explanation vector* $v \in \mathbb{R}^{|\mathscr{L}| \times 1}$. We use the explanation vector v to demonstrate linear algebraic computation of abduction to reach an explanation E starting from an initial vector $v = v(\mathbb{O})$ which is the observation vector (note that we can use the notation $\mathbb{O}$ as a vector without the function notation $v()$ as stated before). At each computation step, we can interpret the meaning of the explanation vector v as: in order to explain $\mathbb{O}$, we have to explain all atoms v_i such that $v[i] > 0$.
An answer of PHCAP is a vector satisfying the following condition:

Definition 7. Answer of a PHCAP[22]**:** The explanation vector v *reaches* an answer E if $v \subseteq \mathbb{H}$. This condition can be written in linear algebra as follows:

$$\theta(v + \mathbb{H}) \leq \theta(\mathbb{H}) \tag{3}$$

where $\mathbb{H}$ is the shorthand of $v(\mathbb{H})$ which is the hypotheses set/vector. θ is a thresholding function mapping an element x of a vector/matrix to 0 if $x < 1$; otherwise map x to 1.

We here mention again Algorithm 1 in [22]. The main idea is built upon the two 1-step abduction for P_{And} (line 5) and P_{Or} (line 19) based on *And*-computable and *Or*-computable conditions. Each 1-step abduction applies on an *explanation vector* starting from the *observation vector* $\mathbb{O}$ until we reach an answer. During the abduction process, the *explanation vector* may "grow" to an *explanation matrix*, denoted by M, as *Or*-rules create new possible branches. Thus, we can abduce explanations by computing matrix multiplication (for *And*-computable matrices), and solving a corresponding MHS problem (for *Or*-computable matrices). Further detailed definitions and proofs of the method are presented in [22].

Algorithm 1. Explanations finding in a vector space

Input: PHCAP consists of a tuple $\langle \mathscr{L}, \mathbb{H}, \mathbb{O}, P \rangle$
Output: Set of explanations $\mathbb{E}$
1: Create an abductive matrix $M_P{}^T$ from P
2: Initialize the *observation matrix* M from $\mathbb{O}$ (obtained directly from the *observation vector* $\mathbb{O}$)
3: $\mathbb{E} = \emptyset$
4: **while True do**
5: $\quad M' = M_P{}^T \cdot M$
6: $\quad M' =$ **consistent**(M')
7: $\quad v\_sum = sum_{col}(M') < 1 - \varepsilon$
8: $\quad M' = M'[v\_sum =$ **False**$]$
9: $\quad$ **if** $M' = M$ **or** $M' = \emptyset$ **then**
10: $\quad\quad v\_ans = \theta(M + \mathbb{H}) \leq \theta(\mathbb{H})$
11: $\quad\quad \mathbb{E} = \mathbb{E} \cup M[v\_ans =$ **True**$]$
12: $\quad\quad$ **return minimal**$(\mathbb{E})$
13: $\quad$ **do**
14: $\quad\quad v\_ans = \theta(M' + \mathbb{H}) \leq \theta(\mathbb{H})$
15: $\quad\quad \mathbb{E} = \mathbb{E} \cup M'[v\_ans =$ **True**$]$
16: $\quad\quad M' = M'[v\_ans =$ **False**$]$
17: $\quad\quad M = M \cup M'[$**not** *Or*-computable$]$
18: $\quad\quad M' = M'[$*Or*-computable$]$
19: $\quad\quad M' = \bigcup_{\forall v \in M'} \bigcup_{\forall s \in \mathbf{MHS}(\mathbb{S}_{(v, P_{Or})})} \left((v \setminus head(P_{Or})) \cup s \right)$
20: $\quad\quad M' =$ **consistent**(M')
21: $\quad$ **while** $M' \neq \emptyset$

Example 1. Consider a PHCAP $\langle \mathscr{L}, \mathbb{H}, \mathbb{O}, P \rangle$ such that:
$\mathscr{L} = \{obs,\ e_1,\ e_2,\ e_3,\ e_4,\ e_5,\ e_6,\ H_1,\ H_2,\ H_3\}$, $\mathbb{O} = \{obs\}$, $\mathbb{H} = \{H_1,\ H_2,\ H_3\}$, and $P = \{obs \leftarrow e_1,\ e_1 \leftarrow e_2 \wedge e_3,\ e_2 \leftarrow e_4 \wedge e_5,\ e_2 \leftarrow e_5 \wedge e_6,\ e_3 \leftarrow e_5,\ e_4 \leftarrow H_1,\ e_5 \leftarrow H_2,\ e_6 \leftarrow H_3\ \}$.

1. The standardized program $P' = \{obs \leftarrow e_1,\ e_1 \leftarrow e_2 \wedge e_3,\ e_2 \leftarrow x_1 \vee x_2,\ x_1 \leftarrow e_4 \wedge e_5,\ x_2 \leftarrow e_5 \wedge e_6,\ e_3 \leftarrow e_5,\ e_4 \leftarrow H_1,\ e_5 \leftarrow H_2,\ e_6 \leftarrow H_3\}$ is represented by the abductive matrix:

$$M_P^T = \begin{array}{c} \\ e_1\\ e_2\\ e_3\\ e_4\\ e_5\\ e_6\\ H_1\\ H_2\\ H_3\\ obs\\ x_1\\ x_2 \end{array} \begin{array}{cccccccccccc} e_1 & e_2 & e_3 & e_4 & e_5 & e_6 & H_1 & H_2 & H_3 & obs & x_1 & x_2 \end{array} \\ \begin{pmatrix} & & & & & & & & & 1.00 & & \\ 0.50 & & & & & & & & & & & \\ 0.50 & & & & & & & & & & & \\ & & & & & & & & & & 0.50 & \\ & & 1.00 & & & & & & & & 0.50 & 0.50 \\ & & & & & & & & & & & 0.50 \\ & & & 1.00 & & & 1.00 & & & & & \\ & & & & 1.00 & & & 1.00 & & & & \\ & & & & & 1.00 & & & 1.00 & & & \\ & & & & & & & & & & & \\ & 1.00 & & & & & & & & & & \\ & 1.00 & & & & & & & & & & \end{pmatrix}$$

2. Iteration 1:
- $M^{(1)} = \theta(M_P^T \cdot M^{(0)})$, where $M^{(0)} = \mathbb{O}$:

$$\begin{array}{c|c} & 0 \\ \hline e_1 & 1.00 \\ e_2 & \\ e_3 & \\ e_4 & \\ e_5 & \\ e_6 & \\ H_1 & \\ H_2 & \\ H_3 & \\ obs & \\ x_1 & \\ x_2 & \end{array} = \begin{array}{c|cccccccccccc} & e_1 & e_2 & e_3 & e_4 & e_5 & e_6 & H_1 & H_2 & H_3 & obs & x_1 & x_2 \\ \hline e_1 & & & & & & & & & & 1.00 & & \\ e_2 & 0.50 & & & & & & & & & & & \\ e_3 & 0.50 & & & & & & & & & & & \\ e_4 & & & & & & & & & & & 0.50 & \\ e_5 & & & 1.00 & & & & & & & & 0.50 & 0.50 \\ e_6 & & & & & & & & & & & & 0.50 \\ H_1 & & & & 1.00 & & & 1.00 & & & & & \\ H_2 & & & & & 1.00 & & & 1.00 & & & & \\ H_3 & & & & & & 1.00 & & & 1.00 & & & \\ obs & & & & & & & & & & & & \\ x_1 & & 1.00 & & & & & & & & & & \\ x_2 & & 1.00 & & & & & & & & & & \end{array} \begin{array}{c|c} & 0 \\ \hline e_1 & \\ e_2 & \\ e_3 & \\ e_4 & \\ e_5 & \\ e_6 & \\ H_1 & \\ H_2 & \\ H_3 & \\ obs & 1.00 \\ x_1 & \\ x_2 & \end{array}$$

3. Iteration 2:
- $M^{(2)} = \theta(M_P^T \cdot M^{(1)})$

$$\begin{array}{c|c} & 0 \\ \hline e_1 & \\ e_2 & 0.50 \\ e_3 & 0.50 \\ e_4 & \\ e_5 & \\ e_6 & \\ H_1 & \\ H_2 & \\ H_3 & \\ obs & \\ x_1 & \\ x_2 & \end{array} = \begin{array}{c|cccccccccccc} & e_1 & e_2 & e_3 & e_4 & e_5 & e_6 & H_1 & H_2 & H_3 & obs & x_1 & x_2 \\ \hline e_1 & & & & & & & & & & 1.00 & & \\ e_2 & 0.50 & & & & & & & & & & & \\ e_3 & 0.50 & & & & & & & & & & & \\ e_4 & & & & & & & & & & & 0.50 & \\ e_5 & & & 1.00 & & & & & & & & 0.50 & 0.50 \\ e_6 & & & & & & & & & & & & 0.50 \\ H_1 & & & & 1.00 & & & 1.00 & & & & & \\ H_2 & & & & & 1.00 & & & 1.00 & & & & \\ H_3 & & & & & & 1.00 & & & 1.00 & & & \\ obs & & & & & & & & & & & & \\ x_1 & & 1.00 & & & & & & & & & & \\ x_2 & & 1.00 & & & & & & & & & & \end{array} \begin{array}{c|c} & 0 \\ \hline e_1 & 1.00 \\ e_2 & \\ e_3 & \\ e_4 & \\ e_5 & \\ e_6 & \\ H_1 & \\ H_2 & \\ H_3 & \\ obs & \\ x_1 & \\ x_2 & \end{array}$$

- Solving MHS: $\{\{x_1, x_2\}, \{e_3\}\}$. MHS solutions: $\{\{e_3, x_1\}, \{e_3, x_2\}\} = M^{(3)}$.

4. Iteration 3:
- $M^{(4)} = \theta(M_P^T \cdot M^{(3)})$

$$\begin{array}{c|cc} & 0 & 1 \\ \hline e_1 & & \\ e_2 & & \\ e_3 & & \\ e_4 & 0.25 & \\ e_5 & 0.75 & 0.75 \\ e_6 & & 0.25 \\ H_1 & & \\ H_2 & & \\ H_3 & & \\ obs & & \\ x_1 & & \\ x_2 & & \end{array} = \begin{array}{c|cccccccccccc} & e_1 & e_2 & e_3 & e_4 & e_5 & e_6 & H_1 & H_2 & H_3 & obs & x_1 & x_2 \\ \hline e_1 & & & & & & & & & & 1.00 & & \\ e_2 & 0.50 & & & & & & & & & & & \\ e_3 & 0.50 & & & & & & & & & & & \\ e_4 & & & & & & & & & & & 0.50 & \\ e_5 & & & 1.00 & & & & & & & & 0.50 & 0.50 \\ e_6 & & & & & & & & & & & & 0.50 \\ H_1 & & & & 1.00 & & & 1.00 & & & & & \\ H_2 & & & & & 1.00 & & & 1.00 & & & & \\ H_3 & & & & & & 1.00 & & & 1.00 & & & \\ obs & & & & & & & & & & & & \\ x_1 & & 1.00 & & & & & & & & & & \\ x_2 & & 1.00 & & & & & & & & & & \end{array} \begin{array}{c|cc} & 0 & 1 \\ \hline e_1 & & \\ e_2 & & \\ e_3 & 0.50 & 0.50 \\ e_4 & & \\ e_5 & & \\ e_6 & & \\ H_1 & & \\ H_2 & & \\ H_3 & & \\ obs & & \\ x_1 & 0.50 & \\ x_2 & & 0.50 \end{array}$$

5. Iteration 4:
   - $M^{(5)} = \theta(M_P^T \cdot M^{(4)})$

$$
\begin{array}{c}
\\ e_1 \\ e_2 \\ e_3 \\ e_4 \\ e_5 \\ e_6 \\ H_1 \\ H_2 \\ H_3 \\ obs \\ x_1 \\ x_2
\end{array}
\begin{pmatrix}
0 & 1 \\
 & \\
 & \\
 & \\
 & \\
 & \\
 & \\
0.50 & \\
0.50 & 0.50 \\
 & 0.50 \\
 & \\
 & \\
 &
\end{pmatrix}
=
\begin{array}{c}
\\ e_1 \\ e_2 \\ e_3 \\ e_4 \\ e_5 \\ e_6 \\ H_1 \\ H_2 \\ H_3 \\ obs \\ x_1 \\ x_2
\end{array}
\begin{pmatrix}
e_1 & e_2 & e_3 & e_4 & e_5 & e_6 & H_1 & H_2 & H_3 & obs & x_1 & x_2 \\
 & & & & & & & & & 1.00 & & \\
0.50 & & & & & & & & & & & \\
0.50 & & & & & & & & & & & \\
 & & & & & & & & & & 0.50 & \\
 & & 1.00 & & & & & & & & 0.50 & 0.50 \\
 & & & & & & & & & & & 0.50 \\
 & & & 1.00 & & & 1.00 & & & & & \\
 & & & & 1.00 & & & 1.00 & & & & \\
 & & & & & 1.00 & & & 1.00 & & & \\
 & & & & & & & & & & & \\
 & 1.00 & & & & & & & & & & \\
 & 1.00 & & & & & & & & & &
\end{pmatrix}
\begin{array}{c}
\\ e_1 \\ e_2 \\ e_3 \\ e_4 \\ e_5 \\ e_6 \\ H_1 \\ H_2 \\ H_3 \\ obs \\ x_1 \\ x_2
\end{array}
\begin{pmatrix}
0 & 1 \\
 & \\
 & \\
 & \\
0.50 & \\
0.50 & 0.50 \\
 & 0.50 \\
 & \\
 & \\
 & \\
 & \\
 & \\
 &
\end{pmatrix}
$$

6. The algorithm stops. Found minimal explanations: $\{ \{H_1, H_2\}, \{H_2, H_3\} \}$.

In solving this problem, Algorithm 1 takes four iterations and a call to the MHS solver. One can notice in Iteration 3 that e_3, appearing in both explanation vectors of $M^{(3)}$, is computed twice. Imagine if an e_3-like node is repeated multiple times, then the computation spending from the second time on is duplicated. To deal with this issue we employ the idea of partial evaluation which is going to be discussed in the next section.

3.2 Partial Evaluation

Now, we define the formal method of partial evaluation in solving PHCAP by adapting the definition of partial evaluation of definite programs in vector spaces in [21].

Definition 8 Partial evaluation in abduction: Let a PHCAP $\langle \mathcal{L}, \mathbb{H}, \mathbb{O}, P \rangle$ where P is a standardized program. For any *And*-rule $r = (h \leftarrow b_1 \wedge \cdots \wedge b_m)$ in P,

- if $body(r)$ contains an atom b_i $(1 \leq i \leq m)$ which is not the head of any rule in P, then remove r.
- otherwise, for each atom $b_i \in body(r)$ $(i = 1, \ldots, m)$, if there is an *And*-rule $b_i \leftarrow B_i$ in P (where B_i is a conjunction of atoms), then replace each b_i in $body(r)$ by the conjunction B_i.

The resulting rule is denoted by $\mathsf{unfold}(r)$. Define

$$\mathsf{peval}(P) = \bigcup_{r \in P_{And}} \mathsf{unfold}(r).$$

$\mathsf{peval}(P)$ is called *partial evaluation* of P.

Example 2 (continue Example 1).

- Let $P' = \{r_1, ..., r_9\}$ where:	- Unfolding rules of P' becomes:	- Then $\mathsf{peval}(P')$ consists of:
$r_1 = (obs \leftarrow e_1)$,	$\mathsf{unfold}(r_1) = (obs \leftarrow e_2 \wedge e_3)$,	$obs \leftarrow e_2 \wedge e_3$,
$r_2 = (e_1 \leftarrow e_2 \wedge e_3)$,	$\mathsf{unfold}(r_2) = (e_1 \leftarrow e_2 \wedge e_5)$,	$e_1 \leftarrow e_2 \wedge e_5$,
$r_3 = (e_2 \leftarrow x_1 \vee x_2)$,	$\mathsf{unfold}(r_3) = r_3$,	$e_2 \leftarrow x_1 \vee x_2$,
$r_4 = (x_1 \leftarrow e_4 \wedge e_5)$,	$\mathsf{unfold}(r_4) = (x_1 \leftarrow H_1 \wedge H_2)$,	$x_1 \leftarrow H_1 \wedge H_2$,
$r_5 = (x_2 \leftarrow e_5 \wedge e_6)$,	$\mathsf{unfold}(r_5) = (x_2 \leftarrow H_2 \wedge H_3)$,	$x_2 \leftarrow H_2 \wedge H_3$,
$r_6 = (e_3 \leftarrow e_5)$,	$\mathsf{unfold}(r_6) = (e_3 \leftarrow H_2)$,	$e_3 \leftarrow H_2$,
$r_7 = (e_4 \leftarrow H_1)$,	$\mathsf{unfold}(r_7) = r_7$,	$e_4 \leftarrow H_1$,
$r_8 = (e_5 \leftarrow H_2)$,	$\mathsf{unfold}(r_8) = r_8$,	$e_5 \leftarrow H_2$,
$r_9 = (e_6 \leftarrow H_3)$.	$\mathsf{unfold}(r_9) = r_9$.	$e_6 \leftarrow H_3$.

We do not consider unfolding rules by *Or*-rules and unfolding *Or*-rules, as in the deduction case considered in [21]. Obviously, $\mathsf{peval}(P) = \mathsf{peval}(P_{And})$ and $\mathsf{peval}(P)$ is a standardized program. The *reduct abductive matrix* $M_P(P^r_{And})^T$ is the representation of P_{And} as presented in Definition 5, therefore, we can base on $M_P(P^r_{And})^T$ to build up the matrix representation of $\mathsf{peval}(P)$.

Example 3 (continue Example 2) .
According to Definition 5 we have the *reduct abductive matrix*:

$$M_P(P'^r_{And})^T = \begin{array}{c|cccccccccccc} & e_1 & e_2 & e_3 & e_4 & e_5 & e_6 & H_1 & H_2 & H_3 & obs & x_1 & x_2 \\ \hline e_1 & & & & & & & & & & 1.00 & & \\ e_2 & 0.50 & & & & & & & & & & & \\ e_3 & 0.50 & & & & & & & a & & & & \\ e_4 & & & & & & & & & & & 0.50 & \\ e_5 & & & 1.00 & & & & & & & & 0.50 & 0.50 \\ e_6 & & & & & & & & & & & & 0.50 \\ H_1 & & & & 1.00 & & & 1.00 & & & & & \\ H_2 & & & & & 1.00 & & & 1.00 & & & & \\ H_3 & & & & & & 1.00 & & & 1.00 & & & \\ obs & & & & & & & & & & & & \\ x_1 & & 1.00 & & & & & & & & & & \\ x_2 & & 1.00 & & & & & & & & & & \end{array}$$

1. $\mathsf{peval}(P')$ can be obtained by computing the power of the *reduct abductive matrix*: $\left(M_P(P'^r_{And})^T\right)^2$, $\left(M_P(P'^r_{And})^T\right)^4$, ... $\left(M_P(P'^r_{And})^T\right)^{2^k}$ where k is the number of peval steps. Here, we reach a fixpoint at $k = 2$.

$$\left(M_P(P'^r_{And})^T\right)^4 = \begin{array}{c|cccccccccccc} & e_1 & e_2 & e_3 & e_4 & e_5 & e_6 & H_1 & H_2 & H_3 & obs & x_1 & x_2 \\ \hline e_1 & & & & & & & & & & & & \\ e_2 & 0.50 & 1.00 & & & & & & & & 0.50 & & \\ e_3 & & & & & & & & & & & & \\ e_4 & & & & & & & & & & & & \\ e_5 & & & & & & & & & & & & \\ e_6 & & & & & & & & & & & & \\ H_1 & & & & 1.00 & & & 1.00 & & & & 0.50 & \\ H_2 & 0.50 & & 1.00 & & 1.00 & & & 1.00 & & 0.50 & 0.50 & 0.50 \\ H_3 & & & & & & 1.00 & & & 1.00 & & & 0.50 \\ obs & & & & & & & & & & & & \\ x_1 & & & & & & & & & & & & \\ x_2 & & & & & & & & & & & & \end{array}$$

We refer to this "stable" matrix as $\mathsf{peval}(P)$ and take it to solve the PHCAP.
2. Iteration 1:
- $M^{(1)} = \theta(\mathsf{peval}(P) \cdot M^{(0)})$, where $M^{(0)} = \mathbb{O}$

$$\begin{array}{c|c} & 0 \\ \hline e_1 & \\ e_2 & 0.50 \\ e_3 & \\ e_4 & \\ e_5 & \\ e_6 & \\ H_1 & \\ H_2 & 0.50 \\ H_3 & \\ obs & \\ x_1 & \\ x_2 & \end{array} = \begin{array}{c|cccccccccccc} & e_1 & e_2 & e_3 & e_4 & e_5 & e_6 & H_1 & H_2 & H_3 & obs & x_1 & x_2 \\ \hline e_1 & & & & & & & & & & & & \\ e_2 & 0.50 & 1.00 & & & & & & & & 0.50 & & \\ e_3 & & & & & & & & & & & & \\ e_4 & & & & & & & & & & & & \\ e_5 & & & & & & & & & & & & \\ e_6 & & & & & & & & & & & & \\ H_1 & & & & 1.00 & & & 1.00 & & & & 0.50 & \\ H_2 & 0.50 & & 1.00 & & 1.00 & & & 1.00 & & 0.50 & 0.50 & 0.50 \\ H_3 & & & & & & 1.00 & & & 1.00 & & & 0.50 \\ obs & & & & & & & & & & & & \\ x_1 & & & & & & & & & & & & \\ x_2 & & & & & & & & & & & & \end{array} \begin{array}{c|c} & 0 \\ \hline e_1 & \\ e_2 & \\ e_3 & \\ e_4 & \\ e_5 & \\ e_6 & \\ H_1 & \\ H_2 & \\ H_3 & \\ obs & 1.00 \\ x_1 & \\ x_2 & \end{array}$$

- Solving MHS: $\{\,\{x_1, x_2\}, \{H_2\}\,\}$. MHS solutions: $\{\,\{H_2, x_2\}, \{H_2, x_1\}\,\} = M^{(2)}$.
3. Iteration 2:
- $M^{(3)} = \theta(\mathsf{peval}(P) \cdot M^{(2)})$

$$\begin{array}{c|cc} & 0 & 1 \\ \hline e_1 & & \\ e_2 & & \\ e_3 & & \\ e_4 & & \\ e_5 & & \\ e_6 & & \\ H_1 & & 0.25 \\ H_2 & 0.75 & 0.75 \\ H_3 & 0.25 & \\ obs & & \\ x_1 & & \\ x_2 & & \end{array} = \begin{array}{c|cccccccccccc} & e_1 & e_2 & e_3 & e_4 & e_5 & e_6 & H_1 & H_2 & H_3 & obs & x_1 & x_2 \\ \hline e_1 & & & & & & & & & & & & \\ e_2 & 0.50 & 1.00 & & & & & & & & 0.50 & & \\ e_3 & & & & & & & & & & & & \\ e_4 & & & & & & & & & & & & \\ e_5 & & & & & & & & & & & & \\ e_6 & & & & & & & & & & & & \\ H_1 & & & & 1.00 & & & 1.00 & & & & 0.50 & \\ H_2 & 0.50 & & 1.00 & & 1.00 & & & 1.00 & & 0.50 & 0.50 & 0.50 \\ H_3 & & & & & & 1.00 & & & 1.00 & & & 0.50 \\ obs & & & & & & & & & & & & \\ x_1 & & & & & & & & & & & & \\ x_2 & & & & & & & & & & & & \end{array} \begin{array}{c|cc} & 0 & 1 \\ \hline e_1 & & \\ e_2 & & \\ e_3 & & \\ e_4 & & \\ e_5 & & \\ e_6 & & \\ H_1 & & \\ H_2 & 0.50 & 0.50 \\ H_3 & & \\ obs & & \\ x_1 & & 0.50 \\ x_2 & 0.50 & \end{array}$$

4. The algorithm stops. Found minimal explanations: $\{\ \{H_1, H_2\}, \{H_2, H_3\}\ \}$.

As we can see in Example 3, partial evaluation precomputes all *And*-nodes and we can just reuse their explanation vectors immediately. The number of iterations is reduced from 4 in Example 1 to 2. Moreover, e_3 is already precomputed to H_2, so we do not need to recompute it twice. Thus, the effect of partial evaluation remarkably boosts the overall performance of Algorithm 1 by reducing the number of needed iterations and also cutting down the cost of redundant computation.

Now let us formalize the partial evaluation step.

Proposition 1. *Let $\langle \mathscr{L}, \mathbb{H}, \mathbb{O}, P \rangle$ be a PHCAP such that P is a standardized program. Let $M_P(P^r_{And})^T$ be the reduct abductive matrix of P, also let v_0 be the vector representing observation of the PHCAP.*
Then $\theta\left(\left(M_P(P^r_{And})^T\right)^2 \cdot v_0\right) = \theta\left(M_P(P^r_{And})^T \cdot \theta\left(M_P(P^r_{And})^T \cdot v_0\right)\right)$

Proof. There are two different cases that we need to consider:

- In case v_0 is an *Or*-computable vector, the matrix multiplication maintains all values of atoms appearing in the heads of *Or*-rules by $M_P(P^r_{And})^T \cdot v_0 = v_0$. This is because we set 1 at the diagonal of the reduct abductive matrix as in Definition 5.[1]
- Suppose that v_0 is an *And*-computable vector. An atom is defined by a single rule since P is a standardized program. Suppose that an atom p_i $(1 \le i \le n)$ is defined by $\frac{1}{m}$ of q_j and q_j is defined by $\frac{1}{l}$ of r_k in $M_P(P^r_{And})^T$. Then p_i is defined by $(\frac{1}{m} \times \frac{1}{l})$ of r_k via q_j in $M_P(P^r_{And})^T$, which is computed by the matrix product $\left(M_P(P^r_{And})^T\right)^2$. This corresponds to the result of abductively unfolding a rule $p_i \leftarrow q_1 \wedge \cdots \wedge q_m$ by a rule $q_j \leftarrow r_1 \wedge \cdots \wedge r_l$ $(1 \le j \le m)$ in P. $\theta(\left(M_P(P^r_{And})^T\right)^2 v_0)$ then represents the results of two consecutive steps of 1-step abduction in P_{And}. And $\left(\left(M_P(P^r_{And})^T\right)^2 \cdot v_0\right)[i] \ge 1$ iff $M_P(P^r_{And})^T \cdot \theta\left(M_P(P^r_{And})^T \cdot v_0\right)[i] \ge 1$ for any $1 \le i \le n$.

Hence, the result holds. □

Partial evaluation has the effect of reducing deduction steps by unfolding rules in advance. Proposition 1 realizes this effect by computing matrix products. Partial evaluation is repeatedly performed as:

$$\mathsf{peval}^0(P) = P \text{ and } \mathsf{peval}^k(P) = \mathsf{peval}(\mathsf{peval}^{k-1}(P)) \ (k \ge 1). \tag{4}$$

The k-step partial evaluation has the effect of realizing 2^k steps of deduction at once. Multiplying an explanation vector and the peval matrix thus realizes an exponential speed-up that has been demonstrated in Example 3.

Proposition 2. *Partial evaluation realized in Proposition 1 has a fixpoint.*

[1] This behavior is unlike the behavior of the previous definition in [22] that we set 0 at the diagonal that will eliminate all values of *Or*-rule head atoms in v_0.

Proof. Note that we assume a program is acyclic. As Algorithm 1 causes no change to atoms in the head of *Or*-rules, one can create a corresponding standardized program containing only *And*-rules. The resulting program, with only *And*-rules, is monotonic so it has a fixpoint for every initial vector. Thus, partial evaluation has a fixpoint. □

Accordingly, incorporating peval to Algorithm 1 is made easy by first finding the reduct abductive matrix and then computing the power of that matrix until we reach a fixpoint. Then we use the output vector to replace the abductive matrix in the Algorithm 1 for computing explanations. The motivation behind this idea is to take advantage of the recent advance in efficient linear algebra routines.

Intuitively speaking, non-zero elements in the reduct abductive matrix represent conjuncts appearing in each rule. By computing the power of this matrix, we assume all *And*-nodes are needed to explain the observation. Then we precompute the explanations for all these nodes. However, the good effect of partial evaluation depends on the graph structure of the PHCAP. If there are many *And*-nodes that just lead to "nothing" or somehow these subgraphs of *And*-rules are not repeated at a certain number of times. Then partial evaluation just does the same job as the normal approach but at a higher cost with computing the power of a matrix. We will evaluate the benefit of partial evaluation in the next section.

4 Experimental Results

In this section, we evaluate the efficiency of partial evaluation based on benchmark datasets that are used in [16, 17, 22]. The characteristics of the benchmark datasets are summarized below. Both dense and sparse formats are considered as the representation of program matrices and abductive matrices in the partial evaluation method.

- **Artificial samples I** (166 problems): deeper but narrower graph structure.
- **Artificial samples II** (117 problems)[2]: deeper and wider graph structure, some problems involve solving a large number of medium-size MHS problems.
- **FMEA samples** (213 problems): shallower but wider graph structure, usually involving a few (but) large-size MHS problems.

For further detailed statistics data, readers should follow the experimental setup in [22].

Additionally, to demonstrate the efficiency of partial evaluation, we do enhancing the benchmark dataset based on the transitive closure problem: $P = \{path(X, Y) \leftarrow edge(X, Y),\ path(X, Y) \leftarrow edge(X, Z) \land path(Z, Y)\}$. First, we generate a PHCAP problem based on the transitive closure of the following single line graph: $edge(1, 2)$, $edge(2, 3)$, $edge(3, 4)$, $edge(4, 5)$, $edge(5, 6)$, $edge(6, 7)$, $edge(7, 8)$, $edge(8, 9)$, $edge(9, 10)$. Then we consider the observation to be $path(1, 10)$, and look for the explanation of it. Obviously, we have to include all the edges of this graph in the explanation and the depth of the corresponding graph or *And*-rules is 10. Next, for each problem instance of the original benchmark, we enumerate rules of the form $e \leftarrow h$, where h is a hypothesis, and append the atom of the observation of the new PHCAP into this rule with a probability of 20%. The resulting problem is expected to have the subgraph of

[2] We excluded the unresolved problem `phcap_140_5_5_5.atms`.

And-rules occur more frequently.

Similar to the experiment setup in [17,22], each method is conducted 10 times with a limited runtime on each PHCAP problem to record the execution time and correctness of the output. The time limit for each run is 20 min, that is, if a solver cannot output the correct output within this limit, 40 min will be penalized to its execution time following PAR-2[3] as used in SAT competitions [12]. Accordingly, for each problem instance, we denote t as the effective solving time, t_{peval} as the time for the partial evaluation step, and t_p as the penalty time. Thus, $t + t_p$ is the total running time. Partial evaluation time t_{peval} and also the extra time for transforming to the standardized format are included in t. We also report t_{peval} separately to give a better insight. All the execution times are reported in Table 1 and Table 2.

The two parts of Table 3 and Table 4 compare the two methods in: the maximum number of explanation vectors $(max(|M|))$, the maximum η_z $(max(\eta_z(M)))$, and the minimum sparsity $min(sparsity(M))$ for each explanation matrix. Finally, max_iter is the number of iterations of the main loop of each method, mhs_calls is the number of MHS problems, and $|\mathbb{E}|$ is the number of correct minimal explanations. For the methods with partial evaluation, we report $peval\_steps$ as the number of partial evaluation steps.

We refer to each method as *Sparse matrix - peval*, *Sparse matrix*, *Dense matrix - peval*, and *Dense matrix* for the linear algebraic method in Algorithm 1 with a sparse representation with partial evaluation, sparse representation without partial evaluation, dense representation with partial evaluation, and dense representation without partial evaluation respectively. Our code is implemented in Python 3.7 using Numpy and Scipy for matrices representation and computation. We also exploit the MHS enumerator provided by PySAT[4] for large-size MHS problems. All the source code and benchmark datasets in this paper will be available on GitHub: https://github.com/nqtuan0192/LinearAlgebraicComputationofAbduction. The computer we perform experiments has the following configurations: CPU: Intel® Xeon® Bronze 3106 @1.70GHz; RAM: 64GB DDR3 @1333MHz; OS: Ubuntu 18.04 LTS 64bit.

4.1 Original Benchmark

Table 1. Detailed execution results for the original benchmark.

Datasets	Artificial samples I (166 problems)			Artificial samples II (117 problems)			FMEA samples (213 problems)		
Algorithms	#solved/ #fastest	$t+t_p$ mean/std	t_{peval} mean/std	#solved/ #fastest	$t+t_p$ mean/std	t_{peval} mean/std	#solved/ #fastest	$t+t_p$ mean/std	t_{peval} mean/std
Sparse matrix - peval	**1,660**	4,243	514	**1,170**	**29,438**	124	**2,130**	**49,481**	84
	89	93	19	246	112	48	726	1,214	4
Sparse matrix	**1,660**	**3,527**	–	**1,170**	35,844	–	2,130	53,553	–
	1,401	29	–	513	62	–	150	1,254	–
Dense matrix - peval	**1,660**	811,841	728,086	**1,170**	140,589	3,599	**2,130**	98,614	25
	13	2,227	31,628	90	1,293	910	1,0007	2,950	3
Dense matrix	**1,660**	27,569	–	**1,170**	205,279	–	**2,130**	131,734	–
	157	183	–	321	1,866	–	247	3,629	–

[3] A PAR-2 score of a solver is defined as the sum of all runtimes for solved instances plus 2 times timeout for each unsolved instance.

[4] https://github.com/pysathq/pysat.

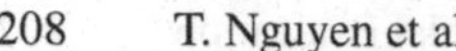

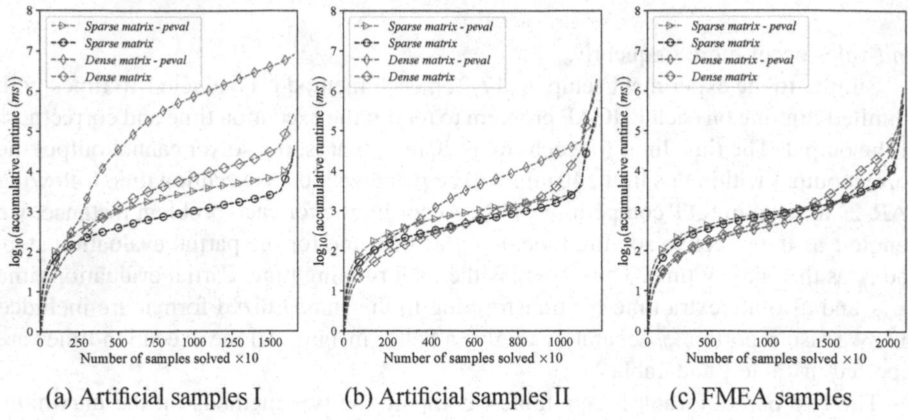

(a) Artificial samples I (b) Artificial samples II (c) FMEA samples

Fig. 1. Effective runtime by the number of solved samples for the original benchmark.

Figure 1 and Table 1 demonstrate the runtime trend and execution time comparison on the original benchmark, while Table 3 gives more detailed information about the sparsity analysis of the dataset in the benchmark. Overall, all algorithms can solve the entire benchmark without any problems. The experiment setup is similar to [22] so readers can compare the data reported in this section with other methods which were reported in [22].

In Artificial samples I, *Sparse matrix* is the fastest algorithm with 1,401 **#fastest** and it finishes the first with 3,527 *ms* on average for each run. *Sparse matrix - peval*, which stands at second place, is slightly slower with average 4,243 *ms* for each run, however, it only is the fastest algorithm in 89 problem instances. A similar trend that the algorithm with partial evaluation is not faster than the original version can be seen with the dense matrix format. In fact, *Dense matrix - peval* is considerably slow in this sample with 811,841 *ms* for each average run, multiple times slower than *Dense matrix*. This could be explained by pointing out that the program matrix size in this dataset is relatively large with **mean** is 2,372.36 as can be seen in the first part of Table 3. In this case, matrix multiplication with the dense format is costly and is not preferable.

In Artificial samples II, *Sparse matrix - peval* is the fastest algorithm with only 29,438 *ms*, while *Sparse matrix* takes 35,844 *ms* for each run on average. However, *Sparse matrix* has higher **#fastest** than *Sparse matrix - peval* that is because many problems in the samples are relatively small. In this dataset, the execution time of *Dense matrix - peval* is significantly improved compared to that of *Dense matrix* with about 25%. In this dataset, the average abductive matrix size is not too large with **mean** is 451.90 while there are multiple branches being created as we see many *mhs_calls*. This condition is favorable for partial evaluation in precomputing multiple branches in advance.

In FMEA samples, a similar trend that partial evaluation significantly improves the original version can be seen in both *Sparse matrix - peval* and *Dense matrix - peval*. *Sparse matrix - peval* again is the fastest algorithm with no doubt, it finishes each run in only about 49,481 *ms*. In spite of that fact, *Dense matrix* is the algorithm with the highest **#fastest** - 1,007. This is because the graph structure of this dataset is shallower, so it produces less complicated matrices that we can consider it is more preferable for dense computation.

4.2 Enhanced Benchmark

Figure 2 and Table 2 demonstrate the runtime trend and execution time comparison on the original benchmark, while Table 4 gives more detailed information about sparsity analysis of the dataset in the benchmark. Overall, the enhanced problems are more difficult than those in the original benchmarks as we see apparently all figures reported in Fig. 2 are higher than that in Fig. 1. However, similar to the original benchmark, all algorithms can solve the entire enhanced benchmark without any problems.

In the enhanced Artificial samples I, with enriched more subgraphs of *And*-rules, now the fastest algorithm is *Sparse matrix - peval* with $12,140 ms$ for each run on average. Interestingly, *Sparse matrix* is the one with the highest **#fastest** $1,389$, although it is not the algorithm that finishes first. *Dense matrix - peval* still cannot catch up with *Dense matrix* even though the execution time of *Dense matrix* now is double what we can see in the previous benchmark. That is because the matrix size is relatively large so we need to increase the depth of embedded subgraphs to see a better effect of partial evaluation with the dense matrix implementation.

In the enhanced Artificial samples II, *Sparse matrix - peval* again takes the first position that it solves in 95,079 ms only for the whole problem samples and being fastest in 254 problem instances. *Sparse matrix* again has the highest **#fastest** 516 but it slower than *Sparse matrix - peval* more than 50% in solving the whole dataset. *Dense matrix - peval* is also faster than *Dense matrix* by more than 50% although there are 323 problems in which *Dense matrix* is the fastest.

Table 2. Detailed execution results for the enhanced benchmark datasets.

Datasets	Artificial samples I (166 problems)			Artificial samples II (117 problems)			FMEA samples (213 problems)		
Algorithms	#solved/ #fastest	$t+t_p$ mean/std	t_{peval} mean/std	#solved/ #fastest	$t+t_p$ mean/std	t_{peval} mean/std	#solved/ #fastest	$t+t_p$ mean/std	t_{peval} mean/std
Sparse matrix - peval	**1,660** 116	**12,140** 124	545 15	**1,170** 254	**95,079** 616	138 4	**2,130** 384	**72,776** 1,103	157 5
Sparse matrix	**1,660** 1,389	16,163 209	– –	**1,170** 516	147,444 1,508	– –	**2,130** 553	74,861 526	– –
Dense matrix - peval	**1,660** 5	869,922 2,434	799,965 58,500	**1,170** 77	380,033 2,228	4,483 688	**2,130** 436	81,837 1,005	103 10
Dense matrix	**1,660** 150	70,365 681	– –	**1,170** 323	613,422 3,651	– –	**2,130** 757	95,996 1,021	– –

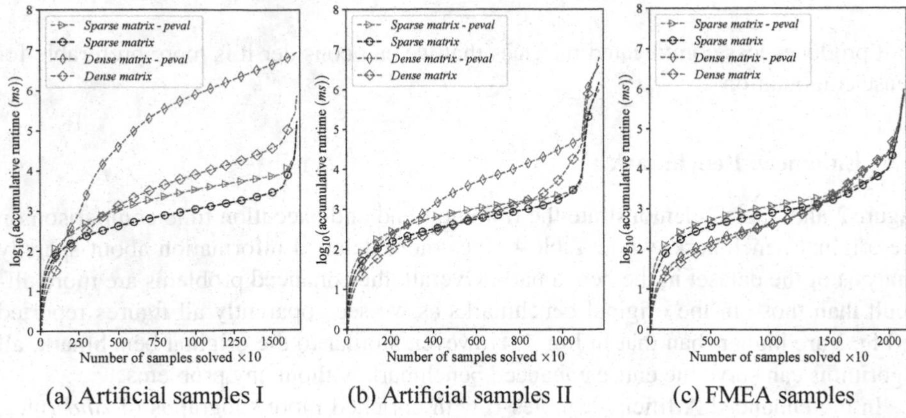

Fig. 2. Effective runtime by the number of solved samples for the enhanced benchmark.

In the enhanced FMEA samples, *Sparse matrix - peval* once again outruns all other algorithms with 72,776*ms* on average for solving the entire dataset. Surprisingly, *Dense matrix* and *Dense matrix - peval* catch up closely with sparse versions that *Dense matrix* has highest **#fastest** with 757. In fact, the shape of the abductive matrices in this dataset is relatively small, and computing these matrices of this size is usually well-optimized. This also can benefit the partial evaluation as we can see *Dense matrix - peval* surpasses *Dense matrix* by more than 12% which is a remarkable improvement.

Discussion: In summary, partial evaluation remarkably improves the linear algebraic approach for abduction. The merit of partial evaluation is that it can be precomputed before abduction steps. Further, once it is computed, we can reuse it repeatedly for different abduction problems. The positive effect can be seen more clearly in case there are multiple subgraphs of *And*-rules exist in the corresponding graph of the PHCAP. In addition, partial evaluation especially boosts the method with sparse representation at a more steady level than with the dense matrix format as reported data for t_{peval} in Table 1 and Table 2.

Table 3. Statistics and sparsity analysis on original benchmark datasets.

Benchmark dataset	Artificial samples I (166 problems)				Artificial samples II (117 problems)				FMEA samples (213 problems)			
Algorithm 1 (without partial evaluation)	**mean**	**std**	**min**	**max**	**mean**	**std**	**min**	**max**	**mean**	**std**	**min**	**max**
$max(\|M\|)$	920.73	10,233.54	1.00	131,413.00	5,245.37	25,961.91	1.00	188,921.00	2,126.49	15,512.54	1.00	154,440.00
$max(\eta_z(M))$	32,162.44	386,905.76	1.00	4,983,288.00	235,884.85	1,138,981.38	1.00	7,302,298.00	43,738.87	334,393.40	1.00	3,459,456.00
$min(sparsity(M))$	0.98	0.05	0.67	1.00	0.94	0.10	0.52	1.00	0.79	0.13	0.46	0.99
max_iter	4.65	5.37	2.00	65.00	7.32	11.46	2.00	91.00	1.94	0.24	1.00	2.00
mhs_calls	5.74	28.22	0.00	349.00	23.15	118.52	0.00	1,208.00	0.93	0.26	0.00	1.00
$\|\mathbb{E}\|$	2.77	5.06	1.00	50.00	3.70	9.62	1.00	63.00	68.89	272.54	1.00	2,288.00
Algorithm 1 (without partial evaluation)	**mean**	**std**	**min**	**max**	**mean**	**std**	**min**	**max**	**mean**	**std**	**min**	**max**
$max(\|M\|)$	627.03	6,479.97	1.00	82,728.00	5,991.66	28,924.93	1.00	188,921.00	1.00	0.00	1.00	1.00
$max(\eta_z(M))$	21,210.33	246,116.37	1.00	3,167,154.00	267,195.89	1,301,308.57	1.00	9,648,741.00	1.00	0.00	1.00	1.00
$min(sparsity(M))$	0.98	0.05	0.67	1.00	0.94	0.09	0.54	1.00	0.97	0.02	0.89	0.99
max_iter	2.87	3.54	1.00	33.00	4.41	6.84	1.00	43.00	1.00	0.00	1.00	1.00
mhs_calls	5.67	27.49	0.00	339.00	30.59	195.41	0.00	2,067.00	0.93	0.26	0.00	1.00
$\|\mathbb{E}\|$	2.77	5.06	1.00	50.00	3.70	9.62	1.00	63.00	68.89	272.54	1.00	2,288.00
$peval\_steps$	3.78	0.95	2.00	5.00	3.71	0.81	2.00	5.00	2.00	0.00	2.00	2.00

Table 4. Statistics and sparsity analysis on benchmark datasets enhanced with transitive closure problem.

Benchmark dataset	Artificial samples I (166 problems)				Artificial samples II (117 problems)				FMEA samples (213 problems)			
Algorithm 1 (without partial evaluation)	**mean**	**std**	**min**	**max**	**mean**	**std**	**min**	**max**	**mean**	**std**	**min**	**max**
$max(\|M\|)$	1,094.92	12,382.96	1.00	159,138.00	7,889.15	34,097.99	1.00	193,943.00	2,918.32	19,190.31	1.00	183,960.00
$max(\eta\_z(M))$	38,796.85	470,699.48	1.00	6,063,138.00	357,363.52	1,587,936.23	1.00	10,607,545.00	59,601.64	410,555.07	1.00	3,886,020.00
$min(sparsity(M))$	0.99	0.03	0.83	1.00	0.96	0.05	0.61	1.00	0.92	0.05	0.8	1.00
max_iter	4.70	5.41	2.00	65.00	7.19	11.34	2.00	91.00	1.97	0.18	1.00	2.00
mhs_calls	5.75	28.22	0.00	349.00	24.83	120.31	0.00	1,208.00	0.93	0.26	0.00	1.00
$\|\mathbb{E}\|$	2.49	5.21	0.00	50.00	4.03	16.72	0.00	168.00	41.46	299.51	0.00	4,000.00
Algorithm 1 (without partial evaluation)	**mean**	**std**	**min**	**max**	**mean**	**std**	**min**	**max**	**mean**	**std**	**min**	**max**
$max(\|M\|)$	727.49	7,677.78	1.00	98,181.00	6,671.67	29,457.12	1.00	193,943.00	1,324.90	8,591.52	1.00	82,440.00
$max(\eta\_z(M))$	25,065.43	294,060.12	1.00	3,785,073.00	277,984.67	1,176,915.48	1.00	6,222,569.00	25,302.40	172,962.09	1.00	1,716,972.00
$min(sparsity(M))$	0.99	0.03	0.83	1.00	0.96	0.06	0.61	1.00	0.94	0.04	0.86	1.00
max_iter	2.91	3.56	1	33.00	4.44	6.93	1.00	48.00	1.66	0.47	1.00	2.00
mhs_calls	5.68	27.49	0.00	339.00	26.59	137.00	0.00	1,391.00	0.93	0.26	0.00	1.00
$\|\mathbb{E}\|$	2.49	5.21	0.00	50.00	4.03	16.72	0.00	168.00	41.46	299.51	0.00	4,000.00
$peval\_steps$	4.20	0.40	4.00	5.00	4.13	0.36	4.00	6.00	4.00	0.00	4.00	4.00

5 Related Work

Propositional abduction has been solved using propositional satisfiability (SAT) techniques in [13], in which a quantified MaxSAT is employed and implicit hitting sets are computed. Another approach to abduction is based on the search for stable models of a logic program [11]. In [25], Saikko et al. [25] have developed a technique to encode the propositional abduction problem as disjunctive logic programming under answer set semantics. Answer set programming has also been employed for first-order Horn abduction in [31], in which all atoms are abduced and weighted abduction is employed.

In terms of linear algebraic computation, Sato et al. [30] developed an approximate computation to abduce relations in Datalog [30], which is a new form of predicate invention in Inductive Logic Programming [20]. They did empirical experiments on linear and recursive cases and indicated that the approach can successfully abduce base relations, but their method cannot compute explanations consisting of possible abducibles.

In this regard, Aspis et al. [2] [2] have proposed a linear algebraic transformation for abduction by exploiting Sakama et al. [27]'s algebraic transformation. Aspis et al. [2] have defined an explanatory operator based on a third-order tensor for computing abduction in Horn propositional programs that simulates deduction through Clark completion for the abductive program [5]. The dimension explosion would arise, unfortunately, and Aspis et al. [2] have not yet reported an empirical work. Aspis et al. [2] propose encoding every single rule as a slice in a third-order tensor then they achieve the growth naturally. Then, they only consider removing columns that are duplicated or inconsistent with the program. According to our analysis, their current method has some points that can be improved to avoid redundant computation. First, they can consider merging all slices of *And*-rules into a single slice to limit the growth of the output matrix. Second, they have to consider incorporating MHS-based elimination strategy, otherwise, their method will waste a lot of computation and resources on explanations that are not minimal.

Nguyen et al. has proposed partial evaluation for computing least models of definite programs [21]. Their method realizes exponential speed-up of fixpoint computation using a program matrix in computing a long chain of *And*-rules. However, computing the least fixpoint of a definite program is very fast with Sparse Matrix-Vector Multiplication (SpMV) [23]. Therefore, the cost of computing the power of the program matrix may only show benefit in a limited number of specific cases. Further, the possibility of applying partial evaluation for model computation in normal logic programs is remaining unanswered in Nguyen et al. 's work [21].

In terms of partial evaluation, Lamma andMello has demonstrated that Assumption based Truth Maintenance System (ATMS) can be considered as the unfolded version of the logic program following bottom-up reasoning mechanism [18]. Our work, on the other hand, could be considered as a linear algebraic version of top-down partial evaluation for abductive programs. In [26], Sakama and Inoue have proposed *abductive partial deduction* with the purpose to preserve the meanings of abductive logic programs [26]. The main idea of this method is that it retains the original clauses together with the unfolded clauses to reserve intermediate atoms which could be used as assumptions [26]. This idea is incorporated in our method already because the matrix representation simply stores every possible clause by nature.

6 Conclusion

We have proposed to improve the linear algebraic approach for abduction by employing partial evaluation. Partial evaluation steps can be realized as the power of the reduct abductive matrix in the language of linear algebra. Its significant enhancement in terms of execution time has been demonstrated using artificial benchmarks and real FMEA-based datasets with both dense and sparse representation, especially more with the sparse format. The performance gain can be more impressive if there are multiple repeated subgraphs of *And*-rules and even more significant if these subgraphs are deeper and deeper. In this case, the benefit of precomputing these subgraphs outweighs the cost of computing the power of the reduct abductive matrix which is considerably expensive.

However, there are many other issues that need to be resolved in future research to realize the full potential of partial evaluation in abduction. If there is a loop in the program, the current method cannot reach a fixpoint. Handling loops and extending the method to work on non-Horn clausal forms is our ongoing work. As we discussed, it may depend on the possibility to derive consequences of clausal theories in a linear algebraic way. Another challenging problem is knowing when to apply partial evaluation and how deep we do unfolding before solving the problem. Even though repeated partial evaluation finishes in finite steps, it is not necessary to perform until an end concerning the cost of the matrix multiplication. An effective prediction of where to stop without sacrificing too much time can significantly improve the overall performance of the linear algebraic method. Moreover, incorporating some efficient pruning techniques or knowing where to zero out in the abductive matrix is also a potential future topic.

Acknowledgements. This work has been supported by JSPS, KAKENHI Grant Numbers JP18H03288 and JP21H04905, and by JST, CREST Grant Number JPMJCR22D3, Japan.

References

1. Apt, K.R., Bezem, M.: Acyclic programs. New Gener. Comput. **9**, 335–364 (1991). https://doi.org/10.1007/BF03037168
2. Aspis, Y., Broda, K., Russo, A.: Tensor-based abduction in Horn propositional programs. In: ILP 2018, CEUR Workshop Proceedings, vol. 2206, pp. 68–75 (2018)
3. Beckman, L., Haraldson, A., Oskarsson, Ö., Sandewall, E.: A partial evaluator, and its use as a programming tool. Artif. Intell. **7**(4), 319–357 (1976). https://doi.org/10.1016/0004-3702(76)90011-4
4. Boutilier, C., Beche, V.: Abduction as belief revision. Artif. Intell. **77**(1), 43–94 (1995). https://doi.org/10.1016/0004-3702(94)00025-V
5. Console, L., Dupré, D.T., Torasso, P.: On the relationship between abduction and deduction. J. Logic Comput. **1**(5), 661–690 (1991). https://doi.org/10.1093/logcom/1.5.661
6. Dai, W.Z., Xu, Q., Yu, Y., Zhou, Z.H.: Bridging machine learning and logical reasoning by abductive learning. In: Proceedings of the 33rd International Conference on Neural Information Processing Systems, Curran Associates Inc., Red Hook, NY, USA (2019)
7. Eiter, T., Gottlob, G.: The complexity of logic-based abduction. J. ACM (JACM) **42**(1), 3–42 (1995). https://doi.org/10.1145/200836.200838
8. Eshghi, K.: Abductive planning with event calculus. In: ICLP/SLP, pp. 562–579 (1988)

9. Futamura, Y.: Partial evaluation of computation process-an approach to a compiler-compiler. High.-Order Symbolic Comput. **12**(4), 381–391 (1999). https://doi.org/10.1023/A:1010095604496.This is an updated and revised version of the previous publication in "Systems, Computers, Control", Volume 25, 1971, pages 45-50
10. Garey, M.R., Johnson, D.S.: Computers and Intractability: a guide to the theory of NP-completeness. Freeman, W.H. (1979). ISBN 0-7167-1044-7
11. Gelfond, M., Lifschitz, V.: The stable model semantics for logic programming. In: ICLP/SLP, vol. 88, pp. 1070–1080 (1988)
12. Heule, M.J., Järvisalo, M., Suda, M.: Sat competition 2018. J. Satisfiability Boolean Model. Comput. **11**(1), 133–154 (2019)
13. Ignatiev, A., Morgado, A., Marques-Silva, J.: Propositional abduction with implicit hitting sets. In: ECAI 2016, Frontiers in Artificial Intelligence and Applications, vol. 285, pp. 1327–1335. IOS Press (2016). https://doi.org/10.3233/978-1-61499-672-9-1327
14. Ignatiev, A., Narodytska, N., Marques-Silva, J.: Abduction-based explanations for machine learning models. In: Proceedings of the AAAI Conference on Artificial Intelligence, vol. 33, pp. 1511–1519 (2019). https://doi.org/10.1609/aaai.v33i01.33011511
15. Josephson, J.R., Josephson, S.G.: Abductive Inference: Computation, Philosophy, Technology. Cambridge University Press, Cambridge (1996)
16. Koitz-Hristov, R., Wotawa, F.: Applying algorithm selection to abductive diagnostic reasoning. Appl. Intell. **48**(11), 3976–3994 (2018). https://doi.org/10.1007/s10489-018-1171-9
17. Koitz-Hristov, R., Wotawa, F.: Faster horn diagnosis - a performance comparison of abductive reasoning algorithms. Appl. Intell. **50**(5), 1558–1572 (2020). https://doi.org/10.1007/s10489-019-01575-5
18. Lamma, E., Mello, P.: A rationalisation of the ATMS in terms of partial evaluation. In: Lau, K.K., Clement, T.P., (eds) Logic Program Synthesis and Transformation, pp. 118–131. Springer, Cham (1993). https://doi.org/10.1007/978-1-4471-3560-9_9
19. Lloyd, J.W., Shepherdson, J.C.: Partial evaluation in logic programming. J. Logic Program. **11**(3–4), 217–242 (1991). https://doi.org/10.1016/0743-1066(91)90027-M
20. Muggleton, S.: Inductive logic programming. New Gener. Comput. **8**(4), 295–318 (1991). https://doi.org/10.1007/BF03037089
21. Nguyen, H.D., Sakama, C., Sato, T., Inoue, K.: An efficient reasoning method on logic programming using partial evaluation in vector spaces. J. Logic Comput. **31**(5), 1298–1316 (2021). https://doi.org/10.1093/logcom/exab010
22. Nguyen, T.Q., Inoue, K., Sakama, C.: Linear algebraic computation of propositional Horn abduction. In: 2021 IEEE 33rd International Conference on Tools with Artificial Intelligence (ICTAI), pp. 240–247. IEEE (2021). https://doi.org/10.1109/ICTAI52525.2021.00040
23. Nguyen, T.Q., Inoue, K., Sakama, C.: Enhancing linear algebraic computation of logic programs using sparse representation. New Gener. Comput. **40**(5), 1–30 (2021). https://doi.org/10.1007/s00354-021-00142-2
24. Rocktäschel, T., Riedel, S.: End-to-end differentiable proving. In: Proceedings of the 31st International Conference on Neural Information Processing Systems, pp. 3791–3803, NIPS 2017, Curran Associates Inc., Red Hook, NY, USA (2017). ISBN 9781510860964
25. Saikko, P., Wallner, J.P., Järvisalo, M.: Implicit hitting set algorithms for reasoning beyond NP. In: KR, pp. 104–113 (2016)
26. Sakama, C., Inoue, K.: The effect of partial deduction in abductive reasoning. In: ICLP, pp. 383–397 (1995)
27. Sakama, C., Inoue, K., Sato, T.: Linear algebraic characterization of logic programs. In: Li, G., Ge, Y., Zhang, Z., Jin, Z., Blumenstein, M. (eds.) KSEM 2017. LNCS (LNAI), vol. 10412, pp. 520–533. Springer, Cham (2017). https://doi.org/10.1007/978-3-319-63558-3_44
28. Sakama, C., Inoue, K., Sato, T.: Logic programming in tensor spaces. Ann. Math. Artif. Intell. **89**(12), 1133–1153 (2021). https://doi.org/10.1007/s10472-021-09767-x

29. Sato, T.: Embedding Tarskian semantics in vector spaces. In: Workshops at the Thirty-First AAAI Conference on Artificial Intelligence (2017)
30. Sato, T., Inoue, K., Sakama, C.: Abducing relations in continuous spaces. In: IJCAI, pp. 1956–1962 (2018). https://doi.org/10.24963/ijcai.2018/270
31. Schüller, P.: Modeling variations of first-order Horn abduction in answer set programming. Fundam. Informaticae **149**(1–2), 159–207 (2016). https://doi.org/10.3233/FI-2016-1446
32. Selman, B., Levesque, H.J.: Abductive and default reasoning: a computational core. In: AAAI, pp. 343–348 (1990)
33. Tamaki, H., Sato, T.: Unfold/fold transformation of logic programs. In: Proceedings of the Second International Conference on Logic Programming, pp. 127–138 (1984)
34. Vasileiou, S.L., Yeoh, W., Son, T.C., Kumar, A., Cashmore, M., Magazzeni, D.: A logic-based explanation generation framework for classical and hybrid planning problems. J. Artif. Intell. Res. **73**, 1473–1534 (2022). https://doi.org/10.1613/jair.1.13431

Using Hybrid Knowledge Bases for Meta-reasoning over OWL 2 QL

Haya Majid Qureshi(✉) and Wolfgang Faber

University of Klagenfurt, Klagenfurt, Austria
{haya.qureshi,wolfgang.faber}@aau.at

Abstract. Metamodeling refers to scenarios in ontologies in which classes and roles can be members of classes or occur in roles. This is a desirable modelling feature in several applications, but allowing it without restrictions is problematic for several reasons, mainly because it causes undecidability. Therefore, practical languages either forbid metamodeling explicitly or treat occurrences of classes as instances to be semantically different from other occurrences, thereby not allowing metamodeling semantically. Several extensions have been proposed to provide metamodeling to some extent. This paper focuses on the extension called Metamodeling Semantics (MS) over OWL 2 QL and the Metamodeling Semantic Entailment Regime (MSER). Reasoning over the latter requires more than an OWL 2 QL reasoner, and in the past a solution that transforms MSER to query answering over Datalog has been proposed. In this paper, we investigate the use of Hybrid Knowledge Bases for MSER. Hybrid Knowledge Bases have rule and ontology components, and we provide translations that keep some portion of the original metamodeling ontology in the ontology part. This is accompanied by a translation of *SPARQL* queries (including metaqueries) to queries over the Hybrid Knowledge Base. The target language is compatible with the hybrid reasoner called *HEXLite-owl-api-plugin*, allowing for an effective reasoning tool over metamodeling ontologies, which we evaluate experimentally.

Keywords: Ontology · Metamodeling · Rules · SPARQL

1 Introduction

Metamodeling helps in specifying conceptual modelling requirements with the notion of meta-classes (for instance, classes that are instances of other classes) and meta-properties (relations between meta-concepts). These notions can be expressed in OWL Full. However, OWL Full is so expressive for metamodeling that it leads to undecidability [17]. OWL 2 DL and its sub-profiles guarantee decidability, but they provide a very restricted form of metamodeling [10] and give no semantic support due to the prevalent Direct Semantics (DS).

Consider an example adapted from [9], concerning the modeling of biological species, stating that all GoldenEagles are Eagles, all Eagles are Birds, and Harry

M. Hanus and D. Inclezan (Eds.): PADL 2023, LNCS 13880, pp. 216–231, 2023.
https://doi.org/10.1007/978-3-031-24841-2_14

is an instance of GoldenEagle, which further can be inferred as an instance of Eagle and Birds. However, in the species domain one can not just express properties of and relationships among species, but also express properties of the species themselves. For example "GoldenEagle is listed in the IUCN Red List of endangered species" states that GoldenEagle as a whole class is an endangered species. Note that this is also not a subclass relation, as Harry is not an endangered species. To formally model this expression, we can declare GoldenEagle to be an instance of new class EndangeredSpecies.

$$Eagle \sqsubseteq Birds, \quad GoldenEagle \sqsubseteq Eagle, \quad GoldenEagle(Harry)$$
$$EndangeredSpecies \sqsubseteq Species, \; EndangeredSpecies(GoldenEagle)$$

Note that the two occurrences of the IRI for GoldenEagle (in a class position and in an individual position) are treated as different objects in the standard direct semantics DS[1], therefore not giving semantic support to punned[2] entities and treating them as independent of each other by reasoners. These restrictions significantly limit meta-querying as well, as the underlying semantics for SPARQL queries over OWL 2 QL is defined by the *Direct Semantic Entailment Regime* [6], which uses *DS*.

To remedy the limitation of metamodeling, Higher-Order Semantics (HOS) was introduced in [13] for OWL 2 QL ontologies and later referred to as Metamodeling Semantics (MS) in [14], which is the terminology that we will adopt in this paper. The interpretation structure of HOS follows the Hilog-style semantics of [2], which allows the elements in the domain to have polymorphic characteristics. Furthermore, to remedy the limitation of metaquerying, the Metamodeling Semantics Entailment Regime (MSER) was proposed in [3], which does allow meta-modeling and meta-querying using SPARQL by reduction from query-answering over OWL 2 QL to Datalog queries.

One observation is that some parts of metamodeling ontologies will not use all the statements. Our basic idea is to use standard ontology reasoners for those parts and use the Datalog rewriting of [3] for the metamodeling portions. To this end, we study the use of hybrid mechanisms of enriching ontologies with rules [16]. Hybrid integration of ontology languages with rule systems has shown significant advances and plays a central role in the development of the Semantic Web [1]. Research has been done on different types for hybrid integration, which can be roughly divided into two types: the first type is homogeneous integration that combines rules and ontologies in the same logical language (e.g., Hybrid MKNF (Minimal Knowledge and Negation as Failure) knowledge bases [18] and DLP (Description Logic Programs) [8]). The second type is heterogeneous integration that keeps the rules and ontologies independent and imposes a strict semantic separation between them (e.g., HEX programs [4]). In this work, we focus on the latter type.

The second observation is that with metamodeling parts inside ontologies, the correctness of the standard reasoner's reasoning is not guaranteed as they

[1] http://www.w3.org/TR/2004/REC-owl-semantics-20040210/.
[2] http://www.w3.org/2007/OWL/wiki/Punning.

follow the *DS*. The aim is to identify the part of the ontology where we can use a standard reasoner with the hope that the standard reasoner is fast enough and the bits that have metamodeling delegate to the MSER rule part. Together with these two observations, our work's motivation is to reflect the idea of efficient query answering in a hybrid knowledge base that addresses the distinct flavour of metamodeling and meta-querying in OWL 2 QL with the current OWL2 reasoner and rule engine.

This paper proposes four hybrid approaches for MSER. The approaches employ different criteria for splitting the input ontology into two parts. All of them also use the query to extract modules of the ontology that are relevant for answering the query. In all cases, the output forms a hybrid knowledge base $\mathcal{K} = (\mathcal{L}, \mathcal{P})$, which can be processed by a hybrid reasoner like *HEXLite-owl-api-plugin*.

The main contributions of this work can be summarised as follows:

- In this work, we continue the line of [3,20] and offer a hybrid approach for MSER evaluation.
- We evaluate the performance of the four hybrid approaches using *HEXLite-owl-api-plugin*.
- In the evaluation, we considered both meta-modeling ontologies and meta-queries.
- We give a detailed theoretical background, and a discussion of the evaluation results.

Section 2 describes the background notions like OWL 2 QL, meta-modeling Semantics (MS), Hybrid Knowledge Bases (HKB) and SPARQL. In Sect. 3, the four approaches are defined. The initial evaluation is presented in Sect. 5. Finally, future directions and related work are discussed in Sect. 6.

2 Preliminaries

This section gives a brief overview of the language and the formalism used in this work.

2.1 OWL 2 QL

This section recalls the syntax of the ontology language OWL 2 QL and the *Metamodeling Semantics* (MS) for OWL 2 QL, as given in [15].

Syntax

We start by recalling some basic elements used for representing knowledge in ontologies: *Concepts*, a set of individuals with common properties, *Individuals*, objects of a domain of discourse, and *Roles*, a set of relations that link individuals. An OWL 2 ontology is a set of axioms that describes the domain of interest. The elements are classified into *literals* and *entities*, where *literals* are values belonging to datatypes and *entities* are the basic ontology elements denoted by *Internationalized Resource Identifiers* (IRI). The notion of the vocabulary V of an OWL 2 QL, constituted by the tuple $V = (V_e, V_c, V_p, V_d, D, V_i, L_{QL})$. In V, V_e is the union of V_c, V_p, V_d, V_i and its elements are called atomic expressions; V_c, V_p, V_d, and V_i are sets of IRIs, denoting, respectively, classes, object properties, data properties, and individuals, L_{QL} denotes the set of literals - characterized as OWL 2 QL datatype maps denoted as DM_{QL} and D is the set of datatypes in OWL 2 QL (including rdfs:Literal). Given a vocabulary V of an ontology $\mathcal{O}$, we denote by Exp the set of well formed expressions over V. For the sake of simplicity we use Description Logic (DL) syntax for denoting expressions in OWL 2 QL. Complex expressions are built over V, for instance, if $e_1, e_2 \in V$ then $\exists e_1.e_2$ is a complex expression. An OWL 2 QL Knowledge Base $\mathcal{O}$ is a pair $\langle \mathcal{T}, \mathcal{A} \rangle$, where $\mathcal{T}$ is the TBox (inclusion axioms) and $\mathcal{A}$ is the ABox (assertional axioms). Sometimes we also let $\mathcal{O}$ denote $\mathcal{T} \cup \mathcal{A}$ for simplicity. OWL 2 QL is a finite set of logical axioms. The axioms allowed in an OWL 2 QL ontology have one of the form: inclusion axioms $e_1 \sqsubseteq e_2$, disjointness axioms $e_1 \sqsubseteq \neg\, e_2$, axioms asserting property i.e., reflexive property $ref(e)$ and irreflexive property $irref(e)$ and assertional axioms i.e., $c(a)$ class assertion, ,$p(a, b)$ object property assertion, and $d(a, b)$ data property assertion. We employ the following naming schemes (possibly adding subscripts if necessary): $\mathfrak{c}$, $\mathfrak{p}$, $\mathfrak{d}$, $\mathfrak{t}$ denote a class, object property, data property and datatype. The above axiom list is divided into TBox axioms (further divided into positive TBox axioms and negative TBox axioms) and ABox axioms. The positive TBox axioms consist of all the inclusion and reflexivity axioms, the negative TBox axioms consist of all the disjointness and irreflexivity axioms and ABox consist of all the assertional axioms. For simplicity, we omit OWL 2 QL axioms that can be expressed by appropriate combinations of the axioms specified in the above axiom list. Also, for simplicity we assume to deal with ontologies containing no data properties.

Meta-Modeling Semantics

The Meta-modeling Semantics (MS) is based on the idea that every entity in V may simultaneously have more than one type, so it can be a class, or an individual, or data property, or an object property or a data type. To formalise this idea, the Meta-modeling Semantics has been defined for OWL 2 QL. In what follows, $\mathbf{P}(S)$ denotes the power set of S. The meta-modeling semantics for $\mathcal{O}$ over V is based on the notion of interpretation, constituted by a tuple $\mathcal{I} = \langle \Delta, \cdot^I, \cdot^C, \cdot^P, \cdot^D, \cdot^T, \cdot^{\mathcal{I}} \rangle$, where

- Δ is the union of the two non-empty disjoint sets: $\Delta = \Delta^o \cup \Delta^v$, where Δ^o is the object domain, and Δ^v is the value domain defined by DM_{QL};
- $\cdot^I : \Delta^o \rightarrow \{True, False\}$ is a total function for each object $o \in \Delta^o$, which indicates whether o is an individual; if $\cdot^C, \cdot^P, \cdot^D, \cdot^T$ are undefined for an o, then we require $o^I = True$, also in other cases, e.g., if o is in the range of $\cdot^C$;
- $\cdot^C : \Delta^o \rightarrow \mathbf{P}(\Delta^o)$ is partial and can assign the extension of a class;
- $\cdot^P : \Delta^o \rightarrow \mathbf{P}(\Delta^o \times \Delta^o)$ is partial and can assign the extension of an object property;

- $\cdot^D : \Delta^o \rightarrow \mathbf{P}(\Delta^o \times \Delta^v)$ is partial and can assign the extension of a data property;
- $\cdot^T : \Delta^o \rightarrow \mathbf{P}(\Delta^v)$ is partial and can assign the extension of a datatype;
- $\cdot^{\mathcal{I}}$ is a function that maps every expression in Exp to Δ^o and every literal to Δ^v.

This allows for a single object o to be simultaneously interpreted as an individual via $\cdot^I$, a class via $\cdot^C$, an object property via $\cdot^P$, a data property via $\cdot^D$, and a data type via $\cdot^T$. For instance, for Example 1, $\cdot^C$, $\cdot^I$ would be defined for *GoldenEagle*, while $\cdot^P$, $\cdot^D$ and $\cdot^T$ would be undefined for it.

The semantics of logical axiom α is defined in accordance with the notion of axiom satisfaction for an MS interpretation $\mathcal{I}$. The complete set of notions is specified in Table 3.B in [15]. Moreover, $\mathcal{I}$ is said to be a model of an ontology $\mathcal{O}$ if it satisfies all axioms of $\mathcal{O}$. Finally, an axiom α is said to be logically implied by $\mathcal{O}$, denoted as $\mathcal{O} \models \alpha$, if it is satisfied by every model of $\mathcal{O}$.

2.2 Hybrid Knowledge Bases

Hybrid Knowledge Bases (HKBs) have been proposed for coupling logic programming (LP) and Description Logic (DL) reasoning on a clear semantic basis. Our approach uses HKBs of the form $\mathcal{K} = \langle \mathcal{O}, \mathcal{P} \rangle$, where $\mathcal{O}$ is an OWL 2 QL knowledge base and $\mathcal{P}$ is a hex program, as defined next.

Hex programs [4] extend answer set programs with external computation sources. We use hex programs with unidirectional external atoms, which import elements from the ontology of an HKB. For a detailed discussion and the semantics of external atoms, we refer to [5]. What we describe here is a simplification of the much more general hex formalism.

Regular atoms are of the form $p(X_1, \ldots, X_n)$ where p is a predicate symbol of arity n and $X_1, \ldots, X_n$ are terms, that is, constants or variables. An external atom is of the form $\&g[X_1, \ldots, X_n](Y_1, \ldots, Y_m)$ where g is an external predicate name g (which in our case interfaces with the ontology), $X_1, \ldots, X_n$ are input terms and $Y_1, \ldots, Y_m$ are output terms.

Next, we define the notion of positive rules that may contain external atoms.

Definition 1. *A hex rule r is of the form*

$$a \leftarrow b_1, \ldots, b_k. \quad k \geq 0$$

where a is regular atom and $b_1, \ldots, b_k$ are regular or external atoms. We refer to a as the head of r, denoted as $H(r)$, while the conjunction $b_1, ..., b_k$ is called the body of r.

We call r ordinary if it does not contain external atoms. A program $\mathcal{P}$ containing only ordinary rules is called a positive program, otherwise a hex program. A hex program is a finite set of rules.

The semantics of hex programs generalizes the answer set semantics. The Herbrand base of $\mathcal{P}$, denoted $HB_{\mathcal{P}}$, is the set of all possible ground versions of atoms and external atoms occurring in $\mathcal{P}$ (obtained by replacing variables with constants). Note that constants are not just those in the standard Herbrand universe (those occuring in $\mathcal{P}$) but also those created by external atoms, which in our case will be IRIs from $\mathcal{O}$. Let the grounding of a rule r be $grd(r)$ and the grounding of program $\mathcal{P}$ be $grd(\mathcal{P}) = \bigcup_{r \in \mathcal{P}} grd(r)$. An interpretation relative to $\mathcal{P}$ is any subset $I \subseteq HB_{\mathcal{P}}$ containing only regular atoms. We write $I \models a$ iff $a \in I$. With every external predicate name $\&g \in G$ we associate an $(n+m+1)$-ary Boolean function $f_{\&g}$ (called oracle function) assigning each tuple $(I, x_1, \ldots, x_n, y_1 \ldots, y_m)$ either 0 or 1, where I is an interpretation and x_i, y_j are constants. We say that $I \models \&g[x_1, \ldots, x_n](y_1, \ldots, y_m)$ iff $f_{\&g}(I, x_1 \ldots, x_n, y_1, \ldots, y_m) = 1$. For a ground rule r, $I \models B(r)$ iff $I \models a$ for all $a \in B(r)$ and $I \models r$ iff $I \models H(r)$ whenever $I \models B(r)$. We say that I is a model of $\mathcal{P}$, denoted $I \models \mathcal{P}$, iff $I \models r$ for all $r \in grd(\mathcal{P})$. The *FLP-reduct* of $\mathcal{P}$ w.r.t I, denoted as $f\mathcal{P}^I$, is the set of all $r \in grd(\mathcal{P})$ such that $I \models B(r)$. An interpretation I is an answer set of $\mathcal{P}$ iff I is a minimal model of $f\mathcal{P}^I$. By $AS(\mathcal{P})$ we denote the set of all answer sets of $\mathcal{P}$. If $\mathcal{K} = \langle \mathcal{O}, \mathcal{P} \rangle$, then we write $AS(\mathcal{K}) = AS(\mathcal{P})$—note that $\mathcal{O}$ is implicitly involved via the external atoms in $\mathcal{P}$. In this paper, $AS(\mathcal{K})$ will always contain exactly one answer set, so we will abuse notation and write $AS(\mathcal{K})$ to denote this unique answer set.

We will also need the notion of query answers of HKBs that contain rules defining a dedicated query predicate q. Given a hybrid knowledge base $\mathcal{K}$ and a query predicate q, let $ANS(q, \mathcal{K})$ denote the set $\{\langle x_1, \ldots, x_n \rangle \mid q(x_1, \ldots, x_n) \in AS(\mathcal{K})\}$.

3 Query Answering Using MSER

We consider SPARQL queries, a W3C standard for querying ontologies. While SPARQL query results can in general either be result sets or RDF graphs, we have restricted ourselves to simple **SELECT** queries, so it is sufficient for our purposes to denote results by set of tuples. For example, consider the following SPARQL query:

```
SELECT ?x ?y ?z WHERE {
        ?x rdf:type ?y.
        ?y rdfs:SubClassOf ?z
    }
```

This query will retrieve all triples $\langle x, y, z \rangle$, where x is a member of class y that is a subclass of z. In general, there will be several variables and there can be multiple matches, so the answers will be sets of tuples of IRIs.

Now, we recall query answering under the Meta-modeling Semantics Entailment Regime (MSER) from [3]. This technique reduces SPARQL query answering over OWL 2 QL ontologies to Datalog query answering. The main idea of this approach is to define (i) a translation function τ mapping OWL 2 QL axioms to datalog facts and (ii) a fixed datalog rule base $\mathcal{R}^{ql}$ that captures inferences in OWL 2 QL reasoning.

The reduction employs a number of predicates, which are used to encode the basic axioms available in OWL 2 QL. This includes both axioms that are explicitly represented in the ontology (added to the Datalog program as facts via τ) and axioms that logically follow. In a sense, this representation is closer to a meta-programming representation than other Datalog embeddings that translate each axiom to a rule.

The function τ transforms an OWL 2 QL assertion α to a fact. For a given ontology $\mathcal{O}$, we will denote the set of facts obtained by applying τ to all of its axioms as $\tau(\mathcal{O})$; it will be composed of two portions $\tau(\mathcal{T})$ and $\tau(\mathcal{A})$, as indicated in Table 1.

Table 1. τ Function

$\tau(\mathcal{O})$	α	$\tau(\alpha)$	α	$\tau(\alpha)$
$\tau(\mathcal{T})$	c1 ⊑ c2	isacCC(c1, c2)	r1 ⊑ ¬ r2	disjrRR(r1,r2)
	c1 ⊑ ∃r2⁻.c2	isacCI(c1,r2,c2)	c1 ⊑ ¬ c2	disjcCC(c1,c2)
	∃r1 ⊑ ∃r2.c2	isacRR(r1,r2,c2)	c1 ⊑ ¬∃r2⁻	disjcCI(c1,r2)
	∃r1⁻ ⊑ c2	isacIC(r1,c2)	∃r1⊑ ¬ c2	disjcRC(r1,c2)
	∃r1⁻ ⊑ ∃r2.c2	isacIR(r1,r2,c2)	$\exists r_1$ ⊑ ¬∃r2	disjcRR(r1,r2)
	∃r1⁻ ⊑ ∃r2⁻.c2	isacII(r1,r2,c2)	∃r1 ⊑ ¬∃r2⁻	disjcRI(r1,r2)
	r1 ⊑ r2	isarRR(r1,r2)	∃r1⁻ ⊑ ¬ c2	disjcIC(r1,c2)
	r1 ⊑ r2⁻	isarRI(r1,r2)	∃r1⁻ ⊑ ¬∃r2	disjcIR(r1,r2)
	c1 ⊑ ∃r2.c2	isacCR(c1,r2,c2)	∃r1⁻ ⊑ ¬∃r2⁻	disjcII(r1,r2)
	∃r1⊑ c2	isacRC(r1,c2)	r1 ⊑ ¬ r2⁻	disjrRI(r1,r2)
	∃r1 ⊑ ∃r2⁻.c2	isacRI(r1,r2,c2)	irref(r)	irrefl(r)
	refl(r)	refl(r)		
$\tau(\mathcal{A})$	c(x)	instc(c,x)	x ≠ y	diff(x,y)
	r(x, y)	instr(r,x,y)		

The fixed program $\mathcal{R}^{ql}$ can be viewed as an encoding of axiom saturation in OWL 2 QL. The full set of rules provided by authors of [3] are reported in the online repository of [20]. We will consider one rule to illustrate the underlying ideas:

$$\text{isacCR(C1,R2,C2)} \leftarrow \text{isacCC(C1,C3), isacCR(C3,R2,C2)}.$$

The above rule encodes the following inference rule:

$$\mathcal{O} \models \text{C1} \sqsubseteq \text{C3}, \mathcal{O} \models \text{C3} \sqsubseteq \exists\text{R2.C2} \Rightarrow \mathcal{O} \models \text{C1} \sqsubseteq \exists\text{R2.C2}$$

Finally, the translation can be extended in order to transform conjunctive SPARQL queries under MS over OWL 2 QL ontologies into a Datalog query. SPARQL queries will be translated to Datalog rules using a transformation τ^q.

τ^q uses τ to translate the triples inside the body of the SPARQL query $\mathcal{Q}$ and adds a fresh datalog predicate q in the head to account for projections. In the following we assume q to be the query predicate created in this way.

For example, the translation of the SPARQL query given earlier will be

$$\text{q(X,Y,Z)} \leftarrow \text{instc(X,Y), isacCC(Y,Z)}.$$

Given an OWL 2 QL ontology $\mathcal{O}$ and a SPARQL query $\mathcal{Q}$, let $ANS(\mathcal{Q}, \mathcal{O})$ denote the answers to $\mathcal{Q}$ over $\mathcal{O}$ under MSER, that is, a set of tuples of IRIs. In the example above, the answers will be a set of triples.

4 MSER Query Answering via Hybrid Knowledge Bases

We propose four variants for answering MSER queries by means of Hybrid Knowledge Bases. We first describe the general approach and then define each of the four variants.

4.1 General Architecture

The general architecture is outlined in Fig. 1. In all cases, the inputs are an OWL 2 QL ontology $\mathcal{O}$ and a SPARQL query $\mathcal{Q}$. We then differentiate between **OntologyFunctions** and **QueryFunctions**. The **OntologyFunctions** achieves two basic tasks: first, the ontology is split into two partitions $\mathcal{O}^{'}$ and $\mathcal{O}^{''}$, then $\tau(\mathcal{O}^{''})$ is produced.

The **QueryFunctions** work mainly on the query. First, a set $\mathcal{N}$ of IRIs is determined for creating *Interface Rules* (IR, simple hex rules), denoted as $\pi(\mathcal{N})$ for importing the extensions of relevant classes and properties from $\mathcal{O}^{'}$. In the simplest case, $\mathcal{N}$, consist of all IRIs in $\mathcal{O}^{'}$, but we also consider isolating those IRIs that are relevant to the query by means of Logic-based Module Extraction (LME) as defined in [11]. Then, τ^q translates $\mathcal{Q}$ into a datalog query $\tau^q(\mathcal{Q})$. Finally, the created hex program components are united (plus the fixed inference rules), yielding the rule part $\mathcal{P} = \mathcal{R}^{ql} \cup \pi(\mathcal{N}) \cup \tau(\mathcal{O}^{''}) \cup \tau^q(\mathcal{Q})$, which together with $\mathcal{O}^{'}$ forms the HKB $\mathcal{K} = \langle \mathcal{O}^{'}, \mathcal{P} \rangle$, for which we then determine $ANS(q, \mathcal{K})$, where q is the query predicate introduced by $\tau^q(\mathcal{Q})$.

4.2 Basic Notions

Before defining the specific variations of our approach, we first define some auxiliary notions. The first definition identifies meta-elements.

Definition 2. *Given an Ontology $\mathcal{O}$, IRIs in $(V_c \cup V_p) \cap V_i$ are meta-elements, i.e., IRIs that occur both as individuals and classes or object properties.*

In our example, *GoldenEagle* is a meta-element. Meta-elements form the basis of our main notion, clashing axioms.

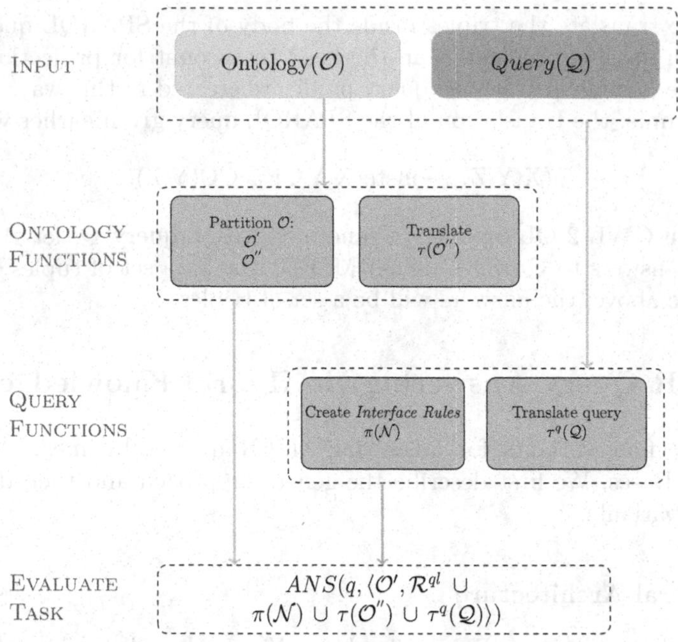

Fig. 1. The overall architecture of hybrid-framework

Definition 3. *Clashing Axioms in $\mathcal{O}$ are axioms that contain meta-elements, denoted as* CA($\mathcal{O}$). *To denote clashing and non-clashing parts in TBox ($\mathcal{T}$) and ABox ($\mathcal{A}$), we write $\mathcal{A}^N = \mathcal{A} \setminus \text{CA}(\mathcal{O})$ as non-clashing ABox, $\mathcal{A}^C = \text{CA}(\mathcal{O}) \cap \mathcal{A}$ as clashing ABox; and likewise $\mathcal{T}^N = \mathcal{T} \setminus \text{CA}(\mathcal{O})$ as non-clashing TBox and $\mathcal{T}^C = \text{CA}(\mathcal{O}) \cap \mathcal{T}$ as clashing TBox.*

The clashing axiom notion allows for splitting $\mathcal{O}$ into two parts and generate $\mathcal{O}'$ without clashing axioms.

We would also like to distinguish between standard queries and meta-queries. A meta-query is an expression consisting of meta-predicates p and meta-variables v, where p can have other predicates as their arguments and v can appear in predicate positions. The simplest form of meta-query is an expression where variables appear in class or property positions also known as *second-order queries.* More interesting forms of meta-queries allow one to extract complex patterns from the ontology, by allowing variables to appear simultaneously in individual object and class or property positions. We will refer to non-meta-queries as standard queries. Moving towards *Interface Rules*, we first define signatures of queries, ontologies, and axioms.

Definition 4. *A* signature $\mathbf{S}(\mathcal{Q})$ *of a SPARQL query $\mathcal{Q}$ is the set of* IRIs *occurring in $\mathcal{Q}$. If no IRIs occur in $\mathcal{Q}$, we define $\mathbf{S}(\mathcal{Q})$ to be the signature of $\mathcal{O}$. Let $\mathbf{S}(\mathcal{O})$ (or $\mathbf{S}(\alpha)$) be the set of atomic classes, atomic roles and individuals that occur in $\mathcal{O}$ (or in axiom α).*

As hinted earlier, we can use $\mathbf{S}(\mathcal{O}')$ for creating interface rules ($\mathcal{O}'$ being the ontology part in the HKB), or use $\mathbf{S}(\mathcal{Q})$ for module extraction via *LME* as defined in [11] for singling out the identifiers relevant to the query, to be imported from the ontology via interface rules. We will denote this signature as $\mathbf{S}(LME(\mathbf{S}(\mathcal{Q}), \mathcal{O}'))$.

We next define the *Interface Rules* for a set of IRIs $\mathcal{N}$.

Definition 5. *For a set a of IRIs $\mathcal{N}$, let $\pi(\mathcal{N})$ denote the hex program containing a rule*

$$instc(C, X) \leftarrow \; \&g[C](X).$$

for each class identifier $C \in \mathcal{N}$, and a rule

$$instr(R, X, Y) \leftarrow \; \&g[R](X, Y).$$

for each property identifier $R \in \mathcal{N}$. Here &g is a shorthand for the external atom that imports the extension of classes or properties from the ontology $\mathcal{O}'$ of our framework.

The rules in $\pi(\mathcal{N})$ are called *Interface Rules* (IR) and serve as the bridge from $\mathcal{O}'$ to $\mathcal{P}$.

4.3 Variants

Now we define the four variants for the ontology functions, and two for the query functions. Since for one ontology function $\mathcal{O}'$ is empty, the two query functions have the same effect, and we therefore arrive at seven different variants for creating the hybrid knowledge bases (HKB).

The difference in the ontology functions is which axioms of $\mathcal{O} = \langle \mathcal{A}, \mathcal{T} \rangle$ stay in $\mathcal{O}'$ and which are in $\mathcal{O}''$, the latter of which is translated to Datalog. We use a simple naming scheme, indicating these two components:

$\mathcal{A}$–$\mathcal{T}$**:** $\mathcal{O}' = \mathcal{A}$, $\mathcal{O}'' = \mathcal{T}$.
NAT–*CAT***:** $\mathcal{O}' = \langle \mathcal{A}^N, \mathcal{T} \rangle$, $\mathcal{O}'' = \langle \mathcal{A}^C, \mathcal{T} \rangle$.
NAT–*CACT***:** $\mathcal{O}' = \langle \mathcal{A}^N, \mathcal{T} \rangle$, $\mathcal{O}'' = \langle \mathcal{A}^C, \mathcal{T}^C \rangle$.
E–*AT***:** $\mathcal{O}' = \emptyset$, $\mathcal{O}'' = \mathcal{O} = \langle \mathcal{A}, \mathcal{T} \rangle$.

E–*AT* serves as a baseline, as it boils down to the Datalog encoding of [3]. *E*–*AT* is also the only variant for which the query functions will not create any Interface Rules, i.e. $\pi(\mathcal{N}) = \emptyset$. We next describe and define each of the ontology functions.

$\mathcal{A}$–$\mathcal{T}$. In this approach, $\mathcal{O}'$ consists only of the ABox $\mathcal{A}$ of $\mathcal{O}$ and the TBox $\mathcal{T}$ of $\mathcal{O}$ is translated into Datalog. Here the main difference to the Datalog encoding of [3] is that the ABox is not translated to facts but stays in the ontology part and can be accessed via Interface Rules.

Definition 6. *Given $\mathcal{O} = \langle \mathcal{A}, \mathcal{T} \rangle$, let the $\mathcal{A}$–$\mathcal{T}$ HKB be $\mathcal{K}^{\mathcal{A}\text{-}\mathcal{T}}(\mathcal{O}) = \langle \mathcal{A}, \mathcal{R}^{ql} \cup \tau(\mathcal{T}) \rangle$.*

NAT–CAT. In this approach, the notion of clashing axioms (cf. Definition 3) is used to separate the ABox into two parts, each of which is combined with the same TBox. One of the resulting ontologies is clash-free and can therefore be reliably treated by a standard ontology reasoner. The other ontology has the meta-assertions and will be treated by the datalog transformation. The link between the two will later be provided by the query functions in the form of Interface Rules.

Definition 7. *Given* $\mathcal{O} = \langle \mathcal{A}, \mathcal{T} \rangle$, *let the NAT–CAT HKB be* $\mathcal{K}^{NAT\text{–}CAT}(\mathcal{O}) = \langle\langle \mathcal{A}^N, \mathcal{T}\rangle, \mathcal{R}^{ql} \cup \tau(\langle \mathcal{A}^C, \mathcal{T}\rangle)\rangle$.

NAT–CACT. This approach is similar to the previous one except that it does not pair the same $\mathcal{T}$ with both $\mathcal{A}^N$ and $\mathcal{A}^C$. Instead, it associates $\mathcal{A}^N$ with the original $\mathcal{T}$ as before, but associates only $\mathcal{T}^C$, the clashing part of $\mathcal{T}$ to $\mathcal{A}^C$, yielding a potentially smaller ontology to be translated to Datalog. Also here, the linking Interface Rules will be added later by the query functions.

Definition 8. *Given* $\mathcal{O} = \langle \mathcal{A}, \mathcal{T} \rangle$, *let the NAT–CACT HKB be* $\mathcal{K}^{NAT\text{–}CACT}(\mathcal{O}) = \langle\langle \mathcal{A}^N, \mathcal{T}\rangle, \mathcal{R}^{ql} \cup \tau(\langle \mathcal{A}^C, \mathcal{T}^C\rangle)\rangle$.

E–AT. This approach is the baseline, has an empty ontology part in the HKB and translates the entire given ontology to Datalog, just like in [3]. Note that here no Interface Rules are necessary.

Definition 9. *Given* $\mathcal{O}$, *let the E–AT HKB be* $\mathcal{K}^{E\text{–}AT}(\mathcal{O}) = \langle \emptyset, \mathcal{R}^{ql} \cup \tau(\mathcal{O})\rangle$.

Next we turn to the query functions. As hinted at earlier, we will consider two versions, which differ in the Interface Rules they create. Both create query rules $\tau^q(\mathcal{Q})$ for the given query, but one (*All*) will create interface rules for all classes and properties in the ontology part of the HKB, while the other (*Mod*) will extract the portion of the ontology relevant to query using *LME* and create Interface Rules only for classes and properties in this module.

For notation, we will overload the $\cup$ operator for HKBs, so we let $\langle \mathcal{O}, \mathcal{P}\rangle \cup \langle \mathcal{O}', \mathcal{P}'\rangle = \langle \mathcal{O} \cup \mathcal{O}', \mathcal{P} \cup \mathcal{P}'\rangle$ and we also let $\langle \mathcal{O}, \mathcal{P}\rangle \cup \mathcal{P}' = \langle \mathcal{O}, \mathcal{P} \cup \mathcal{P}'\rangle$ for ontologies $\mathcal{O}$ and $\mathcal{O}'$ and hex programs $\mathcal{P}$ and $\mathcal{P}'$.

Definition 10. *Given an HKB* $\langle \mathcal{O}, \mathcal{P}\rangle$ *and query* $\mathcal{Q}$, *let the All HKB be defined as* $\mathcal{K}_{All}(\langle \mathcal{O}, \mathcal{P}\rangle, \mathcal{Q}) = \langle \mathcal{O}, \mathcal{P} \cup \tau^q(\mathcal{Q}) \cup \pi(\mathbf{S}(\mathcal{O}))\rangle$.

Definition 11. *Given an HKB* $\langle \mathcal{O}, \mathcal{P}\rangle$ *and query* $\mathcal{Q}$, *let the Mod HKB be defined as* $\mathcal{K}_{All}(\langle \mathcal{O}, \mathcal{P}\rangle, \mathcal{Q}) = \langle \mathcal{O}, \mathcal{P} \cup \tau^q(\mathcal{Q}) \cup \pi(\mathbf{S}(LME(\mathbf{S}(\mathcal{Q}), \mathcal{O})))\rangle$.

We will combine ontology functions and query functions, and instead of $\mathcal{K}_\beta(\mathcal{K}^\alpha(\mathcal{O}), \mathcal{Q})$ we will write $\mathcal{K}^\alpha_\beta(\mathcal{O}, \mathcal{Q})$. We thus get eight combinations, but we will not use $\mathcal{K}^{E\text{–}AT}_{Mod}$, as it unnecessarily introduces Interface Rules. Also note that $\mathcal{K}^{E\text{–}AT}_{All}(\mathcal{O}, \mathcal{Q})$ does not contain any Interface Rules, because the ontology part of $\mathcal{K}^{E\text{–}AT}(\mathcal{O})$ is empty.

4.4 Query Equivalence

We next state that query answering over OWL 2 QL ontologies under MSER is equivalent to query answering over the HKBs obtained using all seven approaches described above.

Theorem 1. *Let $\mathcal{O}$ be a consistent OWL 2 QL ontology, $\mathcal{Q}$ a conjunctive SPARQL query, then for $\mathcal{K}^{\alpha}_{\beta}(\mathcal{O},\mathcal{Q})$, where α is one of $\mathcal{A}-\mathcal{T}$, $\mathcal{NAT}-\mathcal{CAT}$, $\mathcal{NAT}$–$\mathcal{CACT}$, or E–$\mathcal{AT}$, and β is one of All or Mod, it holds that $ANS(\mathcal{Q},\mathcal{O}) = ANS(q,\mathcal{K}^{\alpha}_{\beta}(\mathcal{O},\mathcal{Q}))$, where q is the query predicate introduced by $\tau^{q}(\mathcal{Q})$.*

Proof(sketch). $ANS(\mathcal{Q},\mathcal{O}) = ANS(q,\mathcal{K}^{E\text{–}\mathcal{AT}}_{All}(\mathcal{O},\mathcal{Q}))$ holds because of the results of [3]. $\mathcal{K}^{E\text{–}\mathcal{AT}}_{All}(\mathcal{O},\mathcal{Q})$ has an empty ontology part and the hex program does not contain any external atoms and is equal to the Datalog program defined in [3]. Clearly the answer set of this HKB is equal to the answer set of the Datalog program defined in [3], for which the answer equality has been proved in [3].

$ANS(\mathcal{Q},\mathcal{O}) = ANS(q,\mathcal{K}^{\alpha}_{All}(\mathcal{O},\mathcal{Q}))$ for $\alpha \in \{\mathcal{A}-\mathcal{T}, \mathcal{NAT}-\mathcal{CAT}\}$ holds because the grounding of $\mathcal{K}^{\alpha}_{All}(\mathcal{O},\mathcal{Q})$ is essentially equal to the grounding of $\mathcal{K}^{E\text{–}\mathcal{AT}}_{All}(\mathcal{O},\mathcal{Q})$. This is guaranteed by the external atoms that produce all the facts for ABox axioms that the Datalog transformation creates otherwise. There will be some additional facts stemming from the TBox, but these are also derived via rules in $\mathcal{R}^{ql}$. The answer set of $\mathcal{K}^{\alpha}_{All}(\mathcal{O},\mathcal{Q})$ and $\mathcal{K}^{E\text{–}\mathcal{AT}}_{All}(\mathcal{O},\mathcal{Q})$ is therefore equal, and hence also the answer tuples are equal.

For $ANS(\mathcal{Q},\mathcal{O}) = ANS(q,\mathcal{K}^{\mathcal{NAT}\text{–}\mathcal{CACT}}_{All}(\mathcal{O},\mathcal{Q}))$ we observe that the HKB $\mathcal{K}^{\mathcal{NAT}\text{–}\mathcal{CACT}}_{All}(\mathcal{O},\mathcal{Q})$ still has the same answer set as $\mathcal{K}^{E\text{–}\mathcal{AT}}_{All}(\mathcal{O},\mathcal{Q})$, however now some facts stemming from the TBox portion $\mathcal{T}^{N}$ are now imported via the Inference Rules rather than being derived by Datalog rules.

Concerning $ANS(\mathcal{Q},\mathcal{O}) = ANS(q,\mathcal{K}^{\alpha}_{Mod}(\mathcal{O},\mathcal{Q}))$, the answer set of the HKB $\mathcal{K}^{\alpha}_{Mod}(\mathcal{O},\mathcal{Q})$ is no longer equal to the answer set of the Datalog program defined in [3]. However, we can show that the answer sets restricted to atoms containing the predicate q are equal. This is due to the fact that $LME(\mathbf{S}(\mathcal{Q}),\mathcal{O})$ is a module in the sense of [11] and therefore the conservative extension property holds over $\mathbf{S}(\mathcal{Q})$ as well. So for any axiom α over $\mathbf{S}(\mathcal{Q})$, $\mathcal{O} \models \alpha$ iff $LME(\mathbf{S}(\mathcal{Q}),\mathcal{O}) \models \alpha$. Combining this with the arguments given for *All* shows the claim.

5 Evaluation

In order to see whether the proposed hybrid system is feasible in practice, we conducted two sets of experiments on the widely used Lehigh University Benchmark (LUBM) dataset and on the Making Open Data Effectively USable (MODEUS) Ontologies[3]. The first set of experiments examines the performance of the ontology functions with the query function *Mod*. The second set of experiments considers the performance of the ontology functions except E–$\mathcal{AT}$ with the query

[3] http://www.modeus.uniroma1.it/modeus/node/6.

function *All*. We consider $E{-}\mathcal{AT}$ with the first experiment, even though it does not make use of *Mod*, but it shows the baseline. The second experiment should highlight how much of the gain stems from using *LME*.

The **LUBM** datasets describe a university domain with information like departments, courses, students, and faculty. This dataset comes with 14 queries with different characteristics (low selectivity vs high selectivity, implicit relationships vs explicit relationships, small input vs large input, etc.). We have also considered the meta-queries *mq1*, *mq4*, *mq5*, and *mq10* from [12] as they contain variables in-property positions and are long conjunctive queries. We have also considered two special-case queries *sq1* and *sq2* from [3] to exercise the MSER features and identify the new challenges introduced by the additional expressivity over the ABox queries. Basically, in special-case queries, we check the impact of `DISJOINTWITH` and meta-classes in a query. For this, like in [3], we have introduced a new class named *TypeOfProfessor* and make *FullProfessor*, *AssociateProfessor* and *AssistantProfessor* instances of this new class and also define *FullProfessor*, *AssociateProfessor* and *AssistantProfessor* to be disjoint from each other. Then, in *sq1* we are asking for all those y and z, where y is a professor, z is a type of professor and y is an instance of z. In *sq2*, we have asked for different pairs of professors.

The **MODEUS** ontologies describe the *Italian Public Debt* domain with information like financial liability or financial assets to any given contracts [14]. It comes with 8 queries. These queries are pure meta-queries as they span over several levels of the knowledge base. MODEUS ontologies are meta-modeling ontologies with meta-classes and meta-properties.

We have done the experiments on a Linux batch server, running Ubuntu 20.04.3 LTS (GNU/Linux 5.4.0-88-generic x86_64) on one AMD EPYC 7601 (32-Core CPU), 2.2 GHz, Turbo max. 3.2 GHz. The machine is equipped with 512 GB RAM and a 4TB hard disk. Java applications used OpenJDK 11.0.11 with a maximum heap size of 25 GB. During the course of the evaluation of the proposed variants we have used the time resource limitation as the benchmark setting on our data sets to examine the behavior of different variants. If not otherwise indicated, in both experiments, each benchmark had 3600 min (excluding the $\mathcal{K}$ generation time). For simplicity, we have not included queries that contain data properties in our experiments. We also have included the generation time of the hybrid knowledge base $\mathcal{K}$ including the loading of ontology and query, τ translation, module extraction, generating IR and translating queries. The circle dots in Figs. 2 and 3 explicitly represent the generation time of $\mathcal{K}$ for each query. It takes up to a few seconds depending on the ontology size. All material is available at https://git-ainf.aau.at/Haya.Qureshi/hkb.

We next report the results on LUBM(1) and LUBM(9) consisting of 1 and 9 departments, first with standard queries, followed by meta-queries.

We can observe that performance is satisfactory in Experiment 1. *HEXLite-owl-api-plugin* terminated within the timeout on all queries that we considered, with each of the four variants. With standard queries, $\mathcal{A}{-}\mathcal{T}$ runs generally faster than the other variants. This may seem a bit surprising, especially as $E{-}\mathcal{AT}$ is

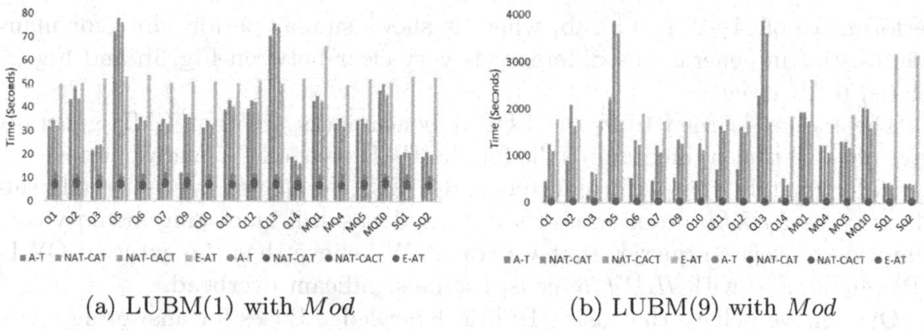

(a) LUBM(1) with *Mod* (b) LUBM(9) with *Mod*

Fig. 2. Experiment 1 with standard and meta queries

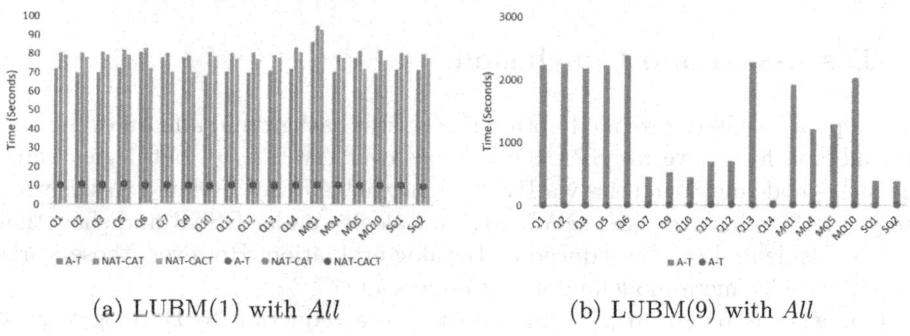

(a) LUBM(1) with *All* (b) LUBM(9) with *All*

Fig. 3. Experiment 2 with standard and meta queries

usually slower. We conjecture that this is due to the LME method, which avoids retrieving some ABox extensions. We also believe that in those cases, in which $E-\mathcal{AT}$ is faster, there is no or little reduction obtained by LME. The other two variants appear to introduce some extra overhead, which does not help for this kind of queries.

Looking at meta-queries, the picture is similar for the smaller LUBM(1) ontology, in which $\mathcal{A}-\mathcal{T}$ was usually fastest. However, the difference to $\mathcal{NAT}-\mathcal{CAT}$ and $\mathcal{NAT}-\mathcal{CACT}$ is comparatively small. $E-\mathcal{AT}$ is significantly slower here.

On the more sizeable LUBM(9) ontology, variant $\mathcal{NAT}-\mathcal{CACT}$ is consistently faster than the others on meta-queries. It appears that the combination of splitting the ontology and LME really pays off for these queries.

On the other hand we can observe from Experiment 2 that the Fig. 3a shows somewhat similar performance as of Fig. 2a in Experiment 1 with only minor differences in time. However in Fig. 3b, it can be seen that only $\mathcal{A}-\mathcal{T}$ was able to perform reasonably on the LUBM(9) ontology and the other two variants had timeouts on all the queries. This behaviour of the $\mathcal{NAT}$–$\mathcal{CAT}$ and $\mathcal{NAT}$–$\mathcal{CACT}$ variants in Experiment 2 on LUBM(9) confirmed our intuition that a lot of the performance gains in Experiment 1 are due to LME. What surprised us was the performance of $\mathcal{A}-\mathcal{T}$ in Fig. 3b, where it shows similar performance for many queries. But in general, the difference is very clear between Fig. 3b and Fig. 2b related to IR usage.

The lessons learned from the LUBM benchmarks are that LME generally pays off, and that a maximum split for the HKB pays off for meta-queries.

Unfortunately, the performance on MODEUS ontologies was not really satisfactory, as *HEXLite* took more than 5 h and was still increasing memory consumption on several queries that we tried. We believe that the external OWL-API plugin used with *HEXLite* causes some significant overhead.

Overall, we believe that using Hybrid Knowledge Bases for answering meta-queries does have good potential, but at the moment seems to be hampered by system performance for larger ontologies.

6 Discussion and Conclusion

In this paper we have provided both a theoretical background and an evaluation of a method for answering SPARQL queries over OWL 2 QL ontologies under the meta-modeling Semantics via Hybrid Knowledge Bases. Related to this work are [7,12,19], which consider OWL 2 QL ontologies under DSER and show that it is decidable and can be reduced to Datalog evaluation. However, those works do not consider meta-modeling or meta-querying.

Our work is based on [3], which presented a reduction to Datalog, argued its correctness, and analysed complexity, and [20], which evaluated performance using various systems supporting Datalog. The main contribution of our paper is that it generalizes these methods from Datalog to Hybrid Knowledge Bases, by translating only parts of the ontology to Datalog and by doing some of the reasoning with ontology tools. Our work also offers several variants of the transformation and evaluates the approach using a hybrid reasoner.

Our future work is to use other hybrid reasoners for answering meta-queries, for example *OntoDLV*[4] or *DLV2 with support for external python atoms*[5]. Also, we intend to expand our work by establishing new variants for transformations to hybrid knowledge bases.

References

1. Antoniou, G., et al.: Combining rules and ontologies: a survey. In: Reasoning on the Web with Rules and Semantics (2005)
2. Chen, W., Kifer, M., Warren, D.S.: HiLog: a foundation for higher-order logic programming. J. Logic Program. **15**(3), 187–230 (1993)
3. Cima, G., De Giacomo, G., Lenzerini, M., Poggi, A.: On the SPARQL metamodeling semantics entailment regime for OWL 2 QL ontologies. In: Proceedings of the 7th International Conference on Web Intelligence, Mining and Semantics, pp. 1–6 (2017)

[4] https://www.dlvsystem.it/dlvsite/ontodlv/.

[5] https://dlv.demacs.unical.it/.

4. Eiter, T., Fink, M., Ianni, G., Krennwallner, T., Redl, C., Schüller, P.: A model building framework for answer set programming with external computations. Theory Pract. Logic Program. **16**(4), 418–464 (2016)
5. Eiter, T., Ianni, G., Schindlauer, R., Tompits, H.: Effective integration of declarative rules with external evaluations for semantic-web reasoning. In: Sure, Y., Domingue, J. (eds.) ESWC 2006. LNCS, vol. 4011, pp. 273–287. Springer, Heidelberg (2006). https://doi.org/10.1007/11762256_22
6. Glimm, B.: Using SPARQL with RDFS and OWL entailment. In: Polleres, A., et al. (eds.) Reasoning Web 2011. LNCS, vol. 6848, pp. 137–201. Springer, Heidelberg (2011). https://doi.org/10.1007/978-3-642-23032-5_3
7. Gottlob, G., Pieris, A.: Beyond SPARQL under OWL 2 QL entailment regime: rules to the rescue. In: Twenty-Fourth International Joint Conference on Artificial Intelligence (2015)
8. Grosof, B.N., Horrocks, I., Volz, R., Decker, S.: Description logic programs: combining logic programs with description logic. In: Proceedings of the 12th International Conference on World Wide Web, pp. 48–57 (2003)
9. Guizzardi, G., Almeida, J.P.A., Guarino, N., de Carvalho, V.A.: Towards an ontological analysis of powertypes. In: JOWO@ IJCAI (2015)
10. Hitzler, P., Krotzsch, M., Rudolph, S.: Foundations of Semantic Web Technologies. CRC Press, Boca Raton (2009)
11. Jiménez-Ruiz, E., Grau, B.C., Sattler, U., Schneider, T., Berlanga, R.: Safe and economic re-use of ontologies: a logic-based methodology and tool support. In: Bechhofer, S., Hauswirth, M., Hoffmann, J., Koubarakis, M. (eds.) ESWC 2008. LNCS, vol. 5021, pp. 185–199. Springer, Heidelberg (2008). https://doi.org/10.1007/978-3-540-68234-9_16
12. Kontchakov, R., Rezk, M., Rodríguez-Muro, M., Xiao, G., Zakharyaschev, M.: Answering SPARQL queries over databases under OWL 2 QL entailment regime. In: Mika, P., et al. (eds.) ISWC 2014. LNCS, vol. 8796, pp. 552–567. Springer, Cham (2014). https://doi.org/10.1007/978-3-319-11964-9_35
13. Lenzerini, M., Lepore, L., Poggi, A.: A higher-order semantics for OWL 2 QL ontologies. In: Description Logics (2015)
14. Lenzerini, M., Lepore, L., Poggi, A.: Metaquerying made practical for OWL 2 QL ontologies. Inf. Syst. **88**, 101294 (2020)
15. Lenzerini, M., Lepore, L., Poggi, A.: Metamodeling and metaquerying in OWL 2 QL. Artif. Intell. **292**, 103432 (2021)
16. Meditskos, G., Bassiliades, N.: Rule-based OWL ontology reasoning systems: implementations, strengths, and weaknesses. In: Handbook of Research on Emerging Rule-Based Languages and Technologies: Open Solutions and Approaches, pp. 124–148. IGI Global (2009)
17. Motik, B.: On the properties of metamodeling in OWL. In: Gil, Y., Motta, E., Benjamins, V.R., Musen, M.A. (eds.) ISWC 2005. LNCS, vol. 3729, pp. 548–562. Springer, Heidelberg (2005). https://doi.org/10.1007/11574620_40
18. Motik, B., Rosati, R.: Reconciling description logics and rules. J. ACM (JACM) **57**(5), 1–62 (2008)
19. Poggi, A.: On the SPARQL direct semantics entailment regime for OWL 2 QL. In: Description Logics (2016)
20. Qureshi, H.M., Faber, W.: An evaluation of meta-reasoning over OWL 2 QL. In: Moschoyiannis, S., Peñaloza, R., Vanthienen, J., Soylu, A., Roman, D. (eds.) RuleML+RR 2021. LNCS, vol. 12851, pp. 218–233. Springer, Cham (2021). https://doi.org/10.1007/978-3-030-91167-6_15

Solving Vehicle Equipment Specification Problems with Answer Set Programming

Raito Takeuchi[1(✉)], Mutsunori Banbara[1], Naoyuki Tamura[2], and Torsten Schaub[3]

[1] Nagoya University, Furo-cho, Chikusa-ku, Nagoya 464-8601, Japan
{takeuchi.raito,banbara}@nagoya-u.jp

[2] Kobe University, Rokko-dai Nada-ku Kobe Hyogo 657-8501, Japan
tamura@kobe-u.ac.jp

[3] Universität Potsdam, August-Bebel-Strasse, 14482 Potsdam, Germany
torsten@cs.uni-potsdam.de

Abstract. We develop an approach to solving mono- and multi-objective vehicle equipment specification problems considering the corporate average fuel economy standard (CAFE problems, in short) in automobile industry. Our approach relies upon Answer Set Programming (ASP). The resulting system *aspcafe* accepts a CAFE instance expressed in the orthogonal variability model format and converts it into ASP facts. In turn, these facts are combined with an ASP encoding for CAFE solving, which can subsequently be solved by any off-the-shelf ASP systems. To show the effectiveness of our approach, we conduct experiments using a benchmark set based on real data provided by a collaborating Japanese automaker.

Keywords: Vehicle equipment specification · Corporate average fuel economy standard · Answer set programming · Multi-objective optimization

1 Introduction

Vehicle equipment specification is the combination of vehicle models and equipments in automobile catalogs. The task of finding such combinations depends on various factors, including laws and regulations of countries or regions, market characteristics, customer preferences, competitors, costs, dependency between equipments, and many others. A great deal of efforts by experienced engineers has been made so far to perform the task. However, the task has become increasingly hard because of the recent globalization of automobile industry. It is therefore important to develop efficient implementations for automatically finding high-quality vehicle equipment specifications.

The vehicle equipment specification problem is generally defined as the task of finding combinations of equipment types and equipment options for each vehicle model, subject to a given set of constraints on variability, dependency, and fuel

M. Hanus and D. Inclezan (Eds.): PADL 2023, LNCS 13880, pp. 232–249, 2023.
https://doi.org/10.1007/978-3-031-24841-2_15

economy. Each equipment type is a different kind of equipment such as transmission and drive type. Each equipment option is a specific option such as CVT, HEV, 2WD, and 4WD. The objective of the problem is to find feasible solutions maximizing the expected sales volume. *The multi-objective vehicle equipment specification problem* involves multiple objective functions that are optimized simultaneously, such as maximizing expected sales volume, minimizing the number of equipment options, etc. The goal of this problem is to find the Pareto front (viz. the set of Pareto optimal solutions [12]).

There is a strong relationship between automobile industry and environmental issues. Fuel economy regulations have been tightened in many countries due to recent international movements against global warming. *The Corporate Average Fuel Economy standard* (CAFE standard; [22,25]) is a regulation for automobiles first enacted by the U.S. Congress in 1975. The CAFE standard has been adopted in the United States and the European Union, and was recently introduced in Japan. The standard requires each automaker to achieve a target for the sales-weighted fuel economy of its entire automobile fleet in each model year. That is, even if a particular kind of vehicle models does not meet it, the automakers can achieve the standards by improving the fuel economy of other models.

Answer Set Programming (ASP; [7,19,23]) is a declarative programming paradigm, combining a rich modeling language with high performance solving capacities. ASP is well suited for modeling combinatorial optimization problems, and has been successfully applied in diverse areas of artificial intelligence and other fields over the last decade [13]. Moreover, recent advances in ASP open up promising directions to make it more applicable to real world problems: preference handling [8,33], constraint ASP [5,6], temporal ASP [9], domain heuristics [11,17], core-guided optimization [1,3], multi-shot ASP solving [18], and many others. Such advances encourage researchers and practitioners to use ASP for solving practical applications. In this paper, we tackle mono- and multi-objective vehicle equipment specification problems based on the CAFE standard as a fuel economy regulation. We refer to them as CAFE problems. The CAFE problems are real-world applications in automobile industry.

This paper describes an approach to solving the CAFE problems based on ASP. The resulting system *aspcafe* reads a CAFE instance expressed in the Orthogonal Variability Model (OVM; [24]) format and converts it into ASP facts. In turn, these facts are combined with an ASP encoding for CAFE solving, which can subsequently be solved by any off-the-shelf ASP systems, in our case *clingo* [15]. We also extend the *aspcafe* encoding to finding Pareto optimal solutions of multi-objective CAFE problems by utilizing the *asprin* system [8], an enhancement of *clingo* with preferences handling. To show the effectiveness of our approach, we conduct experiments using a benchmark set based on real data provided by a collaborating Japanese automaker.

The declarative approach of ASP has several advantages. First, the problems are modeled as first-order logic programs and then solved by general-purpose ASP systems, rather than dedicated implementations. Second, the elaboration

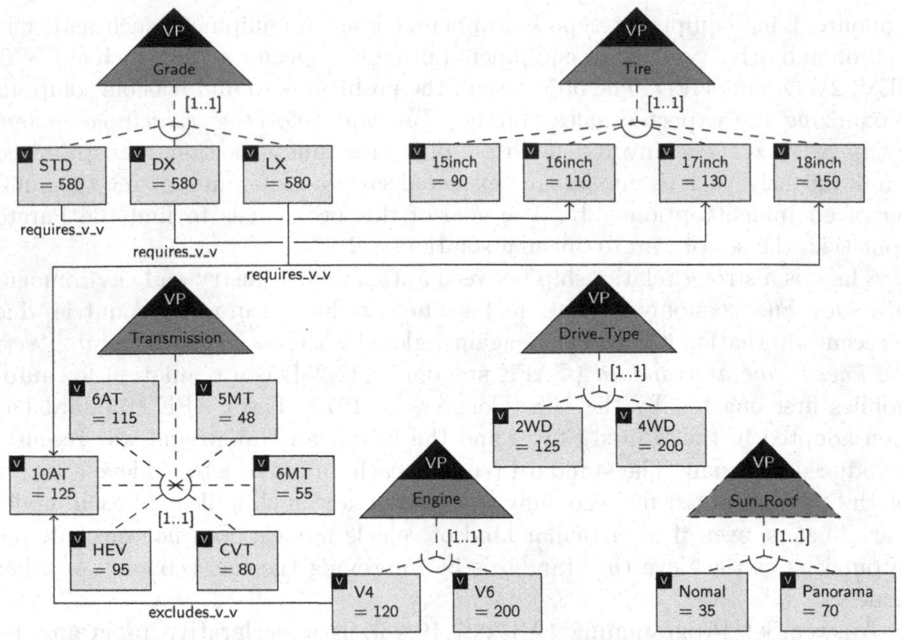

Fig. 1. Example of CAFE problem instance

tolerance of ASP enables domain experts to maintain and modify encodings easily. Finally, it is easy to experiment with the advanced ASP techniques listed above.

From the viewpoint of ASP, the *aspcafe* approach provides insight into advanced modeling techniques for CAFE problems. The main features of *aspcafe* encoding are the following ones: (i) a collection of compact ASP encodings for CAFE solving, (ii) an optimized encoding in view of reducing domain values, (iii) easy extension to multi-objective optimization based on Pareto optimality as well as lexicographic ordering. The *aspcafe* encoding is written in the *clingo* language. We give a brief introduction to ASP's syntax in Sect. 3, but for more details, refer the reader to the literature [15,16].

2 CAFE Problems

As mentioned, CAFE problems are vehicle equipment specification problems based on the CAFE standard. The inputs of a CAFE problem are a problem instance, the number of vehicle models (n), the CAFE standard value (t), a fuel economy curve, and an expected sales volume curve. The problem instance mainly consists of equipment types and equipment options. The CAFE problem is defined as the task of finding combinations of them for each vehicle model, subject to a given set of constraints on variability, dependency, and fuel economy.

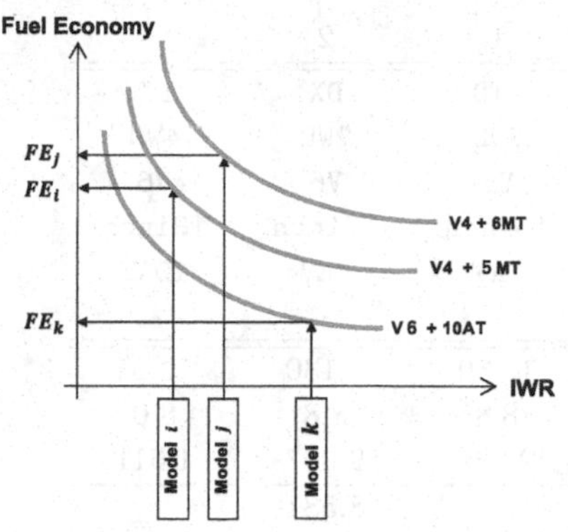

Fig. 2. Correlation curve for fuel economy

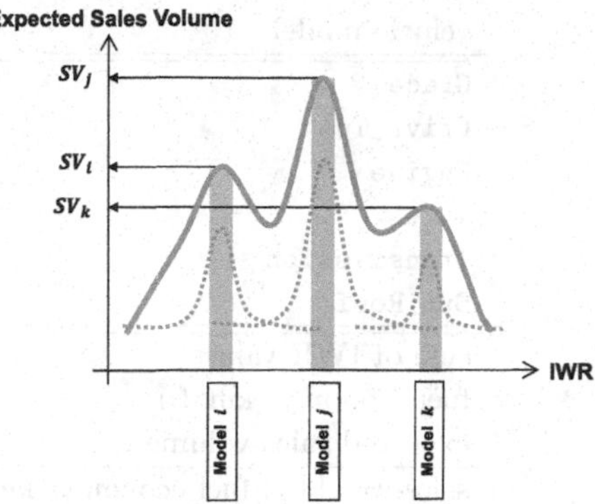

Fig. 3. Distribution curve for expected sales volume

The objective of the problem is to find feasible solutions of maximum expected sales volume.

Figure 1 shows an example of a CAFE problem instance. The example is expressed in the OVM format, a graphical modeling language widely used in software product line engineering [24]. Each triangle node tagged with `VP` represents an equipment type. Mandatory types are represented by solid nodes, while non-mandatory types such as `Sun_Roof` are represented by dashed nodes. Each rectangle node tagged with `V` represents an equipment option. The value assigned to each option indicates the *Inertial Working Rating* (IWR), which intuitively corresponds to its weight. The variability relation between types and options is represented in term of alternative choice (dashed lines) and multiplicity ($[lb..ub]$). For each type, the lb and ub respectively indicate the lower and upper bounds for the number of selectable options. The dependency relation between options is represented by solid arrows, and can either be a requirement dependency (`requires_v_v`) or an exclusion dependency (`excludes_v_v`). This example consists of 6 equipment types, 19 equipment options, and 5 dependency relations. For the multiplicity, every type must have exactly one option. For instance, `Transmission` type can select exactly one of 6 options: `6AT`, `10AT`, `HEV`, `CVT`, `6MT`, and `5MT`. Due to the `excludes_v_v` dependency, `10AT` and `V4` cannot be selected simultaneously.

The constraints of CAFE problem are as follows.

Variability Constraints: For each vehicle model and for each type, the number of options is within the upper and lower bounds (for the multiplicity).

Dependency Constraints: For each vehicle model, the dependency relation between types and options must be satisfied.

Fuel Economy Constraints: Let FE_g and SV_g respectively be the fuel economy and expected sales volume of each vehicle model g ($1 \leq g \leq n$). The

vehicle model	1	2	3
Grade	STD	DX	LX
Drive_Type	2WD	2WD	4WD
Engine	V6	V6	V6
Tire	16inch	17inch	18inch
Transmission	6AT	HEV	10AT
Sun_Roof	-	-	-
sum of IWR values	1,130	1,130	1,255
fuel economy (km/L)	8.8	8.8	8.0
expected sales volume	2,007	2,007	1,511
sales-weighted fuel economy (km/L)		8.581	
sum of expected sales volume		5,525	

Fig. 4. Example of an optimal solution

following CAFE standard must be satisfied[1].

$$\frac{\sum_{g=1}^{n} FE_g \cdot SV_g}{\sum_{g=1}^{n} SV_g} \geq t \tag{1}$$

This constraint expresses that the sales-weighted fuel economy of n models (left-hand side) is greater than or equal to the CAFE standard value t (right-hand side). For each model g, the IWR value of g is the sum of IWR values of options implemented for g. The fuel economy FE_g can be obtained from the IWR value of g through a correlation curve between IWR values and fuel economy as in Fig. 2. In the same way, the expected sales volume SV_g can be obtained through a distribution curve between IWR values and expected sales volume as in Fig. 3, which is a typical distribution curve based on three segments.

Suppose that the number of vehicle models is $n = 3$ and the CAFE standard value is $t = 8.5$ km/L. Figure 4 shows an example of an optimal solution for the problem instance in Fig. 1. The fuel economy of three vehicle models are 8.8, 8.8, and 8.0 km/L in order. Although the vehicle model 3 does not meet this criterion, the overall solution satisfies the CAFE standard, since the sales-weighted fuel economy of the three models (8.581 km/L) is greater than $t = 8.5$. We can also observe that all dependency constraints are satisfied. For example, model 3 implements `18inch` tires and `10AT` transmission, which are both required by `LX` grade.

3 Answer Set Programming

Answer Set Programming (ASP; [7,19,23]) is a declarative programming paradigm widely used in knowledge representation and reasoning. In ASP, prob-

[1] By demand of our collaborating automaker, we use a simple CAFE standard.

lems are modeled as *logic programs* which are finite sets of *rules* of the following form:

```
a0 :- a1,...,am,not am+1,...,not an.
```

where $0 \leq m \leq n$ and each a_i is a propositional *atom* for $0 \leq i \leq n$. The left of ':-' is called *head* and the right is called *body*. The connectives ':-', ',', and 'not' denote 'if', 'conjunction', and 'default negation', respectively. Each rule is terminated by a period '.'. A *literal* is an atom a or not a. Intuitively, the rule means that a_0 must be true if $a_1, \dots, a_m$ are true and if $a_{m+1}, \dots, a_n$ are false. Semantically, a logic program induces a collection of so-called *answer sets*, which are distinguished models of the program based on stable model semantics [19]. ASP rules have two special cases:

```
a0.                                    (fact)
:- a1,...,am,not am+1,...,not an.      (integrity constraint)
```

A rule of empty body is called a *fact*, and we often omit ':-'. A fact is always true and belongs to every answer set. A rule of empty head is called an *integrity constraint*, which means that the conjunction of literals in its body must not hold. Integrity constraints can be used to filter solution candidates.

There have been several language extensions to promote the use of ASP in practice [10]. First, rules with first-order variables are regarded as shorthand for the set of their ground instances. Further language constructs include *conditional literals* and *cardinality constraints*. The former are of the form $\ell_0 : \ell_1, \dots, \ell_m$ and the latter can be written as ℓb $\{c_1; \dots; c_n\}$ ub, where all ℓ_i are literals for $0 \leq i \leq m$, and all c_j are conditional literals for $1 \leq j \leq n$; ℓb and ub indicate lower and upper bounds on the number of satisfied literals in the cardinality constraint. Similarly, *weighted cardinality constraints* are of the form #sum $\{\boldsymbol{t}_1 : \boldsymbol{\ell}_1; \dots; \boldsymbol{t}_n : \boldsymbol{\ell}_n\}$, where all $\boldsymbol{t}_i$ and $\boldsymbol{\ell}_i$ are tuples of terms and literals respectively. Note that the weight refers to the first element of a term tuple. For instance, a useful shortcut expression like `N = #sum {Y,X:a(X,Y)}` binds `N` to the sum of weights `Y` of satisfied literals `a(X,Y)`. In addition to the `#sum` function, *clingo* supports `#min` for the minimum weight and `#max` for the maximum weight. Finally, maximize functions for optimization are expressed as #maximize $\{\boldsymbol{t}_1 : \boldsymbol{\ell}_1; \dots; \boldsymbol{t}_n : \boldsymbol{\ell}_n\}$.

4 The *aspcafe* Approach

In our declarative approach, the *aspcafe* solver accepts a CAFE instance in OVM format and converts it into ASP facts. In turn, these facts are combined with a first-order encoding for CAFE solving, which can subsequently be solved by any general-purpose ASP system such as *clingo* (see Fig. 5).

ASP Fact Format. Listing 1 shows the ASP facts corresponding to the CAFE instance in Fig. 1. The atoms `vp_def/1` and `v_def/3` represent equipment types and equipment options, respectively. For instance, the atom `vp_def("Engine")` represents that `Engine` is an equipment type. The atom

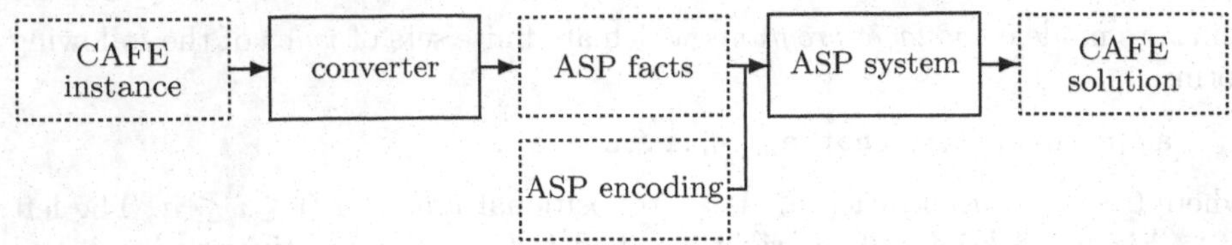

Fig. 5. Architecture of the *aspcafe* solver

```
vp_def("Grade").
v_def("STD", "Grade", 580).  v_def("DX", "Grade", 580).  v_def("LX", "Grade", 580).

vp_def("Drive_Type").
v_def("2WD", "Drive_Type", 125).        v_def("4WD", "Drive_Type", 200).

vp_def("Engine").
v_def("V4", "Engine", 120).             v_def("V6", "Engine", 200).

vp_def("Sun_Roof").
v_def("Nomal",     "Sun_Roof", 35).     v_def("Pnrorama", "Sun_Roof", 70).

vp_def("Tire").
v_def("15_inch_Tire", "Tire",  90).     v_def("16_inch_Tire", "Tire", 110).
v_def("17_inch_Tire","Tire", 130).      v_def("18_inch_Tire", "Tire", 150).

vp_def("Transmission").
v_def("6AT",   "Transmission", 115).    v_def("10AT", "Transmission", 125).
v_def("HEV",   "Transmission",  95).    v_def("CVT",  "Transmission",  80).
v_def("5MT",   "Transmission",  48).    v_def("6MT",  "Transmission",  55).

require_v_v("STD", "16_inch_Tire").     require_v_v("DX",  "17_inch_Tire").
require_v_v("LX",  "18_inch_Tire").     require_v_v("LX",  "10AT").
exclude_v_v("V4",  "10AT").

require_vp("Drive_Type").  require_vp("Engine").        require_vp("Grade").
require_vp("Tire").        require_vp("Transmission").

group(1). group(2). group(3).
```

Listing 1. ASP facts for the CAFE instance in Fig. 1 with the number of models $n = 3$

`v_def("V4","Engine",120)` represents that `V4` is an equipment option of `Engine` type and its IWR value is 120. The atoms `require_v_v/2` and `exclude_v_v/2` represent dependency relation between options. For instance, the atom `require_v_v("LX","10AT")` represents that the `LX` option of `Grade` type requires the `10AT` option of `Transmission` type. In addition, the atom `require_vp/1` represents the mandatory type, and `group/1` represents the vehicle model.

Basic Encoding. Listing 2 shows an ASP encoding for solving CAFE problems. The atom `vp(VP,G)` is intended to represent that the vehicle model `G` implements the equipment type `VP`. The rule in Line 1 introduces the atom `vp(VP,G)` as a solution candidate for each model `G` and each type `VP`. The rule in Line 2 enforces that, for each model `G`, the atom `vp(VP,G)` must hold if `VP` is a mandatory type. The rule in Line 5 represents the variability constraints. The atom `v(V,G)`

```
1  { vp(VP,G) } :- vp_def(VP), group(G).
2  :- not vp(VP,G), require_vp(VP), group(G).

4  % variability constraints
5  1 { v(V,G) : v_def(V,VP,_) } 1 :- vp(VP,G).

7  % fuel economy constraints
8  iwr(S,G) :- S = #sum { IWR,V : v(V,G), v_def(V,_,IWR) }, group(G).
9  fe(FE,G) :- iwr(S,G), fe_map(S,FE).
10 sv(SV,G) :- iwr(S,G), sv_map(S,SV).
11 :- not 0 #sum { (FE-t)*SV,FE,SV,G : fe(FE,G), sv(SV,G) }.

13 % dependency constraints (requirement)
14 :- require_v_v(V1,V2), v(V1,G), not v(V2,G).
15 :- require_v_vp(V,VP), v(V,G), not vp(VP,G).
16 :- require_vp_v(VP,V), vp(VP,G), not v(V,G).
17 :- require_vp_vp(VP1,VP2), vp(VP1,G), not vp(VP2,G).

19 % dependency constraints (exclusion)
20 :- exclude_v_v(V1,V2), v(V1,G), v(V2,G).
21 :- exclude_v_vp(V,VP), v(V,G), vp(VP,G).
22 :- exclude_vp_v(VP,V), vp(VP,G), v(V,G).
23 :- exclude_vp_vp(VP1,VP2), vp(VP1,G), vp(VP2,G).

25 % objective function
26 #maximize { SV,G : sv(SV,G) }.
```

Listing 2. Basic encoding

is intended to represent that the model G implements the option V. This rule enforces that the model G implements exact one option V of type VP if vp(VP,G) holds[2]. The rules in Line 8–11 represent the fuel economy constraints. The atom iwr(S,G) represents that S is the IWR value of vehicle model G, that is, S is the sum of IWR values of options implemented for G. The atom fe(FE,G) represents that the fuel economy of G is FE, which is obtained from the IWR value S of G (iwr(S,G)) through a correlation curve (fe_map(S,FE)) between IWR values and fuel economy. In the same manner, the atom sv(SV,G) represents that the expected sales volume of G is SV. The fuel economy constraint (1) in Sect. 2 is transformed into

$$\sum\nolimits_{g=1}^{n} (FE_g - t) \cdot SV_g \geq 0$$

and is then represented in ASP by a weighted cardinality constraint in Line 11, where the constant t represents the CAFE standard value t. The rules in Line

[2] For simplicity, the number of equipment options is limited to exactly one, but can be easily modified to arbitrary bounds for multiplicity.

```
ub_vp(UB,VP) :- UB = #max { IWR,V : v_def(V,VP,IWR) }, vp_def(VP).
lb_vp(LB,VP) :- LB = #min { IWR,V : v_def(V,VP,IWR) }, vp_def(VP).

ub_iwr(S) :- S = #sum { UB,VP : ub_vp(UB,VP) }.
lb_iwr(S) :- S = #sum { LB,VP : lb_vp(LB,VP), require_vp(VP) }.

iwr(S,G) :- S = #sum { IWR,V : v(V,G), v_def(V,_,IWR) },
            LB <= S, S <= UB, lb_iwr(LB), ub_iwr(UB), group(G).
```

Listing 3. Optimized code in view of reducing the domain values

14–17 represent the requirement constraints on equipment types and options. The rule in Line 14 enforces that, for each vehicle model `G`, `G` implements the option `V2` if `G` implements `V1` and `require_v_v(V1,V2)` holds. The rule in Line 15 enforces that, for each vehicle model `G`, `G` implements the type `VP` if `G` implements the option `V` and `require_v_vp(V,VP)` holds. Conversely, the rule in Line 16 enforces that, for each vehicle model `G`, `G` implements the option `V` if `G` implements the type `VP` and `require_vp_v(VP,V)` holds. The rule in Line 17 enforces that, for each vehicle model `G`, `G` implements the type `VP2` if `G` implements `VP1` and `require_vp_vp(VP1,VP2)` holds. Similarly, the exclusion constraints are represented by integrity constraints in Line 20–23. The objective of the CAFE problem is to maximize the expected sales volume, as defined by the `#maximize` statement in Line 26.

Optimized Encoding. The basic encoding in Listing 2 can concisely represent the constraints of the CAFE problem, but there is still room for improvement. In particular, the rule in Line 8

```
iwr(S,G) :- S = #sum { IWR,V : v(V,G), v_def(V,_,IWR) }, group(G).
```

is very expensive for grounding and solving. Once again, the auxiliary atom `iwr(S,G)` represents that `S` is the sum of IWR values of options implemented for the model `G`. In the rule, the domain of `S` is $0 \leq \mathtt{S} \leq \sum_{j \in V} w_j$, where V is the set of equipment options and w_j is the IWR value of option $j \in V$. However, this domain is quite naive, since the lower bound of zero means no option and the upper bound of $\sum_{j \in V} w_j$ means the total sum of IWR values of all possible options. The grounding process therefore generates a lot of unnecessary grounded rules, which can cause performance degradation. This issue can be resolved by considering, for each type i, the minimum and maximum IWR values of options which i can select, whether i is a mandatory type or not, and the multiplicity of variability constraints. The improved bounds are as follows:

$$\sum_{i \in VP^*} \min_{j \in V_i} w_j \leq \mathtt{S} \leq \sum_{i \in VP} \max_{j \in V_i} w_j$$

where VP is the set of equipment types and VP^* is the set of mandatory equipment types. Each V_i is the set of equipment options that the type $i \in VP$ can select. Listing 3 shows an optimized code for calculating `iwr(S,G)` based on the

```
1 % maximize the expected sales volume
2 #maximize { SV@2,G : sv(SV,G) }.
3
4 % minimize the number of equipment options
5 used_v(V) :- v(V,G).
6 #minimize { 1@1,V : used_v(V) }.
```

Listing 4. Lexicographic optimization of two criteria

improved domain. The optimized encoding is obtained from the basic encoding by replacing the rule in Line 8 with the code in Listing 3. The optimized encoding can be expected to be more effective on large problems than the basic one, since it can reduce the number of rules after grounding.

	Solution 1			Solution 2			Solution 3		
vehicle model	1	2	3	1	2	3	1	2	3
`Grade`	STD	DX	LX	STD	DX	LX	STD	DX	LX
`Drive_Type`	4WD	2WD	4WD	2WD	2WD	4WD	2WD	2WD	4WD
`Engine`	V4	V6	V6	V6	V6	V6	V6	V6	V6
`Tire`	16	17	18	16	17	18	16	17	18
`Transmission`	5MT	HEV	10AT	CVT	HEV	10AT	6AT	HEV	10AT
`Sun_Roof`	Panorama	-	-	Nomal	-	-	-	-	-
sum of IWR values	1,128	1,130	1,255	1,130	1,130	1,255	1,130	1,130	1,255
fuel economy (km/L)	8.9	8.8	8.0	8.8	8.8	8.0	8.8	8.8	8.0
expected sales volume	2,007	2,007	1,511	2,007	2,007	1,511	2,007	2,007	1,511
sales-weighted FE (km/L)	8.6			8.5			8.5		
sum of expected SV	5,525			5,525			5,525		
#options	14			13			12		

Fig. 6. Optimal solutions with the same expected sales volume

Multi-criteria Optimization Based on Lexicographic Ordering. The ASP system *clingo* can deal with lexicographic optimization, which allows for optimizing multiple criteria in the lexicographic order based on their priorities. We here extend the *aspcafe* encoding for two-criteria optimization. One criterion is the maximization of the expected sales volume as before and the other is the minimization of the number of equipment options. The latter aims at reducing the number of production lines as well as at promoting mass production. The extension can be easily done by adding priorities to optimization statements, as can be seen in Listing 4. The `#maximize` statement in Line 2 is the same as the one of the basic encoding (Line 26 in Listing 2), except that the priority `@2` is added. The priority is zero by default, and greater priorities are more significant than smaller ones. The auxiliary atom `used_v(V)` in Line 5 represents

```
% maximize the expected sales volume
#preference (max_sv, more(weight)) { SV,G :: sv(SV,G) }.

% minimize the number of equipment options
used_v(V) :- v(V,G).
#preference (min_op, less(weight)) { 1,V :: used_v(V) }.

% Pareto optimization
#preference (all, pareto) { **max_sv; **min_op }.

% optimization statement
#optimize(all).
```

Listing 5. Pareto optimization of two objective functions

that option `V` is implemented by any vehicle model. Minimizing the number of options is represented by the `#minimize` statement with the priority `@1` in Line 6.

Figure 6 shows all possible optimal solutions of the problem instance in Fig. 1 when the number of vehicle models is $n = 3$ and the CAFE standard value is $t = 8.5$ km/L. In the basic encoding, there exists three optimal solutions with the same expected sales volume (5,525 units), as can be seen in Fig. 6. On the other hand, the extension of basic encoding with two-criteria optimization in Listing 4 produces only Solution 3 as an optimal solution, since the number of options is smaller than the other candidate solutions.

5 Finding Pareto Optimal Solutions of Multi-objective CAFE Problems

We here extend the *aspcafe* encoding to find the Pareto front (viz. the set of Pareto optimal solutions) of multi-objective CAFE Problems. This involves multiple objective functions that are considered separately and optimized simultaneously. This extension can be easily done by utilizing the *asprin* system [8], an enhancement of *clingo* with preference handling. In *asprin*, a preference is defined by the following statement

`#preference(`s`,`t`){`e_1`;...;`e_n`}.`

where s is the preference name, t is the preference type, and e_i is the preference element. *asprin* allows for new user-defined preference types. In addition, it provides some built-in preference types, such as `subset`, `less(weight)`, `more(weight)`, and `pareto`. The preference types `less(weight)` and `more(weight)` are similar to `#minimize` and `#maximize` statements. They deal with sets of preference elements of the form

w`,`t`::`F

	Solution 1			Solution 2		
vehicle model	1	2	3	1	2	3
Grade	STD	DX	LX	STD	DX	LX
Drive_Type	2WD	2WD	4WD	2WD	2WD	4WD
Engine	V6	V6	V6	V6	V6	V6
Tire	16inch	17inch	18inch	16inch	17inch	18inch
Transmission	6AT	HEV	10AT	10AT	HEV	10AT
Sun_Roof	-	-	-	-	-	-
sum of expected SV	5,525			5,475		
#options	12			11		

	Solution 3			Solution 4		
vehicle model	1	2	3	1	2	3
Grade	STD	DX	LX	STD	DX	LX
Drive_Type	2WD	2WD	2WD	2WD	2WD	2WD
Engine	V6	V6	V6	V6	V6	V6
Tire	16inch	17inch	18inch	16inch	17inch	18inch
Transmission	10AT	HEV	10AT	10AT	10AT	10AT
Sun_Roof	-	-	-	-	-	-
sum of expected SV	5,135			4,723		
#options	10			9		

Fig. 7. Pareto optimal solutions that are not dominated each other

where w is an integer, t is a term tuple, and F is a Boolean formula. Preference type `pareto` deals with sets of atoms of the form

`**`s

where s is a preference name. The main feature of *asprin* is that it allows for flexible preference handling by combining multiple preference statements. Finally, an optimization statement

`#optimize(`s`).`

enforces that *asprin* computes answer sets that are optimal with respect to the preference s. For more details of *asprin*, refer the reader to the literature [15].

Extended Encoding. Listing 5 shows an *asprin* encoding for Pareto optimization of two objective functions. The preference statement `max_sv` in Line 2 aims at maximizing the expected sales volume. In a similar way, `min_op` in Line 6 aims at minimizing the number of options. These two preferences are combined according to the `pareto` type in Line 9. And finally, the combined preference `all` is declared subject to optimization in Line 12. The extended encoding is obtained from the basic one by replacing the optimization statement in Line 26 (in Listing 2) with the code in Listing 5.

Figure 7 shows all possible Pareto optimal solutions of the problem instance in Fig. 1 when the number of vehicle models is $n = 3$ and the CAFE standard value is $t = 8.5$ as before. There are four Pareto optimal solutions (Solution 1–4), which are not dominated by each other. For instance, the expected sales volume for Solution 1 is larger than the one of Solution 2, but the number of options for Solution 2 is smaller than the one of Solution 1. Solution 1 is an extreme solution that is the same as Solution 3 in Fig. 6 obtained by lexicographic optimization. As can be seen in Fig. 7, the extended encoding succeeds in finding well-balanced solutions like Solution 2–3, which are not obtained by lexicographic optimization.

6 Experiments

Table 1. Benchmark instances provided by a collaborative Japanese automaker

Instance	#Types	#Options	#Dependency constraints
small	8	21	4
medium	86	229	147
big	315	1,337	0

To evaluate the effectiveness of our declarative approach, we carry out experiments on three benchmark instances based on real data, provided by a collaborating Japanese automaker. Table 1 shows the instance name and the number of equipment types, options, and dependency constraints, from left to right for each instance. Among them, big is very large and consists of 315 types and 1,337 options. Our empirical analysis considers all combinations (15 in total) of three instances and five different CAFE standard values (8.5, 9.0, 9.5, 10.0, and 10.5 km/L). We ran them on a Mac with an Intel Core i7 3.2 GHz processor and 64 GB RAM. We used *clingo*-5.5.2 and *asprin*-3.1.1 for our experiments. We imposed a time limit of 3 h for mono-objective CAFE problems, and of 30 h for multi-objective ones. The number of vehicle models is $n = 3$.

Mono-objective CAFE Problems. We evaluate the *aspcafe* encodings in Sect. 4 for solving mono-objective CAFE problems. At first, Table 2 compares the basic encoding with the optimized encoding. The columns display in order the instance name, the CAFE standard value t in kilometer per liter, and the expected sales volume for each encoding. The better bounds of the last two columns are highlighted in bold. If followed by a superscript '*', these bounds indicate that *aspcafe* proved the optimality of the obtained bounds. That is, the bound '6,021*' in the first row indicates that we found and proved an optimal value 6,021 of the small instance when $t = 8.5$. The symbol 'UNSAT' indicates that we proved the unsatisfiability. The symbol 'N.A' indicates that *clingo* cannot find any feasible solutions within the time limit.

As can be seen in Table 2, the optimized encoding produced the best bounds for 14 combinations and proved that 4 of them are optimal. In contrast, the basic encoding produced the best bounds for 7 combinations. While the number

Table 2. Comparison of different *aspcafe* encodings on the obtained bounds

Instance	CAFE standard value t (km/L)	Expected sales volume	
		Basic encoding	Optimized encoding
small	8.5	**6,021***	**6,021***
	9.0	**5,007***	**5,007***
	9.5	**2,688***	**2,688***
	10.0	**1,318***	**1,318***
	10.5	UNSAT	UNSAT
medium	8.5	6,010	**6,021**
	9.0	**5,595**	**5,595**
	9.5	**3,430**	**3,430**
	10.0	2,245	**2,250**
	10.5	**1,845**	**1,845**
big	8.5	N.A	**3,877**
	9.0	1,038	**4,623**
	9.5	688	**3,121**
	10.0	1,634	**2,100**
	10.5	538	**1,529**
#best bounds		7	14

Table 3. The CPU times of finding optimal solutions for the small instance

Instance	CAFE standard value t (km/L)	CPU time (s)	
		Basic encoding	Optimized encoding
small	8.5	45.547	**33.006**
	9.0	24.687	**12.312**
	9.5	**25.439**	47.863
	10.0	437.215	**0.375**
	10.5	1072.848	**0.072**
average CPU times		321.147	**18.744**

Table 4. Comparison of different *aspcafe* encodings on #constraints after grounding

Instance	CAFE standard value t (km/L)	#Constraints after grounding	
		Basic encoding (ratio)	Optimized encoding (ratio)
small		39,682 (1.00)	**14,356** (0.36)
medium	9.0	65,035 (1.00)	**30,574** (0.47)
big		302,428 (1.00)	**36,516** (0.12)

of optima obtained is identical, the optimized encoding was able to produce much better bounds for all combinations of the `big` instance, than the basic encoding.

Next, Table 3 shows the CPU times of finding optimal solutions (or deciding the unsatisfiability) for each combination of the `small` instance. The better times of the last two columns are highlighted in bold. The table below shows the average CPU time for each encoding. We can observe that the optimized encoding is faster than the basic encoding for all combinations except $t = 9.5$. The optimized encoding is 17 times faster on average than the basic one. In the optimized encoding, the time decreases with increasing CAFE standard value, except for $t = 9.5$. It would be an important future work to investigate the reason why we could not observe the performance advantage of the optimized encoding for $t = 9.5$.

Finally, Table 4 shows the number of constraints after grounding for each instance with $t = 9.0$. The smaller size of the last two columns are highlighted in bold. The numbers in parentheses of the last two columns are the ratio of the basic encoding to the optimized encoding. We can observe that the optimized encoding drastically reduces the number of constraints compared to the basic encoding. In particular, it succeeds in reducing the number of constraints by 88% for the `big` instance. Consequently, our declarative approach scales to large instances, as demonstrated in Table 2 by the fact that the optimized encoding was able to find quality bounds for the `big` instance.

Table 5. Benchmark results of finding Pareto optimal solutions

Instance	CAFE standard value t (km/L)	#Feasible solutions	#Pareto optimal solutions	CPU time (s)
`small`	8.5	41,217	8*	34.697
	9.0	3,961	5*	1,102.318
	9.5	374	1*	93,336.660
	10.0	28	1*	1.959
	10.5	0	0	0.295

Multi-objective CAFE problems. We evaluate the extended encoding in Sect. 5 for solving multi-objective CAFE problems. Table 5 shows the results of finding the Pareto front, that is, the set of Pareto optimal solutions. The columns display in order the instance name, the CAFE standard value t in kilometer per liter, the number of feasible solutions, the number of Pareto optimal solutions, and the CPU time of finding the Pareto front. The numbers followed by a superscript '*', indicate that *asprin* was able to enumerate all Pareto optimal solutions. That is, the first row indicates that we found the Pareto front of the `small` instance with $t = 8.5$ in 34 s, which consists of 8 Pareto optimal solutions. As can be seen in Table 5, the extended encoding was able to find the

Pareto front of all combinations of the `small` instance. On the other hand, we met the difficulty of finding Pareto optimal solutions for both `medium` and `big` instances. This shows a limitation of our approach at present. To overcome this issue, it is worth implementing an advanced technique of finding the Pareto front based on P-minimal model generation [30].

7 Conclusion

We proposed an ASP-based approach to solving mono- and multi-objective CAFE problems, which are real-world applications in automobile industry. The resulting system *aspcafe*[3] accepts a problem instance in OVM format and converts it into ASP facts. In turn, these facts are combined with an ASP encoding in Sect. 4–5 for CAFE problem solving, which is subsequently solved by *clingo* and *asprin*. The main features of our declarative approach are as follows:

Expressiveness: The collection of *aspcafe* encodings confirmed that ASP's modeling language is expressive enough to compactly express the constraints of CAFE problems.

Extensibility: We extended the *aspcafe* encoding to find the set of Pareto optimal solutions of multi-objective CAFE problems. The extension is easily done by utilizing *asprin*'s preference handling framework.

Efficiency: The *aspcafe*'s declarative approach scales to real-world instances, as demonstrated by the fact that the optimized encoding was able to deal with large instances regarded as hopeless by a collaborating automaker.

Our declarative approach can be extended to a wide range of regulations and constraints, such as the ZEV (Zero-Emission Vehicle) regulation. We will investigate this possibility, and the results will be applied to more practical vehicle equipment specification problems.

ASP has been applied to several problems in product configuration [4,14,26, 31,32]. Perhaps, the most relevant works are approaches to automobile product configuration based on SAT, MaxSAT, and Constraint Programming [2,21,27–29,34–36]. Most of them focus on detecting inconsistencies of product configuration data with customer preferences. From the viewpoint of ASP, Gençay et al. have proposed an ASP-based approach to validating and analyzing automobile product configuration on the Renault car benchmark [20]. They have used cautious and brave reasoning for four applications to discover groups of product configuration that are strongly related to each other. Although the aim of this study is different from those relevant works, investigating synergistic effects would be interesting future work.

References

1. Alviano, M., Dodaro, C.: Anytime answer set optimization via unsatisfiable core shrinking. Theory Pract. Logic Program. **16**(5–6), 533–551 (2016)

[3] All source code is available from https://github.com/banbaralab/aspcafe.

2. Amilhastre, J., Fargier, H., Marquis, P.: Consistency restoration and explanations in dynamic CSPs application to configuration. Artif. Intell. **135**(1–2), 199–234 (2002)
3. Andres, B., Kaufmann, B., Matheis, O., Schaub, T.: Unsatisfiability-based optimization in clasp. In: Dovier, A., Santos Costa, V. (eds.) Technical Communications of the Twenty-Eighth International Conference on Logic Programming (ICLP 2012), vol. 17, pp. 212–221. LIPIcs (2012)
4. Aschinger, M., Drescher, C., Gottlob, G., Vollmer, H.: Loco - A logic for configuration problems. ACM Trans. Comput. Logic **15**(3), 20:1–20:25 (2014)
5. Balduccini, M., Lierler, Y.: Integration schemas for constraint answer set programming: a case study. Theory Pract. Logic Program. **13**(4-5-Online-Supplement) (2013)
6. Banbara, M., Kaufmann, B., Ostrowski, M., Schaub, T.: Clingcon: the next generation. Theory Pract. Logic Program. **17**(4), 408–461 (2017)
7. Baral, C.: Knowledge Representation, Reasoning and Declarative Problem Solving. Cambridge University Press, Cambridge (2003)
8. Brewka, G., Delgrande, J., Romero, J., Schaub, T.: asprin: customizing answer set preferences without a headache. In: Bonet, B., Koenig, S. (eds.) Proceedings of the Twenty-Ninth National Conference on Artificial Intelligence (AAAI 2015), pp. 1467–1474. AAAI Press (2015)
9. Cabalar, P., Kaminski, R., Schaub, T., Schuhmann, A.: Temporal answer set programming on finite traces. Theory Pract. Logic Program. **18**(3–4), 406–420 (2018)
10. Calimeri, F., et al.: ASP-Core-2: input language format (2012)
11. Dodaro, C., Gasteiger, P., Leone, N., Musitsch, B., Ricca, F., Schekotihin, K.: Combining answer set programming and domain heuristics for solving hard industrial problems. Theory Pract. Logic Program. **16**(5–6), 653–669 (2016)
12. Ehrgott, M.: Multicriteria Optimization. Springer, Heidelberg (2005)
13. Erdem, E., Gelfond, M., Leone, N.: Applications of ASP. AI Mag. **37**(3), 53–68 (2016)
14. Friedrich, G., Ryabokon, A., Falkner, A.A., Haselböck, A., Schenner, G., Schreiner, H.: (Re)configuration using answer set programming. In: Shchekotykhin, K.M., Jannach, D., Zanker, M. (eds.) Proceedings of the IJCAI 2011 Workshop on Configuration. CEUR Workshop Proceedings, vol. 755. CEUR-WS.org (2011)
15. Gebser, M., et al.: Potassco User Guide. second edition. University of Potsdam (2015). http://potassco.org
16. Gebser, M., Kaminski, R., Kaufmann, B., Schaub, T.: Answer Set Solving in Practice. Morgan and Claypool Publishers, San Rafael (2012)
17. Gebser, M., Kaufmann, B., Otero, R., Romero, J., Schaub, T., Wanko, P.: Domain-specific heuristics in answer set programming. In: desJardins, M., Littman, M. (eds.) Proceedings of the Twenty-Seventh National Conference on Artificial Intelligence (AAAI 2013), pp. 350–356. AAAI Press (2013)
18. Gebser, M., Kaminski, R., Kaufmann, B., Schaub, T.: Multi-shot ASP solving with clingo. Theory Pract. Logic Program. **19**(1), 27–82 (2019)
19. Gelfond, M., Lifschitz, V.: The stable model semantics for logic programming. In: Kowalski, R., Bowen, K. (eds.) Proceedings of the Fifth International Conference and Symposium of Logic Programming (ICLP 1988), pp. 1070–1080. MIT Press (1988)
20. Gençay, E., Schüller, P., Erdem, E.: Applications of non-monotonic reasoning to automotive product configuration using answer set programming. J. Intell. Manuf. **30**(3), 1407–1422 (2019)

21. Küchlin, W., Sinz, C.: Proving consistency assertions for automotive product data management. J. Autom. Reason. **24**(1/2), 145–163 (2000)
22. National Highway Traffic Safety Administration: Corporate average fuel economy. https://www.nhtsa.gov/laws-regulations/corporate-average-fuel-economy
23. Niemelä, I.: Logic programs with stable model semantics as a constraint programming paradigm. Ann. Math. Artif. Intell. **25**(3–4), 241–273 (1999)
24. Pohl, K., Böckle, G., van der Linden, F.: Software Product Line Engineering: Foundations. Principles and Techniques. Springer, Heidelberg (2005)
25. Shiau, C.S.N., Michalek, J.J., Hendrickson, C.T.: A structural analysis of vehicle design responses to corporate average fuel economy policy. Transp. Res. Part A: Policy Pract. **43**(9–10), 814–828 (2009)
26. Simons, P., Niemelä, I., Soininen, T.: Extending and implementing the stable model semantics. Artif. Intell. **138**(1–2), 181–234 (2002)
27. Sinz, C., Blochinger, W., Küchlin, W.: PaSAT - parallel SAT-checking with lemma exchange: implementation and applications. Electron. Notes Discrete Math. **9**, 205–216 (2001)
28. Sinz, C., Kaiser, A., Küchlin, W.: Detection of inconsistencies in complex product configuration data using extended propositional SAT-checking. In: Proceedings of the Fourteenth International Florida Artificial Intelligence Research Society Conference, pp. 645–649. AAAI Press (2001)
29. Sinz, C., Kaiser, A., Küchlin, W.: Formal methods for the validation of automotive product configuration data. Artif. Intell. Eng. Des. Anal. Manuf. **17**(1), 75–97 (2003)
30. Soh, T., Banbara, M., Tamura, N., Le Berre, D.: Solving multiobjective discrete optimization problems with propositional minimal model generation. In: Beck, J.C. (ed.) CP 2017. LNCS, vol. 10416, pp. 596–614. Springer, Cham (2017). https://doi.org/10.1007/978-3-319-66158-2_38
31. Soininen, T., Niemelä, I., Tiihonen, J., Sulonen, R.: Representing configuration knowledge with weight constraint rules. In: Provetti, A., Son, T.C. (eds.) Proceedings of the AAAI Spring 2001 Symposium on Answer Set Programming, pp. 195–201. AAAI Press (2001)
32. Tiihonen, J., Heiskala, M., Anderson, A., Soininen, T.: Wecotin - A practical logic-based sales configurator. AI Commun. **26**(1), 99–131 (2013)
33. Wakaki, T., Inoue, K., Sakama, C., Nitta, K.: Computing preferred answer sets in answer set programming. In: Vardi, M.Y., Voronkov, A. (eds.) LPAR 2003. LNCS (LNAI), vol. 2850, pp. 259–273. Springer, Heidelberg (2003). https://doi.org/10.1007/978-3-540-39813-4_18
34. Walter, R., Felfernig, A., Küchlin, W.: Constraint-based and SAT-based diagnosis of automotive configuration problems. J. Intell. Inf. Syst. **49**(1), 87–118 (2017)
35. Walter, R., Küchlin, W.: Remax - A MaxSAT aided product (re-)configurator. In: Proceedings of the Sixteenth International Configuration Workshop. pp. 59–66. CEUR Workshop Proceedings, CEUR-WS.org (2014)
36. Walter, R., Zengler, C., Küchlin, W.: Applications of MaxSAT in automotive configuration. In: Proceedings of the Fifteenth International Configuration Workshop. pp. 21–28. CEUR Workshop Proceedings, CEUR-WS.org (2013)

UAV Compliance Checking Using Answer Set Programming and Minimal Explanations Towards Compliance (Application Paper)

Sarat Chandra Varanasi[1,2](✉), Baoluo Meng[1], Christopher Alexander[1], and Szabolcs Borgyos[1]

[1] General Electric Research, Niskayuna, NY, USA
{saratchandra.varanasi,baoluo.meng,christopher.alexander, szabolcs.borgyos}@ge.com

[2] The University of Texas at Dallas, Richardson, TX, USA
sxv153030@utdallas.edu

Abstract. We describe a continuation of prior work on automated compliance checking process for Unmanned Aerial Vehicles using Answer Set Programming. We describe a new algorithm to perform minimal explanations for offending compliance rules. This explanation is also performed for predicate answer set programs and the paper provides an extension to the algorithm that supported only propositional answer set programs. This improvement increases the expressivity of rules that can be captured in the compliance checking process. We take advantage of the goal-directed execution and constraint-solving capabilities of the s(CASP) engine in order to both compliance check the rules and compute the minimal explanations for violating rules. We further aim to map more rules from the AMA safety code into ASP.

Keywords: Answer set programming · Automated flight readiness approval · Minimal explanation computation

1 Introduction

Unmanned Aerial Vehicles (UAV) are ubiquitous today with widespread applications ranging from recreational uses to serious military operations. Due to the increase in usage of drones, UAV Traffic Management (UTM) is an important problem for the aviation industry [16]. A key component of UTM is the pre-flight check that involves adhering to guidelines that are necessary for safe traffic. There are several documents and guideline that recommend the safe usage and flight operation of UAVs ranging from recreational to commercial use. One such code for recreational use is provided by the Academy of Model Aeronautics (AMA), named the AMA safety code. The AMA safety code consists of several chapters and sections varying from general flight operation guides to usage of

M. Hanus and D. Inclezan (Eds.): PADL 2023, LNCS 13880, pp. 250–260, 2023.
https://doi.org/10.1007/978-3-031-24841-2_16

specific components such as radio control [1]. These rules are written in English and are interpreted (possibly ambiguously) by model aircraft operators. Further, these rules constitute the pre-flight check, an important step that a flyer has to self-regulate and adhere. Often, a manual self-compliance check of these rules results in misinterpretation and delays operations. Hence, the capture of these natural language rules in Answer Set Programming facilitates an unambiguous and automated approach towards pre-flight compliance checks. Prior work has been done in the capture of the AMA Safety code into Answer Set Programming [19,20]. Our prior work captured the AMA safety code into propositional answer set program while also providing minimal explanation towards compliance for offending pre-flight conditions. We extend the technique to support predicate answer set programs with constraints over real numbers. This enables numerical pre-flight conditions to be captured precisely and reasoned upon, thus providing a natural minimal explanations for offending conditions than possible with using propositional answer set programs. We illustrate our technique with an example in the AMA Safety code.

The rest of the paper is organized is follows: We first provide a background of Answer Set Programming along with Goal-Directed ASP and its support for Predicate Answer Set Programs. We then briefly discuss the AMA Safety Code and its capture in Propositional ASP. We then discuss the limitations of using a purely propositional approach to capture the rules and how they can be mitigated using Predicate ASP. We also state our assumptions for Predicate Answer Set Programs to capture AMA Safety Code followed by an extension of minimal explanation computation algorithm from prior work. Finally, we conclude with potential future works.

2 Background

2.1 Answer Set Programming

Answer Set Programming (ASP) is a widely popular formalism used to perform knowledge representation and reasoning and with many applications ranging from commonsense reasoning, modeling planning and cyber physical system domains to natural language question answering [4–6,8,11,13]. An ASP program models knowledge in the form of rules of the form $p \leftarrow q_1, .., q_n, not\ r_1, , .., not\ r_m$, where $p, q_1, ..q_n, r_1, ..r_m$ are all propositions. The following reading is to be given to the rules: "Infer p to be true if it is known that q_1 through q_n are true and r_1 through r_m are not known to be true. Also, never assert both p and the opposite of p, where the opposite of p can be either *not* p or the classical negation $-p$. Note that, this paper does not require the use of $-p$. This simple informal semantics gives enough power to perform commonsense reasoning and also perform non-monotonic reasoning whereby inferences can be revised in light of new information. The formal semantics for ASP is often termed the stable model semantics. More details about stable model semantics and the expressivity of ASP can be found elsewhere [11]. A model of an ASP program is the set of satisfiable assignments to propositions found in the rules of the program. A model of an ASP program is also termed as an Answer Set and hence the eponymous paradigm. Further the stable model semantics allows for multiple answer sets for a single program.

2.2 Goal-Directed Answer Set Programming

Fundamentally, the stable model semantics of ASP is provided for propositional programs where all the rules state facts or implications about propositions. However, for knowledge and reasoning to be expressive predicates are necessary. Hence, in practice, predicate answer set programs are often propositionalized through a process of grounding and solved for their answer sets. ASP Solvers such as Clingo [10] solve ASP programs with a huge number of rules and also perform plan synthesis for industrial scale problems. Nonetheless, it is easy to make the grounding process non-terminating by either introducing a function symbol in the predicate arguments or by assuming a dense real-valued domain. As an example, consider the rule, `p(f(X)) :- not q(X)`. Grounding based solvers such as Clingo cannot solve for uninterpreted functions without assuming a domain for the variable `X`. Also, consider a rule written in the style of a constraint-logic program as: `p(X) :- X #> 2, X #<= 3`. This rule form assumes a dense real-valued domain for `X` and results in an uncountable grounding by definition. The goal-directed ASP system s(CASP) [2] is designed to address these challenges in solving Predicate Answer Set Programs and has demonstrated several applications in modeling cyber physical systems that involve dense real-valued time [18] and along with other commonsense reasoning applications [7]. However, s(CASP) essential provides a query support to ASP through its operation semantics. At a high level, s(CASP) provides a Prolog-like interface to ASP. A user can typically issue a query $?-p(\bar{X})$, (where $\bar{X}$ is used to denote a set of variables $X_1, X_2, .., X_k$) and see the bindings for $\bar{X}$ along with a justification for $p(\bar{X})$. This justification for $p(\bar{X})$ is also called the partial model for $p(\bar{X})$. s(CASP) does not propositionalize the program but rather explores only part of the knowledge base that is relevant to the issued query. A salient feature of s(CASP) is its ability to perform constraint-solving over reals that precisely enable the previously mentioned applications in CPS modeling [18]. More recently, s(CASP) has been used to model requirements of altitude alerting system from the avionics domain [12]. Note that, there exist numerical constraint solving capabilities in other ASP solvers such as EZCSP [3] and *clingcon* [14]. However, they lack an operational semantics of ASP. The operational semantics of s(CASP) enables us to directly query $?-$ *not* $p(\bar{X})$ and derive the set of conditions that precisely contribute to *not* $p(\bar{X})$.

2.3 Abductive Reasoning in ASP

ASP allows for the modeling of abductive reasoning. This enables one to make assumptions about the truth of certain propositions termed abducibles and make inferences based on the assumptions. In ASP, usually, the directive `#abducible p` means that the proposition `p` may either be true or false in the knowledge base and inferences can be drawn based on whether `p` is indeed true or false respectively. The `#abducible` directive is really a shorthand for even loops through negation in ASP. That is, for an abducible `p`, two rules that are circular in nature are used, namely, `p :- not neg_p.` and `neg_p :- not p.`, where `neg_p` is an invented proposition that is not referred elsewhere in the knowledge base.

2.4 AMA Safety Code for Model Aircraft Operators in ASP

The Academy of Model Aeronautics has a set of guidelines for pre-flight check to ensure safe operation of UAVs [1,19,20]. The rules can be directly translated into ASP. For example, the rules look as follows:

- (Rule 1 in General AMA Chapter) I will not fly a model aircraft in a careless or reckless manner
- (Rule 7 in General AMA Chapter) I will not fly a model aircraft whose weight exceeds 55 pounds, unless certified through AMA large program
- (Rule from Large AMA Chapter) A large turbine model airplane (LTMA 1) is restricted to a maximum speed of 200 mph.

Each rule from the AMA code is written as a potential violation. For example, consider the rule regarding a model aircraft weighing more than 55 pounds and not being certified through the AMA Large Program. Its equivalent mapping to ASP can be written as: `violation :- weight_greater_than_55, not ama_large_certified.` All the rules are written in terms of *defaults* and *exceptions*. In such case, the rules will be of the form, `some_violation :- default, not exception.` The mapping of the general AMA rules into ASP can be found elsewhere [18,19].

2.5 Minimal Explanation for Violating Rules

We briefly discuss the prior approach used to compute minimal explanation towards compliance for offending rules. A set of input conditions needed for the pre-flight check are provided by the user using a web interface [19,20] as part of a questionnaire. Every response provided by the user in the questionnaire corresponds to a propositional literal depicting an input condition and is added as a fact to the knowledge base. Subsequently, the ASP program modeling the AMA code is checked for any violation based on the supplied questionnaire responses. If there is a violation, the user is shown a justification tree for that violation. To compute minimal explanation for compliance, the set of all possible input conditions are abduced by s(CASP) in the back-end and the compliant models are returned. From the set of all possible compliant models, the smallest model that retains as many as the original questionnaire responses is picked. This model represents the minimal set of conditions that need to be flipped from the original questionnaire response, in order to progress towards compliance. In other words, the smallest model whose input conditions intersect the maximum with the original questionnaire responses is selected and returned to the user. In case of Rule 7, the minimal explanation (according to the criteria just mentioned) will be to get certified through the AMA Large Program when the aircraft weights above 55 pounds [18,19].

3 Extending Computing Minimal Explanation to Predicate ASP

Consider the Rule 7 from AMA General Chapter, the violation can be written in Predicate ASP as:

```
violation :- weight(W), W #> 55, not ama_large_certified.
```

Limitations to Using Propositional ASP Programs. It is plain that propositional ASP programs are less expressive than predicate ASP programs. However, there are more issues. When complex numerical constraints are involved, we require the CLP(R) solving capabilities to perform reasoning over numeric domains. s(CASP) will simplify the arithmetic constraints (as much as possible). This is clearly not possible with propositional ASP programs. Further, the explanation provided by s(CASP) will use default natural language terms for the arithmetic $<, >, \leq$ and $\geq$ operators directly.

Symbolic Constraints Provided by s(CASP). Due to CLP(R) capabilities, s(CASP) can return a set of arithmetic constraints whose variables are not ground. This capability exist in Prolog with CLP(R) but not for negated queries. This capability is illustrated by a short example. Consider the following rule `p(X) :- X #> 2, X #< 3`. Submitting the query `?- p(X)` to either Prolog or s(CASP) would return the binding for `X` as $2 < X < 3$. However, the query `?- not(p(X))` returns no bindings due to the negation-as-failure semantics of `not` operator in Prolog. However, due to support for disunification and the ability to produce bindings for negated queries, s(CASP) returns two answers for `?- not p(X)`: one with $X \leq 2$ and $X \geq 3$.

Abducing Input Conditions Involved in Predicates. In case of the propositional ASP program, `#abducible weight_greater_than_55` would be used to determine a condition where there might not be a violation. That is, we use the abductive reasoning capabilities of ASP to derive a model where the input conditions do not cause any violation. When the input conditions are all propositional, the abductive reasoning is simple as illustrated in our prior work [19,20]. If the input conditions involved are predicates (such as `weight(X)`), we need to perform similar abdutcive reasoning to derive models that do not cause a violation. However, s(CASP) has no direct support for declaring `weight(X)` as an abducible. Therefore, we explicitly identify conditions that are first-order predicates and use the even loop for abduction as follows:

```
weight(X) :- not neg_weight(X).
neg_weight(X) :- not weight(X).
```

Corresponding to predicate `weight(X)`, another predicate `neg_weight(X)` is invented and ensured that does not name clash with any other predicate already provided by the user. Note that both `weight(X)` and `neg_weight(X)` are involved in an even loop. The invented predicate `neg_weight(X)` is solely used for the purpose of generating two models: one model in which `weight(X)` is true and another mode where `weight(X)` is false. This is equivalent to the abductive reasoning performed using the `#abducible`. In general a predicate `p(X1, X2, ..., Xk)` which is part of the user input (from questionnaire), has the even loop as seen below:

```
p(X1, X2, ..., Xk) :- not neg_p(X1, X2, ..., Xk).
neg_p(X1, X2, ..., Xk) :- not p(X1, X2, ..., Xk).
```

Finally, the original violation predicate needs to re-written as follows:

```
violation(W) :- weight(W), W #> 55, not ama_large_certified.
```

Note that the violation has an argument `W`. The variable `W` is moved into the head of the violation rule to simplify the s(CASP) execution. Otherwise, `W` would exist merely in the body of the rule causing s(CASP) to excessively check perform consistency checks. More details about head and body variables in s(CASP) programs can be found elsewhere [2] and not directly related to the discussion.

Assumptions Involved and Final Algorithms. Finally, we state all the assumptions involved in the computation of minimal explanation when predicates are included in the ASP program. We re-iterate the assumptions used in our prior work [18,19] for the sake of a coherent presentation.

1. The compliance rules are encoded as an answer set program.
2. The rules are a set of violation definitions, that is, every violation from the compliance rules is numbered by the rule number found in the compliance document. That is, if *rule n* states a violation, then the ASP program would have a rule with head $violation_n$.
3. For a given answer set program Π, $Models(\Pi)$ denotes the set of all answer sets of Π
4. For a given answer set program Π, $Models_q(\Pi)$ denotes the set of all partial models for query $? - q$.
5. A response to a question in the questionnaire corresponds to an input literal fact. For example, if a user selects "Yes" for *Was the operator under influence of alcohol or drugs?*, then the input literal fact `alcohol_drug_influence` is added to Π. If the user selects "No", then the literal fact is not added. Additionally, for an input field that is based on a predicate (such as `weight(W)`), upon an input say 50, the fact `weight(50).` is added to the knowledge base.
6. The literals (including predicates) are also referred to as conditions in the discussion below.
7. There are no circular rules in the knowledge base except the ones involving abducible rules.

Let C denote the set of conditions from a questionnaire. Conditions are the facts collected by the questionnaire. Let Π denote the ASP program encoding compliance rules. Let $C_{violate}$ denote the set of conditions that enable some violation $violation_i$. That is, $violation_i \in PartialModel$ such that $PartialModel \in Models(\Pi \cup C_{violate})$. A simple approach to find the conditions towards compliance is to query $? - not\ violate_i$ to s(CASP) and gather its partial stable models. However, this will not guarantee that the conditions will be found. For example, if $violate_i \leftarrow not\ c$, then the dual for $violate_i$ will be $not\ violate_i \leftarrow c$. However, directly querying for $? - not\ violate_i$ will cause a failure as c needs to be populated from the questionnaire. To address this, c can be treated as an abducible. Then, s(CASP) will abduce one world in which c would be true, and hence find support for $? - not\ violate_i$. Therefore, every condition $c \in C$ would be treated as an abducible and then $? - not\ violate_i$ can be queried. The partial model $C_{\neg violate}$ whose literals have the least difference to the literals of $C_{violate}$ is computed. The truth values of literals from $C_{\neg violate}$ represent the minimal set of conditions to achieve compliance, while retaining as may decisions selected by the user. Additionally, input conditions based on predicates have explicit even loops as mentioned in the previous section. Further, if a rule contains an input based on a predicate, then all the variables involved in the predicate are pushed to the head of the rule.

Algorithm 1. Compute minimal explanation to enable compliance of $violation_i$

Require: $violation_i \in PartialModel \wedge PartialModel \in Models(\Pi)$
Ensure: Compute $C_{\neg violation_i}$ where $not\ violation_i \in Models(\Pi)$
  for all *propositions* $c \in C$ **do**
    $\Pi \leftarrow \Pi \cup \{\#abducible\ c\}$
  end for
  for all *predicates* $p(\bar{X}) \in C$ **do**
    $\Pi \leftarrow \Pi \cup \{p(\bar{X}) \leftarrow not\ neg\_p(\bar{X})\}$
    $\Pi \leftarrow \Pi \cup \{neg\_p(\bar{X}) \leftarrow not\ p(\bar{X})\}$
  end for
  for all $violation_j \in \Pi$ s.t. $violation_i$ contains a predicate $p(\bar{X})$ **do**
    Replace $violation_j$ with $violation_j(\bar{X})$ in Π
  end for
  $PartialModels \leftarrow Models_{(not\ violation_i(\bar{X})}(\Pi)$ ▷ For no loss of generality we can assume $violation_i$ is replaced with $violation(\bar{X})$ in Π
  Find $model \in PartialModels$
    such that $model \cap C_{violate} \neq \emptyset \wedge |model \setminus C_{violate}|$ is smallest
    return *model*
  If no $model \in PartialModels$ exists
    such that $model \cap C_{violate} \neq \emptyset$
    return smallest $model \in PartialModel$

As an illustration, the original violation for Rule 7 and its re-written `not violation(W)` is shown using the SWISH Prolog and s(CASP) interface [21], in

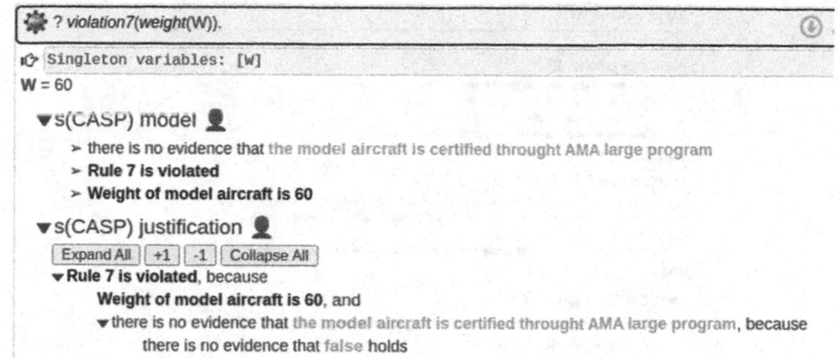

Fig. 1. English justification tree for violation of Rule 7 from AMA Safety Code

Figs. 1 and 2. Note that the justifications and the s(CASP) models are rendered in English. Our algorithm would pick the model corresponding to $W > 55$, from Fig. 2, in order to retain as many as user inputs as possible. Because the user inputs a model aircraft weight greater than 55 pounds, s(CASP) recommends the user to go through the AMA large program.

Discussion: Note that we have used the set intersection ($\cap$) operator to find a model that retains some of the conditions from the offending model $C_{violate}$. This intersection is the literal intersection operator when considering only propositional ASP programs. For predicate ASP programs, the constraints produced by *not* $violate(\bar{X})$ are checked for consistency with what the user has provided for $\bar{X}$ in the questionnaire. For example, consider, violation_i :- condition1(X), X #> 50, condition2(Y) #< 100 and that the user input is <X,Y> = <60, 110>, then, Algorithm 1 will retain the value of Y = 110 as it does not cause a violation. Yet, there exists a violation due to X = 60. Therefore, the intersection operator from Algorithm 1 actually involves checking that a compliant partial model, which requires that Y #>= 100, is consistent with the value provided by the user, namely, Y = 110. By the same process, another compliant partial model, which requires that X #<= 50, Y #< 100, is ruled out, because Y #< 100 is not consistent with the value the user provided, namely, Y = 110.

Implementation Details. The entire application for compliance checking is implemented as a web application. The questionnaire is presented on the front-end whereas the ASP reasoning is performed on the back-end web by exchanging answer sets and justifications through a REST[1] service. The REST service in-turn interacts with the backend SWI Prolog based s(CASP) application [21]. The application using only propositional answer set programs is available [17]. More details about the user interface and the justifications returned by our application can be found elsewhere [19,20].

[1] REST stands for Representation State Transfer that typically provides a HTTP endpoint and is widely used in today's web applications.

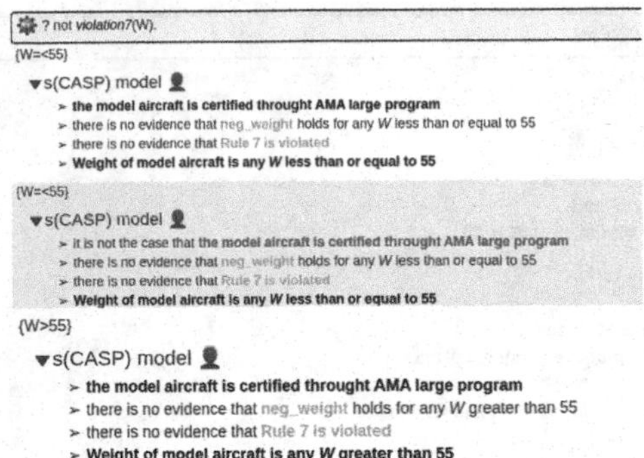

Fig. 2. s(CASP) returns 2 models for $W \leq 55$ and 1 model for $W > 55$

4 Conclusion and Future Work

We have described an extension of algorithm for computation of minimal explanation of compliance when the underlying knowledge base has predicates as opposed to having only propositions. We have illustrated this with an example using one of the rules in AMA Safety Code. The technique can be generalized to arbitrary rule texts from different domains. The only assumption is that the rules should not describe arithmetic constraints that are beyond the scope of CLP(R). Future work would involve integrating more expressive non-linear arithmetic solvers *(dReal)* [9] with s(CASP). Because *dReal* solver is SMT-based, a translation of non-linear arithmetic constraints encountered by s(CASP) into the SMT-LIB format and back to unifying s(CASP) variables is necessary. There exist tools such as *KeYmaera* that prove theorems about dynamics of hybrid systems often involving nonlinear real-valued arithmetic. In order to handle non-linear arithmetic, *KeYmaera* still relies on the constraint solving capabilites of SAT solvers [15]. Our work solely focuses on improving the support for constraint solving over a set of non-linear constraints as opposed to proving theorems involving dynamics of systems involving non-linear arithmetic. Combining expressive solvers with the power of s(CASP) is part of future work. Another interesting direction for future work is to consider automatically translating text in English into ASP rules. Recent works in natural language understanding have performed the translation of English text into ASP programs for the SQuAD and bAbI data sets [4–6]. Eventually, we would like to capture the entire AMA Safety Code and also FAA Part 107 regulations (P107) that concerns commercial licensing for small unmanned aircraft systems (sUAS). The expected turnaround time for a manual approval for P107 from FAA is currently 90 days and automating P107 compliance checking will certainly speed up the whole process.

References

1. Ama safety code (2021). https://www.modelaircraft.org/sites/default/files/documents/100.pdf
2. Arias, J., Carro, M., Salazar, E., Marple, K., Gupta, G.: Constraint answer set programming without grounding. Theory Pract. Logic Program. **18**(3–4), 337–354 (2018)
3. Balduccini, M., Lierler, Y.: Constraint answer set solver EZCSP and why integration schemas matter. Theory Pract. Logic Program. **17**(4), 462–515 (2017)
4. Basu, K., Shakerin, F., Gupta, G.: AQuA: ASP-based visual question answering. In: Komendantskaya, E., Liu, Y.A. (eds.) PADL 2020. LNCS, vol. 12007, pp. 57–72. Springer, Cham (2020). https://doi.org/10.1007/978-3-030-39197-3_4
5. Basu, K., Varanasi, S., Shakerin, F., Arias, J., Gupta, G.: Knowledge-driven natural language understanding of English text and its applications. In: Proceedings of the AAAI Conference on Artificial Intelligence, vol. 35, pp. 12554–12563 (2021)
6. Basu, K., Varanasi, S.C., Shakerin, F., Gupta, G.: Square: semantics-based question answering and reasoning engine. arXiv preprint arXiv:2009.10239 (2020)
7. Chen, Z., Marple, K., Salazar, E., Gupta, G., Tamil, L.: A physician advisory system for chronic heart failure management based on knowledge patterns. Theory Pract. Logic Program. **16**(5–6), 604–618 (2016)
8. Erdem, E., Gelfond, M., Leone, N.: Applications of answer set programming. AI Mag. **37**(3), 53–68 (2016)
9. Gao, S., Kong, S., Clarke, E.M.: dReal: an SMT solver for nonlinear theories over the reals. In: Bonacina, M.P. (ed.) CADE 2013. LNCS (LNAI), vol. 7898, pp. 208–214. Springer, Heidelberg (2013). https://doi.org/10.1007/978-3-642-38574-2_14
10. Gebser, M., et al.: The potsdam answer set solving collection 5.0. KI-Künstliche Intell. **32**(2), 181–182 (2018)
11. Gelfond, M., Kahl, Y.: Knowledge Representation, Reasoning, and the Design of Intelligent Agents: The Answer-Set Programming Approach. Cambridge University Press, Cambridge (2014)
12. Hall, B., et al.: Knowledge-assisted reasoning of model-augmented system requirements with event calculus and goal-directed answer set programming. arXiv preprint arXiv:2109.04634 (2021)
13. Nguyen, T.H., Bundas, M., Son, T.C., Balduccini, M., Garwood, K.C., Griffor, E.R.: Specifying and reasoning about CPS through the lens of the NIST CPS framework. Theory Pract. Log. Program. 1–41 (2022)
14. Ostrowski, M., Schaub, T.: ASP modulo CSP: The clingcon system. Theory Pract. Logic Program. **12**(4–5), 485–503 (2012)
15. Quesel, J.D., Mitsch, S., Loos, S., Aréchiga, N., Platzer, A.: How to model and prove hybrid systems with KeYmaera: a tutorial on safety. Int. J. Softw. Tools Technol. Transf. **18**(1), 67–91 (2016)
16. Rumba, R., Nikitenko, A.: The wild west of drones: a review on autonomous-UAV traffic-management. In: 2020 International Conference on Unmanned Aircraft Systems (ICUAS), pp. 1317–1322. IEEE (2020)
17. Varanasi, S.C.: Flight readiness ASP (2022). https://github.com/ge-high-assurance/flight-readiness-asp
18. Varanasi, S.C., Arias, J., Salazar, E., Li, F., Basu, K., Gupta, G.: Modeling and verification of real-time systems with the event calculus and s(CASP). In: Cheney, J., Perri, S. (eds.) PADL 2022. LNCS, vol. 13165, pp. 181–190. Springer, Cham (2022). https://doi.org/10.1007/978-3-030-94479-7_12

19. Varanasi, S.C., Meng, B., Alexander, C., Borgyos, S., Hall, B.: Automating UAV flight readiness approval using goal-directed answer set programming. arXiv preprint arXiv:2208.12199 (2022)
20. Varanasi, S.C., Meng, B., Alexander, C., Borgyos, S., Hall, B.: Unmanned aerial vehicle compliance checking using goal-directed answer set programming. In: Goal-Directed Execution of Answer Set Programs (2022)
21. Wielemaker, J., Arias, J., Gupta, G.: s(CASP) for SWI-prolog. In: Arias, J., et al. (eds.) Proceedings of the International Conference on Logic Programming 2021 Workshops, Goal-Directed Execution of Answer Set Programs, Co-located with the 37th International Conference on Logic Programming (ICLP 2021), Porto, Portugal (Virtual), 20–21 September 2021. CEUR Workshop Proceedings, vol. 2970. CEUR-WS.org (2021)

Jury-Trial Story Construction and Analysis Using Goal-Directed Answer Set Programming

Zesheng Xu[1], Joaquín Arias[2(✉)], Elmer Salazar[1], Zhuo Chen[1], Sarat Chandra Varanasi[1], Kinjal Basu[1], and Gopal Gupta[1]

[1] The University of Texas at Dallas, Richardson, USA
{zesheng.xu,sarat-chandra.varanasi,kinjal.basu,gupta}@utdallas.edu
[2] CETINIA, Universidad Rey Juan Carlos, Madrid, Spain
joaquin.arias@urjc.es

Abstract. Answer Set Programming (ASP) is a well known paradigm for knowledge representation and for automating commonsense reasoning. Query-driven implementations of Predicate ASP, e.g., the s(CASP) system, permit top-down execution of answer set programs without the need for grounding them. In this paper we consider the problem of analyzing a jury trial for the crime of homicide. We demonstrate how knowledge related to reaching a guilty/not-guilty verdict can be modeled in ASP, and how s(CASP) can be used to analyze the situations under various assumptions. Detailed justification can be provided for each possible verdict, thanks to s(CASP). The goal, and the main contribution, of this paper is to show how a complex situation—whose analysis has been traditionally considered to be within the purview of only humans—can be automatically analyzed using a query-driven predicate ASP system such as s(CASP). For illustration, we use the well-known fictional case of Commonwealth of Massachusetts v. Johnson used in the cognitive science literature. Frank Johnson and Alan Caldwell had a quarrel, which was later followed by Frank Johnson stabbing Alan Caldwell. However, this seemingly simple case was complex in many ways: the lack of critical information led to different verdicts from different jurors. This paper attempts to formalize the logic behind each story construction using s(CASP) and expanding it to similar situations.

Keywords: Answer set programming · Commonsense reasoning · Explanation-based decision making

1 Introduction

Jury decision making has been extensively studied in criminology and sociology literature [8,13]. In particular, the explanation-based decision making model of

We thank the anonymous reviewers for suggestions for improvement. Authors are partially supported by MICINN projects TED2021-131295B-C33 (VAE), PID2021-123673OB-C32 (COSASS), and RTI2018-095390-B-C33 (InEDGEMobility MCI-U/AEI/FEDER, UE), US NSF Grants (IIS 1910131, IIP 1916206), and US DoD.

M. Hanus and D. Inclezan (Eds.): PADL 2023, LNCS 13880, pp. 261–278, 2023.
https://doi.org/10.1007/978-3-031-24841-2_17

Pennington and Hastie [23,24] is widely used for understanding juror behavior. In this paper we use ASP and the s(CASP) system to emulate and analyze jury decision making. According to Pennington and Hastie's model, decision makers construct a causal model to explain available facts. The decision maker simultaneously learns, creates, or discovers a set of alternative decisions. A decision is reached when the "causal model of the evidence is successfully matched to an alternative in the choice set" [23,24]. Constructing a causal model to explain available facts can be thought of as *story construction* where a story is essentially a causal chain. We model this causal chain as a set of ASP rules. Various scenarios can be represented as a set of ASP facts [5]. Alternative verdicts (guilty, innocent, etc.) correspond to queries. These queries can be checked for entailment against the ASP facts and rules assembled together. The jury-decision process can thus be formally represented and mechanized. Counterfactual reasoning can also be performed within the ASP framework to examine alternative scenarios.

Answer Set Programming (ASP) [7,10,16] is a well-known programming paradigm for representing knowledge as well as solving combinatorial problems. ASP can be effectively used for automating commonsense reasoning [17]. Traditionally, answer set programs have been executed by first *grounding* them—turning predicate answer set programs into propositional ones–and then using SAT solver technology to compute models (called answer sets) of the transformed program [15]. Grounding leads to associations encapsulated in variables being lost, which makes justification of computed answers extremely hard. Recently, goal-directed implementations of predicate ASP such as s(CASP) have been proposed that execute predicate answer set programs in a goal-directed manner [4,20]. Programs need not be grounded before being executed on s(CASP). Query-driven evaluation implies that a proof-tree is readily generated for the query which serves as its justification [2]. We use the s(CASP) system as our execution engine for analyzing the jury-decision process. Note that there are other proposals for goal-directed implementation of ASP [9,11,14], however, they can only handle propositional answer set programs. The s(CASP) system can execute predicate answer set programs directly (no grounding is needed).

The main contribution of this paper is to show that complex situations such as jury trials can be modeled with ASP and jury decisions automatically computed using the s(CASP) system based on the story construction model of Pennington and Hastie. Thus, commonsense reasoning applications of the most complex type can be modeled using s(CASP) [3,6]. Detailed explanations are also provided by s(CASP) for any given decision reached. The s(CASP) system also allows for counterfactual situations to be considered and reasoned about. We assume that the reader is familiar with ASP and s(CASP). An introduction on ASP can be found in Gelfond and Kahl's book [17] while a fairly detailed overview of the s(CASP) system can be found elsewhere [3,4].

1.1 Motivation and Background

Our goal is to show that complex commonsense knowledge can be represented in ASP and that the s(CASP) query-driven predicate ASP system can be used for querying this knowledge. Rather than performing the story construction manually and then use the story to derive the verdict manually, we formalize the story and automatically check whether it entails a certain verdict. We illustrate our framework using the case of a jury trial considered in social sciences research [24].

Note that commonsense knowledge can be emulated using (i) default rules, (ii) integrity constraints, and (iii) multiple possible worlds [17,18]. Default rules are used for jumping to a conclusion in the absence of exceptions, e.g., a bird normally flies, unless it's a penguin. Thus, if we are told that Tweety is a bird, we jump to the conclusion that Tweety flies. Later, if we are told that Tweety is a penguin, we withdraw the conclusion that Tweety can fly. Default rules with exceptions represent an elaboration tolerant way of representing knowledge [7].

```
flies(X) :- bird(X), not abnormal_bird(X).
abnormal_bird(X) :- penguin(X).
```

Integrity constraints allow us to express impossible situations and invariants. For example, a person cannot sit and stand at the same time.

```
false :- sit(X), stand(X).
```

Finally, multiple possible worlds allow us to construct alternative universes that may have some of the parts common but other parts inconsistent. For example, the cartoon world of children's books has a lot in common with the real world (e.g., birds can fly in both worlds), yet in the former birds can talk like humans but in the latter they cannot.

We make use of default rules, integrity constraints and multiple worlds to represent commonsense knowledge pertaining to the jury trial.

We apply our techniques to the explanation-based decision making framework of Pennington and Hastie, to automatically explain how a verdict follows from a constructed story. Pennington and Hastie use taxonomies proposed by Collins [22] regarding inference. They tested their jury decision making framework by conducting experiments that replicates the simulated trial of Commonwealth of Massachusetts v. Johnson with test human subjects who acted as jurors. Their work intended to delineate the roles of inference procedures in explanation-based decision processes to illustrate the pervasiveness and variety of inferences in decision making. They proposed a framework which detailed how human utilized story construction using known facts and world knowledge to justify their final decision. Their emphasis is on how possible scenarios can be captured through their framework. Our interest is in automating their framework using ASP and the s(CASP) system. Further, we plan to expand the scenarios within the given structure to construct counterfactual situations with s(CASP).

1.2 Method

Our primary aim is to demonstrate the power of s(CASP) for representing commonsense knowledge needed in story construction. In the explanation-based decision making model, a story is a collection of causal chains. Each causal chain can be broken down and represented with ASP rules.

Pennington and Hastie have broken down the story construction process of a jury into trial knowledge, world knowledge of similar events, and knowledge about story structures. Trial knowledge (e.g., Jack and Jill are a married couple) can be simply input into the s(CASP) system as fact statements. The world knowledge corresponds to all the commonsense knowledge pertaining to the case (e.g., a person may kill another person in self defence). Knowledge about story structure, likewise, is also represented as commonsense knowledge (e.g., intent to kill arises before the actual murder). The commonsense knowledge has to be modeled manually. However, once it has been coded, it can be reused. After we have the facts and commonsense knowledge, we can execute the queries representing the various verdicts in the s(CASP) system and show that indeed the query is entailed by the commonsense knowledge. As the set of facts changes, the corresponding verdict can be computed.

We assume that the reader is familiar with ASP, commonsense reasoning in ASP, and the s(CASP) system. Details can be found in [10,17,18], and [4], respectively. The s(CASP) system is available on Github as well as under SWI Prolog, including its web-based version called SWISH [26].

2 Automating Jury Decision-making

We illustrate the automation of Pennington and Hastie's explanation based decision making model using s(CASP) through a murder trial case. This is the hypothetical case of commonwealth of Massachusetts v. Johnson used by Pennington and Hastie in their work [23,24]. The story of the case is as follows: *the defendant, Frank Johnson, had quarreled in a bar with the victim, Alan Caldwell, who threatened him with a razor. Later in the same evening, they went out and quarreled. During the fight, Frank Johnson stabbed Alan Caldwell, resulting in Alan Caldwell's death. Prior to the fight, at the bar, Alan Caldwell's girlfriend, Sandra Lee, had asked Frank Johnson for a ride to the race track the next morning.* There are a number of disputed points in this case:

- Who initiated the later fight that resulted in the death of Alan Caldwell?
- Did Sandra Lee's action of asking Frank Johnson for a ride initiate the chain of events?
- Did Frank Johnson hold the knife with the intent to murder Alan Caldwell or was it for self-defense?
- Did Frank Johnson have the knife on him or did he retrieve it from home?
- What were the emotional states of Frank Johnson and Alan Caldwell as the events unfolded?
- How much do the emotional states of both parties influence the case?

There are also more details such as testimonies from Frank Johnson and other witnesses. Such testimonies provided even more degrees of detail that further complicate the case because they often imply biases in narration, which can further impact the jury's story construction process. After the jury is presented with the information about the case, the jurors will need to come up with a verdict and explain how they reached their conclusion. This is the point at which the authors of the original paper [24] apply their framework trying to capture the jurors' story construction verdict-reaching process. Our goal is to completely automate this reasoning.

There are two parts to modeling the scenario: modeling the facts and modeling the rules that capture the world knowledge and the reasoning behind story construction. Indisputable details and facts from the trial are represented as s(CASP) facts, while the rest of knowledge is represented as s(CASP) rules. During the trial, many new pieces of information come to light, often describing a key factor with elaborate details and explanations for cause and effect. However, as shown in Pennington and Hastie's study, the juries often pick up on only a few significant details to justify their model, thus determining what to translate into our model also poses some difficulties. Ours is an automated solution, so we represent knowledge that is available before or during the trial.

2.1 Modeling Trial Knowledge

Information that is established in the trial knowledge is extracted and represented as a set of facts.

```
1 person(frank_johnson).
2 person(alan_caldwell).
3 person(sandra_lee).
```

Facts 1, 2, and 3 model the participants of the scenario.

```
4 defendant(frank_johnson).
5 victim(alan_caldwell).
6 dead(alan_caldwell).
```

Facts 4 and 5 model the respective role of Frank Johnson and Alan Caldwell in this trial. Fact 6 represents that Alan Caldwell died from the fight.

```
7 friend(sandra_lee,alan_caldwell).
8 friend(alan_caldwell,sandra_lee).
```

Facts 7 and 8 model the (reflexive) friendship relation between Sandra Lee and Alan Caldwell.

```
9  in_relationship(sandra_lee,alan_caldwell).
10 in_relationship(alan_caldwell,sandra_lee).
```

Facts 9 and 10 model that Sandra Lee and Alan Caldwell are in a relationship.

```
11 has_weapon(frank_johnson).
12 has_weapon(alan_caldwell).
```

Facts 11 and 12 model that Frank Johnson and Alan Caldwell have a weapon.

```
13 primary_conflict(frank_johnson, alan_caldwell).
14 primary_conflict(alan_caldwell, frank_johnson).
```

Facts 13 and 14 model that Frank Johnson and Alan Caldwell are the main characters in the fight.

```
15 clashed_earlier(alan_caldwell, frank_johnson).
16 clashed_earlier(frank_johnson, alan_caldwell).
```

Facts 15 and 16 model that Frank Johnson and Alan Caldwell had clashed earlier (bar fight).

```
17 had_weapon_earlier(alan_caldwell).
```

Fact 17 models that Alan Caldwell has a weapon at the earlier bar fight.

```
18 threatened_earlier(alan_caldwell, frank_johnson).
```

Fact 18 models that Alan Caldwell threatened Frank Johnson earlier.

```
19 close_contact(sandra_lee,frank_johnson).
20 close_contact(frank_johnson,sandra_lee).
```

Facts 19 and 20 model the close contact between Sandra Lee and Frank Johnson (Sandra Lee asking Frank Johnson for a ride). For simplicity, reflexive relationships have been explicitly modeled (e.g., `friend/2` relation). Note that we define predicates with descriptive names, rather than represent knowledge in some generic manner (e.g., using Neo-Davidsonian Logic [12]). Meaning is thus encapsulated in the predicate name. Encapsulating meaning in predicate names is fine since we hand-code all the knowledge explicitly.

2.2 Modeling World Knowledge

Modeling commonsense knowledge about the world is the harder part, however, the task is facilitated by ASP. Our model is based upon the jurors' thought process in the Commonwealth of Massachusetts v. Johnson case but can be expanded for other scenarios. The following is our encoding for story construction:

```
21 major_prior_clash(X,Y) :-
22       person(X), person(Y), X \= Y, clashed_earlier(X,Y), had_weapon_earlier(X).
23 major_prior_clash(X,Y) :-
24       person(X), person(Y), X \= Y, clashed_earlier(Y,X), had_weapon_earlier(Y).
25 minor_prior_clash(X,Y) :-
26       person(X), person(Y), X \= Y, clashed_earlier(Y,X), not n_minor_clash(X,Y).
27 n_minor_clash(X,Y) :- had_weapon_earlier(X).
28 n_minor_clash(X,Y) :- had_weapon_earlier(Y).
29 n_minor_clash(X,Y) :- threatened_earlier(X,Y).
30 n_minor_clash(X,Y) :- threatened_earlier(Y,X).
31 no_prior_conflict(X,Y) :-
32      person(X), person(Y), X \= Y, not clashed_earlier(X,Y).
```

Lines 21 through 32 define and model different levels of conflict between two parties (X and Y) that they had prior to the main conflict (the bar fight). A conflict is major if X and Y clashed earlier (`clashed_earlier(X,Y)`) and one of them(X or Y) had a weapon then (`had_weapon_earlier(X)`). A conflict is minor if X and Y clashed earlier (`clashed_earlier(X,Y)`), and neither of them (X or Y) had weapon earlier (`had_weapon_earlier(X)`) and we cannot prove that it was *not* a minor conflict (`n_minor_clash(X,Y)`). There was not a conflict between X and Y if we cannot prove that they had clashed earlier (`not clashed_earlier(X,Y)`). Note that `n_minor_clash(X,Y)` acts as an exception to `minor_prior_clash(X,Y)` rule; it checks if persons X and Y threatened each other or one of them had any weapon without repeating `minor_prior_clash(X, Y)` clause too many times. Next we model the notion of jealousy.

```
33 jealous(X,Y) :- person(X), person(Y), person(Z), Y \= X, Y \= Z, X \= Z,
34       in_relationship(X,Z), close_contact(Y,Z), not ab_jealous(X,Y).
35 jealous(X,Y):- person(X), person(Y), X \= Y, not rich(X), rich(Y).
36 ab_jealous(X,Y) :- forgiving_person(X), not n_ab_jealous(X,Y).
37 ab_jealous(X,Y) :- person(X), person(Y), person(Z),Y \= X, Y \= Z, X \= Z,
38       in_relationship(X,Z), close_contact(Y,Z), open_relationship(Y,Z).
39 ab_jealous(X,Y) :- person(X), person(Y), person(Z),Y \= X, Y \= Z, X \= Z,
40       in_relationship(X,Z), close_contact(Y,Z), oblivious(X).
```

Lines 33–40 define and model jealousy that was proposed in the original Pennington and Hasti paper. X can be reasoned to be jealous of Y (`jealous(X, Y)`) if: (i) Y had close contact with Z, and Z is X's girlfriend, or (ii) Y is richer than X; unless exceptions apply.

Lines 36–40 model exceptions to the rules defining `ab_jealous(X,Y)`. Line 36 states that if X is a forgiving person, he might forgive person Y. Lines 37–38 state that X would not be jealous at Y for close relationship with his girlfriend Z if X and Z are in an open relationship (`open_relationship(Y,Z)`). Lines 39–40 state that X would not be jealous of Y for close relationship with his girlfriend Z if X is an oblivious person and might not realize that they are having an affair.

```
41 angry(X,Y):- jealous(X,Y), X \= Y.
42 angry(X,Y):-
43       person(X),person(Y), X \= Y, clashed_earlier(X,Y), not ab_angry(X,Y).
44 angry(X,Y):-
45       person(X), person(Y), X \= Y, mentally_unstable(X), not ab_angry(X,Y).
46 ab_angry(X,Y):- not mentally_unstable(X), forgiving_person(X),
47      minor_prior_clash(X,Y), not n_ab_angry(X,Y).
48 ab_angry(X,Y):- not mentally_unstable(X), friend(X,Y),
49      minor_prior_clash(X,Y), not n_ab_angry(X,Y).
50 n_ab_angry(X,Y):- not ab_angry(X,Y).
```

Lines 41–50 define and model the reason behind how X can be angry at Y (note that all the definitions are specific to the case):

- Line 41: X is jealous at Y (`jealous(X,Y)`).
- Lines 42–43: They had prior clashes (`clashed_earlier(X,Y)`).
- Lines 44–45: X is mentally unstable.

unless there are proven exceptions preventing the conclusion from being drawn (`ab_angry(X,Y)`). Line 43 does not have exceptions due to `jealous(X,Y)` already having its own exception rules (`ab_jealous(X,Y)`), thus an exception is not needed here. Lines 46–49 reason that X is not angry at Y if X is a forgiving person (`forgiving_person(X)`) or X is friend with Y (`friend(X,Y)`) and they only had minor prior clash (`minor_prior_clash(X,Y)`).

Lines 46–49 model the exception that prevents X from being angry at Y. Do note they both have the `not mentally_unstable(X)` condition, preventing it from being applied to rule in lines 44–45. We currently do not have any exceptions to the rule in line 41, but it can be added later.

Lines 46–50 create an even loop over negation [17], which will create *abducibles* [18] for generating two scenarios: one in which X is angry at Y and another where X is not angry at Y. The reason behind this logic is that X has a chance of still being angry at Y despite the exception.

```
51 scared(X,Y) :-
52     person(X), person(Y), X \= Y, threatened_earlier(Y,X), not ab_scared(X,Y).
53 scared(X,Y) :- person(X), person(Y), X \= Y, had_weapon_earlier(Y),
54     clashed_earlier(X,Y), not has_weapon(X), not ab_scared(X,Y).
55 scared(X,Y) :-
56     person(X), person(Y), X \= Y, mentally_unstable(Y), not ab_scared(X,Y).
57 ab_scared(X,Y) :-
58     not mentally_unstable(Y), minor_prior_clash(X,Y), forgiving_person(X).
```

Rules 51–58 define and model the reason behind how X can be scared of Y. X can be reasoned to be scared of Y if:

- Lines 51–52: Y has threatened X earlier (`threatened_earlier(Y,X)`).
- Lines 53–54: They had prior clashes (`clashed_earlier(X,Y)`), and Y had a weapon (`had_weapon_earlier(Y)`) while X does not (`not has_weapon(X)`).
- Lines 55–56: Y is mentally unstable.

unless there are proven exceptions preventing the reasoning (`ab_scared(X,Y)`).

Lines 57–58 reason that X is not scared of Y if X is a forgiving person (`forgiving_person(X)`) and they only had minor prior clash (`minor_prior_clash(X,Y)`). The `not mentally_unstable(X)` literal, prevents it from being applied to the rule in lines 55–56.

```
59 initiator(X) :- not exist_an_initiator, angry(X,Y), clashed_earlier(X,Y),
60                 threatened_earlier(X,Y).
61 initiator(X) :- not exist_an_initiator, angry(X,Y),
62                 clashed_earlier(X,Y).
63 initiator(X) :- not exist_an_initiator, angry(X,Y), no_prior_clash(X,Y).
64 initiator(X) :- not exist_an_initiator, defendant(X), victim(Y).
65 initiator(X) :- known_initiator_of_conflict(X).
66 initiator(X) :- not exist_an_initiator, potential_initiator_of_clash(X).
67 exist_an_initiator :- person(Z), known_initiator_of_conflict(Z).
68 potential_initiator_of_clash(X) :- angry(X,Y), has_weapon(X),
69     primary_conflict(X,Y), X \= Y, not n_potential_initiator_of_clash(X).
70 n_potential_initiator_of_clash(X) :- not potential_initiator_of_clash(X).
```

Lines 59–70 reason if X is the initiator, i.e., did X starts the fight that lead to the death of Alan Caldwell? In other words, did X have the motive to start the conflict? X can be argued to be the initiator of this case if:

- Lines 59–60: X had a prior clash with Y (`clashed_earlier(X,Y)`) and X had threatened Y (`threatened_earlier(X,Y)`).
- Lines 61–62: X had a prior clash with Y (`clashed_earlier(X,Y)`) and X can be argued to be angry at Y.

- Line 63: X can be argued to be angry at Y (`angry(X,Y)`) and there was no evidence of prior conflict between them (`no_prior_clash(X,Y)`).
- Line 64: By default, the defendant X (`defendant(X)`) is the initiator of the case and there exists a victim Y (`victim(Y)`).
- Line 65: X is the initiator if it is a known fact (`known_initiator_of_conflict(X)`).
- Line 66: This rule creates an abducible if we do not clearly know who initiated the conflict.

Lines 67–70 all share a common exception: `not exist_an_initiator`. It prevents any further reasoning that one can be the initiator of the case when there is a known initiator.

Line 67 is the exception that will be true if there exist any person who is a known initiator of the conflict.

Lines 68–70 create an even loop used in line 46 to create an abducible. Note that it will only create such an abducible to argue that X is the potential initiator of the case if X is angry at Y (`angry(X,Y)`), X has a weapon (`has_weapon(X)`), and X and Y are involved in the primary conflict (`primary_conflict(X,Y)`).

```
71 provoker(X):- initiator(X), had_weapon_earlier(X), threatened_earlier(X,Y).
72 provoker(X):- initiator(X), retrieved_weapon(X), angry(X,Y).
```

Lines 71–72 of the code define and reason how X is the provoker in this case. X can be argued to be the provoker if:

- Line 71: X can be proven to be the initiator of the conflict (`initiator(X)`), and he had a weapon earlier (`had_weapon_earlier(X)`), and threatened another person Y with it (`threatened_earlier(X,Y)`), who is other party of the conflict (`primary_conflict(X,Y)`).
- Line 72: X can be proven to be the initiator of the conflict (`initiator(X)`), if he retrieved a weapon from elsewhere (`retrieved_weapon(X)`), and he is angry at Y (`angry(X,Y)`), where Y is the other party in the conflict (`primary_conflict(X,Y)`).

This provoker rule-set seems redundant with the initiator rules. However, it is not, because person X can be argued to be an initiator but not the provoker of the case. Initiator rules only reason if person X is the initiator of the chain of events, but not necessarily the one for the primary conflict. For example, Alan Caldwell and Frank Johnson can be argued to be the initiators of the conflict, but we might not be able to argue they are the provoker of the primary conflict.

Currently, the initiator and provoker rules specifically model the Commonwealth of Massachusetts v. Johnson case. For example, they do not capture the possibility where one party might not have a weapon but still is the provoker in the scenario. Such situations can be covered by adding more rules.

```
retrieved_weapon(X) :- person(X), has_weapon(X), not had_weapon_earlier(X),
    not n_retrieved_weapon(X).
n_retrieved_weapon(X) :- not retrieved_weapon(X).
```

Lines 73–75 create an abducible on how X obtained his weapon if he was not known to have a weapon prior to the primary conflict (`has_weapon(X),not had_weapon_earlier(X)`). This is because the source of his weapon is unknown, so there are 2 possibilities proposed in Pennington and Hastie's paper:

- He had it on him but was concealed.
- He returned home and retrieved it.

Surely, there are many more possibilities for the source of his weapon (e.g., someone handed him the weapon), but we ignore them as they are not considered in the original case. More rules can be added to represent these possibilities.

2.3 Verdict Representation

Next, we discuss how potential verdicts are modeled.

```
76 murderer(X):- defendant(X), provoker(X), victim(Y), dead(Y).
77 self_defense(X):- defendant(X), provoker(Y), Y \= X, scared(X,Y).
```

Lines 76–77 models how juries can reach a verdict, as proposed in Pennington and Hastie's paper [24]. It can be argued that X committed murder (`murderer(X)`) if X is the defendant (`defendant(X)`) of the case, he has been proven to be a provoker (`provoker(X)`), and there exists a victim Y who died from this conflict (`victim(Y),dead(Y)`). It can also be argued that X acted in self-defense (`self_defense(X)`) if X is defendant (`defendant(X)`) of the case, and it can be proven that there exists another provoker Y (`Y = X,provoker(Y)`) and X can be reasoned to be scared of Y (`scared(X,Y)`), thus X felt his life was threatened.

```
78 verdict(X, murder) :- murderer(X).
79 verdict(X, self_defense) :- self_defense(X).
80 verdict(X, hung_jury) :- not murderer(X), not self_defense(X).
```

Lines 78–80 correspond to determining the verdict, where `verdict(Person, Verdict)` is the top level query. Line 78 will return a murder verdict for X if X can be proven to have committed murder (`murderer(X)`). Line 79 will return self-defense verdict if X can be proven to have acted in self-defense (`self_defense(X)`). Line 80 will return "hung_jury" verdict if we could not prove that X committed murder or acted in self-defense (`not murderer(X), not self_defense(X)`).

```
81 ?- verdict(Person, Verdict).
```

Line 81 queries the program. It will return the result based on known facts, world knowledge, and story construction logic. Since `Person` and `Verdict` are variables, this query will explore all possible scenarios our program entails.

Table 1. Run-time in ms. for the different scenarios

	Version A				Version B	Version C
Model number	1	3	4	5	1	1
Run time (ms)	39.85	11.16	22.37	4.08	64.66	4258.76

3 Evaluation of Different Verdicts and Scenarios

After executing the query `?- verdict(Person,Verdict)`, we will get different (partial) answers set based on facts and rules included in the program. The s(CASP) system will provide justification –essentially a proof tree– for each of them. We next discuss a few possible verdicts and scenarios that led to them.

Table 1 shows run-times (in milliseconds), and answer set number, for generating: (i) four verdicts given the facts and rules described so far, and (ii) two additional versions, including counterfactual reasoning, described in Sect. 4.

Frank Johnson is the Murderer and Proven to be the Initiator: Figure 1 shows the justification,[1] of the first answer set, where Frank Johnson is charged with murder. We are able to prove that he committed murder by proving that he is a provoker (`provoker(frank_johnson)`) through rule 71. Because we can prove that he had a conflict with Alan Caldwell (`clashed_earlier(frank_johnson,alan_caldwell)`), and he is not friends with Alan Caldwell (`not friend(frank_johnson,alan_caldwell)`) (thus ignoring the rest of rule 44), we can argue he was angry with Alan Caldwell, making him an initiator (`initiator(frank_johnson)`) through rule 61. Further, we also argued he retrieved his weapon from his home (`retrieved_weapon(frank_johnson),has_weapon(frank_johnson)`), and since he was angry, we can reason that he intended to escalate the conflict and prepared for it, thus proving he committed murder through rule 76.

Frank Johnson is the Murderer and Assumed Initiator: Figure 2 shows the justification (for the third answer) when Frank Johnson is also charged with murder (assuming he is the initiator). In this scenario, we proved him to be a provoker using a different rule to prove he is an initiator using the abducible (lines 67–69). Here we can argue for him to be the initiator based on the assumption that because he had a weapon (`has_weapon(frank_johnson)`), and can be reasoned to be angry at the other party in the conflict (`primary_conflict(frank_johnson,alan_caldwell)`, (`angry(frank_johnson,alan_caldwell)`)), thus proving he committed murder through rule in line 76.

[1] This justification is generated by invoking `scasp --tree --short --human murder.pl` and after including the directive `#show` which allows to select the predicates to appear. Note: in the following justifications the negated literals have been omitted by adding the flag `--pos`.

```
JUSTIFICATION_TREE:
'verdict' holds (for frank_johnson, and murder), because
   'murder' holds (for frank_johnson), because
      'defendant' holds (for frank_johnson), and
      'provoker' holds (for frank_johnson), because
        'initiator' holds (for frank_johnson), because
           there is no evidence that 'exist_an_initiator' holds, because
              there is no evidence that 'known_initiator_of_conflict' holds (for alan_caldwell), and
              there is no evidence that 'known_initiator_of_conflict' holds (for frank_johnson), and
              there is no evidence that 'known_initiator_of_conflict' holds (for sandra_lee).
           'angry' holds (for frank_johnson, and alan_caldwell), because
              'clashed_earlier' holds (for frank_johnson, and alan_caldwell), and
              there is no evidence that 'ab_angry' holds (for frank_johnson, and alan_caldwell), because
                 there is no evidence that 'friend' holds (for frank_johnson, and alan_caldwell).
           'clashed_earlier' holds (for frank_johnson, and alan_caldwell), justified above.
        'retrieved_weapon' holds (for frank_johnson), because
           'has_weapon' holds (for frank_johnson), and
           there is no evidence that 'had_weapon_earlier' holds (for frank_johnson), and
           it is assumed that 'retrieved_weapon' holds (for frank_johnson).
        'angry' holds (for frank_johnson, and alan_caldwell), because
           'clashed_earlier' holds (for frank_johnson, and alan_caldwell), justified above, and
           there is no evidence that 'ab_angry' holds (for frank_johnson, and alan_caldwell), justified above.
      'victim' holds (for alan_caldwell), and
      'dead' holds (for alan_caldwell).
```

Fig. 1. Murder Verdict through proving him as provoker

```
JUSTIFICATION_TREE:
'verdict' holds (for frank_johnson, and murder), because
   'murder' holds (for frank_johnson), because
      'defendant' holds (for frank_johnson), and
      'provoker' holds (for frank_johnson), because
         'initiator' holds (for frank_johnson), because
            'potential_initiator_of_clash' holds (for frank_johnson), because
               'angry' holds (for frank_johnson, and alan_caldwell), because
                  'clashed_earlier' holds (for frank_johnson, and alan_caldwell).
               'has_weapon' holds (for frank_johnson), and
               'primary_conflict' holds (for frank_johnson, and alan_caldwell), and
               it is assumed that 'potential_initiator_of_clash' holds (for frank_johnson).
         'retrieved_weapon' holds (for frank_johnson), because
            'has_weapon' holds (for frank_johnson), justified above, and
            it is assumed that 'retrieved_weapon' holds (for frank_johnson).
         'angry' holds (for frank_johnson, and alan_caldwell), because
            'clashed_earlier' holds (for frank_johnson, and alan_caldwell), justified above.
      'victim' holds (for alan_caldwell), and
      'dead' holds (for alan_caldwell).
```

Fig. 2. Murder Verdict through proving him as provoker via abducible

Frank Johnson Acted in Self-defense Because Alan Caldwell Threatened Him: Figure 3 shows the justification (for the fourth answer) where we reason that Frank Johnson acted in self-defense. This reasoning depends on proving Alan Caldwell to be the initiator of this conflict through rule in line 59 where he threatened Frank Johnson earlier (`threatened_earlier(alan_caldwell, frank_johnson)`) with weapon (`had_weapon_earlier(alan_caldwell)`). And, because he had threatened Frank Johnson earlier, we can argue that he was the initiator of the chain of events, thus proving that Frank Johnson acted in self-defense, through rule in line 77.

Frank Johnson Acted in Self-defense Because Alan Caldwell was the Initiator Due to Jealousy: Finally, Fig. 4 shows the justification (for the fifth answer) where Frank Johnson is reasoned to have acted in self-defense. The reasoning rests on proving Alan Caldwell to be the initiator of this conflict through

```
JUSTIFICATION_TREE:
'verdict' holds (for frank_johnson, and self_defense), because
   'self_defense' holds (for frank_johnson), because
      'defendant' holds (for frank_johnson), and
      'provoker' holds (for alan_caldwell), because
        'initiator' holds (for alan_caldwell), because
           'angry' holds (for alan_caldwell, and frank_johnson), because
              'jealous' holds (for alan_caldwell, and frank_johnson), because
                'in_relationship' holds (for alan_caldwell, and sandra_lee), and
                'close_contact' holds (for frank_johnson, and sandra_lee), and
                'in_relationship' holds (for alan_caldwell, and sandra_lee), justified above, and
                'close_contact' holds (for frank_johnson, and sandra_lee), justified above.
           'clashed_earlier' holds (for alan_caldwell, and frank_johnson), and
           'threatened_earlier' holds (for alan_caldwell, and frank_johnson).
        'had_weapon_earlier' holds (for alan_caldwell), and
        'threatened_earlier' holds (for alan_caldwell, and frank_johnson).
      'scared' holds (for frank_johnson, and alan_caldwell), because
        'threatened_earlier' holds (for alan_caldwell, and frank_johnson), justified above, and
        'clashed_earlier' holds (for alan_caldwell, and frank_johnson), justified above, and
        'had_weapon_earlier' holds (for alan_caldwell), justified above.
```

Fig. 3. Innocent Verdict because he was threatened

```
JUSTIFICATION_TREE:
'verdict' holds (for frank_johnson, and self_defense), because
   'self_defense' holds (for frank_johnson), because
      'defendant' holds (for frank_johnson), and
      'provoker' holds (for alan_caldwell), because
        'initiator' holds (for alan_caldwell), because
           'angry' holds (for alan_caldwell, and frank_johnson), because
              'jealous' holds (for alan_caldwell, and frank_johnson), because
                'in_relationship' holds (for alan_caldwell, and sandra_lee), and
                'close_contact' holds (for frank_johnson, and sandra_lee), and
                'in_relationship' holds (for alan_caldwell, and sandra_lee), justified above, and
                'close_contact' holds (for frank_johnson, and sandra_lee), justified above.
           'clashed_earlier' holds (for alan_caldwell, and frank_johnson), and
           'threatened_earlier' holds (for alan_caldwell, and frank_johnson).
        'had_weapon_earlier' holds (for alan_caldwell), and
        'threatened_earlier' holds (for alan_caldwell, and frank_johnson).
      'scared' holds (for frank_johnson, and alan_caldwell), because
        'threatened_earlier' holds (for alan_caldwell, and frank_johnson), justified above, and
        'clashed_earlier' holds (for alan_caldwell, and frank_johnson), justified above, and
        'threatened_earlier' holds (for alan_caldwell, and frank_johnson), justified above.
```

Fig. 4. Innocent verdict via proving Alan Caldwell as initiator due to jealousy

rules in lines 59, 41, and 13. We argued that he is angry toward Frank Johnson because he was jealous (`jealous(alan_caldwell,frank_johnson)`) of his close contact (`close_contact(frank_johnson,sandra_lee)`) with Sandra Lee who is his girlfriend (`in_relationship(alan_caldwell,sandra_lee)`). Because he had threatened Frank Johnson earlier (`threatened_earlier(alan_caldwell, frank_johnson)`), that makes him both the initiator (`initiator(alan_caldwell)`) and provoker (`provoker(alan_caldwell)`) of this case. This allows us to reason that Frank Johnson was scared of him (`scared(frank_johnson, alan_caldwell)`), letting us prove that Frank Johnson acted in self-defense through rule in line 77.

```
'verdict' holds (for frank_johnson, and murder), because
   'murder' holds (for frank_johnson), because
      'defendant' holds (for frank_johnson), and
      'provoker' holds (for frank_johnson), because
        'initiator' holds (for frank_johnson), because
           'angry' holds (for frank_johnson, and alan_caldwell), because
              'clashed_earlier' holds (for frank_johnson, and alan_caldwell), and
              'forgiving_person' holds (for frank_johnson), and
              'clashed_earlier' holds (for alan_caldwell, and frank_johnson), and
              'had_weapon_earlier' holds (for alan_caldwell), and
              'friend' holds (for frank_johnson, and alan_caldwell).
           'clashed_earlier' holds (for frank_johnson, and alan_caldwell), justified above.
        'retrieved_weapon' holds (for frank_johnson), because
           'has_weapon' holds (for frank_johnson), and
           it is assumed that 'retrieved_weapon' holds (for frank_johnson).
        'angry' holds (for frank_johnson, and alan_caldwell), because
           'clashed_earlier' holds (for frank_johnson, and alan_caldwell), justified above.
      'victim' holds (for alan_caldwell), and
      'dead' holds (for alan_caldwell).
```

Fig. 5. Possible scenario if they were friends

```
JUSTIFICATION_TREE:
'verdict' holds (for frank_johnson, and murder), because
   'murder' holds (for frank_johnson), because
      'defendant' holds (for frank_johnson), and
      'provoker' holds (for frank_johnson), because
        'initiator' holds (for frank_johnson), because
           'defendant' holds (for frank_johnson), justified above, and
           'victim' holds (for alan_caldwell).
        'retrieved_weapon' holds (for frank_johnson), because
           'has_weapon' holds (for frank_johnson), and
           it is assumed that 'retrieved_weapon' holds (for frank_johnson).
        'angry' holds (for frank_johnson, and alan_caldwell), because
           'friend' holds (for frank_johnson, and joe_smith), and
           'angry' holds (for joe_smith, and alan_caldwell), because
              'clashed_earlier' holds (for joe_smith, and alan_caldwell).
      'victim' holds (for alan_caldwell), and
      'dead' holds (for alan_caldwell).
```

Fig. 6. Introducing a new person who is connected with both characters

4 Counterfactual Reasoning

Since we use ASP and the flexible framework of the s(CASP) execution engine, we can explore many more options by considering additional facts and constructing different scenarios using the same story-construction structure.

Version B: The following is a scenario if Frank Johnson and Alan Caldwell were found out to actually be friends with each other, and Frank Johnson was a forgiving person. To simulate this, we simply add the following facts:

```
forgiving_person(frank_johnson).
friend(alan_caldwell, frank_johnson).
friend(frank_johnson, alan_caldwell).
```

Figure 5 shows a justification tree where Frank Johnson is still charged with murder. However, the verdict relies on a different line of reasoning and thus a different model is constructed: Despite being friends with Alan Caldwell and being a forgiving person, Frank Johnson found Alan Caldwell's threat to him to be unforgivable, got angry at him, and murdered his friend in cold blood.

Version C: We can even take this a step further. In this newer scenario, Frank Johnson and Alan Caldwell never had the earlier conflict. Instead, we introduce his friend, Joe Smith, who had a conflict with Alan Caldwell and threatened him earlier. To simulate this, we have to make a few changes. First, we commented out facts modeling Alan and Frank's bar fight in lines 13–18 of the trial knowledge section. Then we added the following additional knowledge as facts:

```
person(joe_smith).
friend(frank_johnson, joe_smith).
friend(joe_smith,frank_johnson).
clashed_earlier(joe_smith,alan_caldwell).
threatened_earlier(joe_smith,alan_caldwell).
```

and to accommodate this we added two new rules to define possible ways how one person can be angry at another:

```
angry(X,Y):- person(X), person(Y), person(Z), X \= Y, Y \= Z, Z \= X,
     friend(X,Z), angry(Z,Y).
angry(X,Y):- person(X), person(Y), person(Z), X \= Y, Y \= Z, Z \= X,
     friend(X,Z), angry(Y,Z).
```

In this model, we argued Frank Johnson is still be to charged with murder (see justification tree in Fig. 6). Through our newly added rules, we argued that he is angry at Alan Caldwell because his friend Joe Smith is angry at Alan Caldwell. What is interesting in this proof tree is that a large portion of the tree is used to prove that Frank Johnson was angry at Alan Caldwell. Because in our newly introduced rules above, X is angry at Y, if his friend Z is angry at Y, or Y is angry at Z. This shows how this newly added dimension of "friend" is able to increase the complexity significantly.

5 Discussion

Our modeling of the jury decision-making process does not take into account any laws that may be applicable. This is primarily because Pennington and Hastie's formulation only deals with jury deliberations pertaining to determining a guilty or not guilty verdict. If any laws are applicable, then they can be modeled as well. For example, if an evidence is illegally obtained, then it must be excluded. This can be modeled by turning facts pertaining to evidence into default rules (evidence ought to be considered unless it is tainted).

Note also that the knowledge base we have developed to model various concepts such as angry, scared, jealous, etc., should be reusable, i.e., it should be possible to reuse them in determining a verdict for another (different) case. This should indeed be possible as the knowledge (as ASP rules) has been represented in quite generic manner.

In our modeling of the jury decision-making process, we have modeled time in a very simple manner. It's possible to refine our system to model time more accurately. For this we have to resort to the event calculus [3]. In a story, situations unfold over time, which can be modeled in the event calculus, including inertia. The s(CASP) system can perform reasoning over real numbers using constraints, which can be used to faithfully model time. With more accurate modeling, we can reason about future possibilities as well as past history. For instance, to explain that Alan Caldwell is still alive at time point τ in some scenario, modeling with the event calculus in s(CASP) will allow us to get a list of all possible ways of his murder that did not happen before time τ. We leave the event calculus modeling using s(CASP) for future research.

Finally, with respect to related work, logic programming and ASP have been applied to automating legal reasoning [1,19,21,25], however, in our case, we are more interested in modeling jury deliberations. To the best of our knowledge, ours is the first attempt to use ASP for this purpose. With respect to generating justification for a given atom in the answer set—a prominent feature of the s(CASP) system—there has been considerable work (see the paper by Arias et al. for a detailed discussion [2]). However, most of these methods for generating a justification are designed for propositional answer set programs. In contrast, s(CASP) provides justification for predicate answer set programs.

6 Conclusion

In this paper we showed how ASP and the s(CASP) system can be used to represent knowledge and reason about a jury trial based on Pennington and Hastie's explanation based decision making model. Pennington and Hastie's model is based on story construction that we represent in ASP. We use the goal-directed s(CASP) ASP system to study various verdicts that can be reached under various assumptions, as well as add counterfactual information to represent alternative situations. The s(CASP) system and its query-driven nature is crucial to the study. The s(CASP) system can be viewed as Prolog extended with stable-model semantics-based negation.

Our work shows that ASP/s(CASP) can represent the story construction logic behind jury decisions quite elegantly. Our model was able to capture the schematic of the process combining external facts with knowledge of familiar cases to come up with reasonable stories from which verdicts can be derived. Pennington and Hastie's original paper regards the problem of reaching a verdict using the story construction process as a difficult problem. We show that ASP and s(CASP) can be used to formalize this process. Our work also shows how ASP and s(CASP) can be used to model complex human thought processes [18].

References

1. Aravanis, T.I., Demiris, K., Peppas, P.: Legal reasoning in answer set programming. In: IEEE ICTAI 2018, pp. 302–306. IEEE (2018)
2. Arias, J., Carro, M., Chen, Z., Gupta, G.: Justifications for goal-directed constraint answer set programming. In: Proceedings 36th International Conference on Logic Programming (Technical Communications). EPTCS, vol. 325, pp. 59–72. Open Publishing Association (2020). https://doi.org/10.4204/EPTCS.325.12

3. Arias, J., Carro, M., Chen, Z., Gupta, G.: Modeling and reasoning in event calculus using goal-directed constraint answer set programming. Theory Pract. Logic Program. **22**(1), 51–80 (2022). https://doi.org/10.1017/S1471068421000156
4. Arias, J., Carro, M., Salazar, E., Marple, K., Gupta, G.: Constraint answer set programming without grounding. Theory Pract. Logic Program. **18**(3–4), 337–354 (2018). https://doi.org/10.1017/S1471068418000285
5. Arias, J., Moreno-Rebato, M., Rodriguez-García, J.A., Ossowski, S.: Modeling administrative discretion using goal-directed answer set programming. In: Alba, E., et al. (eds.) CAEPIA 2021. LNCS (LNAI), vol. 12882, pp. 258–267. Springer, Cham (2021). https://doi.org/10.1007/978-3-030-85713-4_25
6. Arias, J., Törmä, S., Carro, M., Gupta, G.: Building information modeling using constraint logic programming. Theory Pract. Logic Program. **22**(5), 723–738 (2022). https://doi.org/10.1017/S1471068422000138
7. Baral, C.: Knowledge Representation, Reasoning and Declarative Problem Solving. Cambridge University Press, Cambridge (2003)
8. Bennett, W., Feldman, M.: Reconstructing Reality in the Courtroom. Rutgers University Press, New Brunswick (1981)
9. Bonatti, P.A., Pontelli, E., Son, T.C.: Credulous resolution for answer set programming. In: Proceedings of the AAAI 2008, pp. 418–423. AAAI Press (2008)
10. Brewka, G., Eiter, T., Truszczynski, M.: Answer set programming at a glance. Commun. ACM **54**(12), 92–103 (2011)
11. Costantini, S., Formisano, A.: Query answering in resource-based answer set semantics. Theory Pract. Log. Program. **16**(5–6), 619–635 (2016)
12. Davidson, D.: Inquiries into Truth and Interpretation: Philosophical Essays, vol. 2. Oxford University Press, Oxford (2001)
13. Elwork, E., Sales, B., Alfini, J.: Juridic decisions: in ignorance of the law or in light of it? Cogn. Sci. **1**, 163–189 (1977)
14. Gebser, M., Gharib, M., Mercer, R.E., Schaub, T.: Monotonic answer set programming. J. Log. Comput. **19**(4), 539–564 (2009)
15. Gebser, M., Kaminski, R., Kaufmann, B., Schaub, T.: Clingo = ASP + control: preliminary report. arXiv 1405.3694 (2014). https://arxiv.org/abs/1405.3694
16. Gebser, M., Kaufmann, B., Kaminski, R., Ostrowski, M., Schaub, T., Schneider, M.: Potassco: the Potsdam answer set solving collection. AI Commun. **24**(2), 107–124 (2011)
17. Gelfond, M., Kahl, Y.: Knowledge Representation, Reasoning, and the Design of Intelligent Agents: The Answer-Set Programming Approach. Cambridge University Press, Cambridge (2014)
18. Gupta, G., et al.: Tutorial: automating commonsense reasoning. In: ICLP Workshops. CEUR Workshop Proceedings, vol. 3193. CEUR-WS.org (2022)
19. Kowalski, R., Datoo, A.: Logical English meets legal English for swaps and derivatives. Artif. Intell. Law **30**(2), 163–197 (2022)
20. Marple, K., Salazar, E., Gupta, G.: Computing stable models of normal logic programs without grounding. arXiv preprint arXiv:1709.00501 (2017)
21. Morris, J.: Constraint answer set programming as a tool to improve legislative drafting: a rules as code experiment. In: ICAIL, pp. 262–263. ACM (2021)
22. Pennington, N., Hastie, R.: The logic of plausible reasoning: a core theory. Cogn. Sci. **13**, 1–49 (1989)
23. Pennington, N., Hastie, R.: Explaining the evidence: tests of the story model for juror decision making. J. Pers. Soc. Psychol. **62**(2), 189 (1992)
24. Pennington, N., Hastie, R.: Reasoning in explanation-based decision making. Cognition **49**, 123–163 (1993)

25. Satoh, K., et al.: PROLEG: an implementation of the presupposed ultimate fact theory of Japanese civil code by PROLOG technology. In: Onada, T., Bekki, D., McCready, E. (eds.) JSAI-isAI 2010. LNCS (LNAI), vol. 6797, pp. 153–164. Springer, Heidelberg (2011). https://doi.org/10.1007/978-3-642-25655-4_14
26. Wielemaker, J., Arias, J., Gupta, G.: s(CASP) for SWI-Prolog. In: Proceedings of the 37th ICLP 2021 Workshops, vol. 2970. CEUR-WS.org (2021). http://ceur-ws.org/Vol-2970/gdeinvited4.pdf

Pruning Redundancy in Answer Set Optimization Applied to Preventive Maintenance Scheduling

Anssi Yli-Jyrä(✉), Masood Feyzbakhsh Rankooh, and Tomi Janhunen

Tampere University, Tampere, Finland
{anssi.yli-jyra,masood.feyzbakhshrankooh,tomi.janhunen}@tuni.fi

Abstract. Multi-component machines deployed, e.g., in paper and steel industries, have complex physical and functional dependencies between their components. This profoundly affects how they are maintained and motivates the use of logic-based optimization methods for scheduling preventive maintenance actions. Recently, an abstraction of maintenance costs, called miscoverage, has been proposed as an objective function for the preventive maintenance scheduling (PMS) of such machines. Since the minimization of miscoverage has turned out to be a computationally demanding task, the current paper studies ways to improve its efficiency. Given different answer set optimization encodings of the PMS problem, we motivate constraints that prune away some sub-optimal and otherwise redundant or invalid schedules from the search space. Our experimental results show that these constraints may enable up to ten-fold speed-ups in scheduling times, thus pushing the frontier of practically solvable PMS problem instances to longer timelines and larger machines.

1 Introduction

Multi-component machines deployed, e.g., in paper and steel industries, have complex physical and functional dependencies between their components forming an entire production line. Moreover, the machinery should be kept constantly in operation to maximize production. These aspects profoundly affect the ways in which such machines can be maintained in the first place. Besides this, further constraints emerge from resources available for carrying out particular maintenance actions; see, e.g., [4] for taking such features into account. There are two main-stream *maintenance policies*, viz. *corrective* and *preventive*. The former is failure-driven and typically supersedes the latter that is complementary by nature and aims to ensure the full operationality of a machine in the long run.

The multitude of concerns, as briefed above, calls for highly flexible scheduling methods. Such potential is offered by logic-based methods supporting optimization, including *constraint optimization problems* (COPs) [13, S. 7.4], *mixed-integer programming* (MIP) [13, S. 15.4], *maximum satisfiability* (MaxSAT) [11], and *answer set optimization* (ASO) [14]. Since answer set programming [3] is known to be well-suited for solving scheduling and resource allocation problems

M. Hanus and D. Inclezan (Eds.): PADL 2023, LNCS 13880, pp. 279–294, 2023.
https://doi.org/10.1007/978-3-031-24841-2_18

[2,5,7,12], this paper continues the development of ASO-based encodings [17] for solving *preventive maintenance scheduling* (PMS) problems. Unlike the conventional scheduling that aims at minimization of the production time (makespan), maintenance scheduling is complementary when it tries to keep the production line operational as long as possible by rationalizing the maintenance actions and minimizing the downtimes during which the machine is unavailable for use, either due to a scheduled maintenance break or due to a machine failure. Therefore, PMS is not, at least directly, a special case of job-shop scheduling but an interesting problem of its own.

The previous ASO encodings [17] abstract away from the details of maintenance costs and aim at *covering* the timeline of each component by coordinated maintenance breaks, and any service actions target at the component-specific recommended maintenance intervals. Delays in the maintenance give rise to *under-coverage* while servicing too often denotes *over-coverage.* Based on this correspondence, the minimization of the respective objective function penalizes for overall *miscoverage.* Since the machine is assumed to be inoperable during service breaks, the breaks form a central resource to be utilized maximally for maintenance, as far as excessive servicing is avoidable. For the same reason, components cannot be maintained independently of each other and, in addition, we assume the omission of all components serviceable during the normal operation of the machine. The upper bound of the number of maintenance breaks is a part of the definition for feasible schedules, but the parameter can be minimized by tightening it as long as the miscoverage of the optimal schedule remains the same or at an acceptable level.

The existing ASO-based encodings from [17] can be understood as *golden designs* when solving PMS problem instances: they provide the baseline for the performance of the solution methods, and the stable optimums that are correct with respect to the underlying formal definition. Since the global minimization of miscoverage tends to be computationally demanding in the numbers of both service breaks and components, the current paper studies ways to improve the efficiency of encodings by incorporating constraints that prune away sub-optimal and otherwise redundant or invalid schedules from the search space. In addition to expressing some known properties of optimal solutions to remove sub-optimal solutions directly, such pruning constraints may also be used to break *symmetries* present in the search space. Symmetries have been studied in the context of constraint programming [10] as well as ASP [6]. Symmetry breaking constraints are often incorporated by programmers themselves, but there is also recent work aiming at their automated detection [15,16]. Regardless how such constraints are devised, they intensify optimization by removing both (i) equally good solutions that are uninteresting modulo symmetries and (ii) invalid (partial) solutions.

Our experimental results indicate that pruning constraints may enable up to ten-fold speed-ups in scheduling times, thus pushing the frontier of practically solvable PMS problem instances to longer timelines and larger machines. The main contributions of our work are as follows:

- The characterization of miscoverage-based PMS [17] in terms of new constraints that prune the search space for optimal schedules.
- The respective correctness arguments for the pruning constraints, indicating that at least one globally optimal schedule will be preserved.
- Experimental analysis revealing the effects of the pruning constraints on the performance of the ASP systems in the computation of optimal schedules.

In this way, we demonstrate the extensibility of the original framework with additional constraints; the focus being on performance improvement whereas other constraints affecting the quality of schedules are left as future work.

The plot for this article is as follows. In Sect. 2, we recall the formal definitions of multi-component machines and their preventive maintenance schedules, including the coverage-based objective functions that are relevant for optimization. The respective baseline encodings [17] of preventive maintenance scheduling (PMS) problems are summarized in Sect. 3. Then, we are ready to present novel constraints that can be used to prune the search space for optimal schedules in various respects. Besides the proofs of correctness, we encode these constraints in terms of rules that are compatible with the baseline encodings, enabling straightforward combinations. The experimental part in Sect. 5 investigates the effect of the new constraints on the performance of ASP systems when solving PMS problems optimally. Section 6 concludes this work.

2 Definitions

In this section, we recall the preventive maintenance scheduling problem from [17]. The definitions abstract away from the complications of practical maintenance on purpose, ignoring such aspects as the shutdown/restarting costs, the duration of maintenance, the availability of maintenance resources, and the feasible combinations of maintenance actions. The abstracted problem is founded on the notions of a multi-component machine (Definition 1) and its preventive maintenance schedule (Definition 2). These tightly related concepts are explained as follows.

Definition 1. *A multi-component machine* $\mathcal{M} = \langle C, \iota, \rho \rangle$ *comprises of a set of* components C, *an* initial lifetime *function* $\iota : C \to \mathbb{N}$, *and a* recommended maintenance interval *function* $\rho : C \to \mathbb{N}$, *satisfying* $\iota(c) < \rho(c)$ *for each* $c \in C$.

The failure rate of each component c will start to grow rapidly when the time passed since the most recent maintenance action of the component reaches the recommended maintenance interval $\rho(c) > 0$. The initial lifetime $\iota(c)$ of a component c aims to capture, e.g., situations where the component has been maintained just a moment before the schedule starts. In that case, the component's lifetime at the beginning of the schedule is just a single time step less than the recommended maintenance interval, i.e., $\iota(c) + 1 = \rho(c)$. But it is also possible that the component is older, in which case its initial lifetime is even smaller. Thus we have the constraint $\iota(c) < \rho(c)$. If $\iota(c) = 0$, the recommendation is that c is serviced as soon as possible.

Definition 2. *A* preventive maintenance schedule *(PMS, or* schedule *for short) for a multi-component machine* $\mathcal{M}$ *is a quadruple* $S = \langle h, A, \ell, b \rangle$ *where*

- $h \in \mathbb{N} \backslash \{0\}$ – *the last time step called the* horizon,
- $A : C \times \{1, \ldots, h\} \to \{0, 1\}$ – service selection function *over the components and the time steps (implying the set of maintenance breaks* $B = \{t \mid c \in C,\ 1 \le t \le h,\ \text{and } A(c,t) = 1\}$ *for the whole machine),*
- $\ell \in \{1, \ldots, h\}$ – *the* upper bound for the last break time, $\max B \cup \{1\}$.
- $b \in \{0, \ldots, \ell\}$ – *the* upper bound for the number of maintenance breaks, $|B|$,

For simplicity, we assume that $\iota(c) < h$ for all $c \in C$, because otherwise the component c need not be scheduled for maintenance. The schedule will impose a machine-wide *break* $t \in B$ at those time steps where at least one component is maintained, but it cannot break the run of the machine for maintenance more than b times or after the time step ℓ. For each component $c \in C$, we define $B_c = \{t \mid A(c,t) = 1\}$ as the set of time steps $1 \le t \le h$ when c is maintained by some *maintenance action.* The effects of all scheduled maintenance actions are uniform in the following sense: if a maintenance action is applied to a component c at time step t, the component c becomes immediately as good as new, meaning that its next due time for maintenance is as far away as the recommended maintenance interval $\rho(c)$ suggests.

The purpose of a schedule is to determine the maintenance plan for a given *timeline* $1..h$ and to keep the machine in a healthy state during the whole timeline. To support this goal, the maintenance times B_c of each component in the PMS try to *cover* the whole timeline of each component c in the following sense:

1. The initial lifetime $\iota(c)$ of a component c *covers* the time steps $1..\iota(c)$ once.
2. The maintenance of a component c at time step t *covers* the time steps $t... \min\{h, t + \rho(c) - 1\}$ once.
3. A particular time step can be covered more than once in relation to a component c, if c is maintained during its initial lifetime or c is subjected to two or more maintenance actions closer than $\rho(c)$ time steps to one another.

The covering of the component-wise timelines cannot be done just by maximizing the maintenance of components at every possible maintenance break but it has to be done optimally, with respect to some cost function. Our work focuses on a cost function based on *miscoverage* $mc(\mathcal{M}, S)$. The fully general form of this function was originally given in [17], but here we confine ourselves to a specific version that is practically sufficient. The specific version divides into three components:

1. The *super-coverage* $sc(\mathcal{M}, S)$. For reasons explained in [17, Lemma 3], we immediately ignore all schedules where a component is maintained in such a way that a time step is covered more than twice. This property of the schedule corresponds to a single, but unnecessary cost that makes ignored schedules bad enough to be completely thrown away during the optimization.

Listing 1. A PMS problem instance

```
comp(1,5,2).    comp(3,7,0).   comp(5,9,0).    comp(7,5,4).
comp(2,10,0).   comp(4,4,3).   comp(6,11,2).   comp(8,8,0).
```

2. The *under-coverage* $uc(\mathcal{M}, S)$ of schedules is a measure that amounts to the total number of component-wise time steps neither covered by initial lifetime nor by the set of time steps when each components is maintained.
3. The *over-coverage* $oc(\mathcal{M}, S)$, the dual notion for the under-coverage, is, in its full generality, a complex concept. The current work replaces it with a proxy, the *(specific) over-coverage* $oc_2(\mathcal{M}, S)$. It amounts to the total number of component-wise time steps covered exactly twice.

The sum of the super-coverage, under-coverage, and the (specific) over-coverage costs gives us the (specific) *miscoverage function* $mc_2(\mathcal{M}, S) = sc(\mathcal{M}, S)+uc(\mathcal{M}, S)+oc_2(\mathcal{M}, S)$ for schedules S, the only cost function studied in this paper. By Lemma 3 in [17], the optimal schedules under the specific miscoverage function coincide with those under the (general) miscoverage function. Schedules with non-zero super-coverage cannot be optimal and they are, thus, excluded from the search space. Under this exclusion, the specific and general miscoverage coincide: $mc_2(\mathcal{M}, S) = mc(\mathcal{M}, S)$.

Definition 3. *Let f be a cost function mapping multi-component machines and their schedules to integer-valued costs. An* optimizing PMS problem P_f *is then a function problem whose inputs consists of a multi-component machine $\mathcal{M} = \langle C, \iota, \rho\rangle$ and a triple $\langle h, \ell, b\rangle$ of scheduling parameters. The* solution *to problem P_f is a schedule $S = \langle h, A, \ell, b\rangle$ that minimizes the value $f(\mathcal{M}, S)$.*

Definition 4. *The MISCOVERAGE PMS problem is an optimizing PMS problem that uses the miscoverage $mc(\mathcal{M}, S)$ as the cost for each schedule S.*

3 An ASO-Based Implementation

In what follows, we give a baseline ASP encoding called *Elevator* for the MISCOVERAGE PMS problem in the language fragment of the GRINGO grounder as described by Gebser et al. in [9]. The Elevator encoding is one of the four encodings introduced and published in connection to [17]. In addition, we announce a new encoding, called Mixed. Since all the five baseline ASP encodings will be used later in our experiments, we give brief characterizations of them.

- Elevator is an AI Planning -style encoding with a generic set of *states* $0, \ldots, n$ for each component subject to restriction $n = 2$. The states indicate various degrees of coverage over the time steps.
- *2-Level* is a related encoding for states 1 and 2 only represented by separate predicates. This encoding is already explained in detail in [17].

Listing 2. PMS encoding: counting maintenance coverage (Elevator)

```
1  time(0..h). comp(C) :- comp(C,_,_). val(0..2).
2  { break(T): time(T), T>0, T<=l } <= b.
3  1 <= { serv(C,T): comp(C) } :- break(T).
4
5  inc(C,T)      :- serv(C,T).
6  dec(C,T+R)    :- serv(C,T), comp(C,R,_), time(T+R).
7  dec(C,L+1)    :- comp(C,R,L), 0<L, time(L+1).
8
9  scnt(C,0,0) :- comp(C,R,0).
10 scnt(C,0,1) :- comp(C,R,L), 0<L.
11
12 scnt(C,T,V+1) :- inc(C,T), not dec(C,T), scnt(C,T-1,V),
13                  comp(C), time(T), val(V), val(V+1).
14 scnt(C,T,V-1) :- not inc(C,T), dec(C,T), scnt(C,T-1,V),
15                  comp(C), time(T), val(V), val(V-1).
16 scnt(C,T,V)   :- scnt(C,T-1,V), not inc(C,T), not dec(C,T),
17                  comp(C), time(T), val(V), time(T-1).
18 scnt(C,T,V)   :- scnt(C,T-1,V), inc(C,T), dec(C,T),
19                  comp(C), time(T), val(V), time(T-1).
20 #minimize { 1@1,C,T: scnt(C,T,0), T>0;
21             1@1,C,T: scnt(C,T,2);       }.
```

- *Compact* is a compacted version of the 2-Level encoding written with one predicate for both states.
- *1-Level* is a transition-less encoding to recognize the occurrence of the state combinations $\{1, 2\}$ and $\{2\}$ from which all three states can be derived.
- *Mixed* is a synthesis of a 1-Level and 2-Level encodings. The state 1 is computed without transitions, whereas the encoding for the state 2 uses a transition from state 1.

In the ASP framework, a machine is encoded in terms of the predicate `comp/3`. For example, the atom `comp(6,11,2)` specifies that $\iota(c_6) = 2$ and $\rho(c_6) = 11$ for the component c_6. Listing 1 has an 8-component machine, with recommended maintenance intervals of the components in the range 4–11, and the initial lifetimes in the range 0–4. The scheduling parameters h, b, and ℓ are encoded as ASP constants `h`, `b`, and `l`, and the maintenance actions $\langle c, t\rangle$ with $A(c, t) = 1$ as atoms `serv`(c, t). In addition, the experiments of the current paper assume that $\ell = h$, although ℓ can also be set freely in the problem encoding.

The encoding is inspired by approaches to AI Planning where we have a set of states, a set of actions, and state transitions enforced by the actions. In Line 1, we define the timeline using `time/1` as a domain predicate, extract the identities of the components and restrict, using the predicate `val/1`, the component-wise state-space based on three states that encode 0-, 1-, and 2-fold coverage for each component and each time step. As to be argued later, this restriction is sufficient to implement miscoverage in the sense of Sect. 2. In Line 2, we select with predicate `break/1` at most `b` time steps for maintenance breaks that occur no later than the time step `l` (the parameter ℓ). Then, for each break, at least one component is selected for maintenance using the predicate `serv/2` in Line 3. These definitions induce the search space of all feasible schedules.

The rest of the encoding describes the miscoverage of the schedule and optimization based on the resulting cost function. We have defined two *derived* action predicates, `inc/2` and `dec/2`, taking the component `C` and the time step `T` as arguments. These actions can co-occur or be both absent, so that the true set of actions is the powerset of {`inc`,`dec`} for each `C` and `T`. The increment action (`inc` in Line 5) encodes the change where the maintenance of a component `C` at a time step `T` covers a set of time steps starting from `T`, possibly incrementing the state counter that indicates how many times the time step is being covered. The decrement action (`dec` in Lines 6–7) indicates that either a recommended maintenance interval or the initial lifetime has just ended and the respective state counter must be decremented to reflect the required change in the count. Note that there is no chance that coverage due to the initial lifetime and coverage due to some maintenance action could end simultaneously.

The service state of the machine at each time step is encoded with the (functional) predicate `scnt(C,T,N)` mapping the component `C` and the time step `T` to the state `N` ranging from `0` to `2`, as restricted by the `val` predicate. The state `0` indicates that the time step of the component is not covered at all. The state `1` indicates that the time step is covered exactly once, while the state `2` indicates a double coverage. In Lines 9–10, the state of the component at the moment before the beginning of the timeline is declared to be either `0` or `1`, depending on the initial lifetime of the component.

Lines ranging from 12 to 19 define state transitions that depend on the combination of the elementary actions `inc` and `dec` at each time step 1...`h`. From the state of the previous time step `T-1`, the component moves deterministically to one of the target states, depending on the combination of the elementary actions at the current time step `T`. The set of actions {`inc`} is treated by Line 12, {`dec`} in Line 14, {} in Line 16, and {`inc`, `dec`} in Line 18. These four cases, respectively, correspond to increasing, decreasing, and (the last two) maintaining coverage through inertial transitions that do not change the state of the component. As a combined effect of the initialization of the state (either `0` or `1`) and the subsequent transitions, the `scnt/3` predicate will map each component and each time step to one of the three states. Only the states of the time steps 1...`h` matter when in comes to the computation of miscoverage, i.e., the sum of the time steps where a component's state is either `0` or `2`. This sum is to be minimized by Lines 20–21.

4 Pruning Constraints and Correctness Proofs

In this section, we present new constraints that can be used to prune the search space for optimal schedules. To this end, let $\mathcal{M} = \langle C, \iota, \rho \rangle$ be a multi-component machine and $S = \langle h, A, \ell, b \rangle$ a PMS for $\mathcal{M}$. For every $c \in C$, recall the set of time steps B_c at which preventive maintenance actions are used to service component c according to S. For each component c, and time step $1 \leq i \leq h$, we define a function analogous to the `scnt/3` predicate of Listing 2 as

$$cnt_S(c,i) = min\Big(\big|\{j \in B_c | j \leq i < j + \rho(c)\}\big| + \big|\{1 | \iota(c) > 0, i \leq \iota(c)\}\big|, 2 \Big). \quad (1)$$

The miscoverage $mc(S, \mathcal{M})$ can be alternatively computed as

$$mc(S, \mathcal{M}) = \big|\{\langle c, i \rangle | c \in C, 1 \leq i \leq h,\ cnt_S(c,i) \neq 1\}\big|. \quad (2)$$

For $k \geq 1$ and $1 \leq t_1 < t_2 \leq h+1$, we also define

$$\sigma(S, c, t_1, t_2, k) = \left|\{i | t_1 \leq i < t_2,\ cnt_S(c, i) = k\}\right|. \tag{3}$$

We now define six properties that can be used for pruning the search space.

Definition 5. *Let $\mathcal{M} = \langle C, \iota, \rho \rangle$ be a multi-component machine, $S = \langle h, A, \ell, b \rangle$ be a solution to the MISCOVERAGE PMS problem with $\mathcal{M}$ as the input, and $t \in B$ be a scheduled maintenance break of S. The schedule S is deemed*

- *lagging at t iff for all components $c \in C$ we have $cnt_S(c, t-1) = 0$;*
- *congested at t iff for all components $c \in C$, we have $cnt_S(c, t) = 2$;*
- *under-serving for a component $c \in C$ at t iff $A(c, t) = 0$ and $\sigma(S, c, t, t', 0) > \sigma(S, c, t, t', 1)$, where $t' = min(t + \rho(c), h)$;*
- *over-serving for a component $c \in C$ at t iff $A(c, t) = 1$ and $\sigma(S, c, t, t', 2) \geq \sigma(S, c, t, t', 1)$, where $t' = min(t + \rho(c), h)$;*
- *under-tight for a component $c \in C$ at t iff $cnt_S(c, t) = 0$;*
- *over-tight for a component $c \in C$ at t iff $cnt_S(c, t-1) = 2$.*

For the sake of simplicity, Definition 5 assumes that ℓ is equal to h. However, if this assumption does not hold, it suffices to require t of Definition 5 to be strictly smaller than ℓ. In what follows, we prove that at least one globally optimal schedule will be preserved after pruning all schedules that have any of the properties of Definition 5. Let $\mathcal{M} = \langle C, \iota, \rho \rangle$ be a multi-component machine, $\langle h, \ell, b \rangle$ be a triple of scheduling parameters, and $\mathcal{P}$ be the MISCOVERAGE PMS problem given $\mathcal{M}$ and $\langle h, \ell, b \rangle$ as the input. For every solution $S = \langle h, A, \ell, b \rangle$ to $\mathcal{P}$, we define a measure $\|\cdot\|_{\mathcal{M}}$ for *service delay* by setting

$$\|S\|_{\mathcal{M}} = \sum_{c \in C,\ 1 \leq t \leq h,\ A(c,t)=1} t. \tag{4}$$

Lemma 1 shows that congested schedules are redundant and can be eliminated by delaying service actions without increasing miscoverage. In other words, any congested schedule is either symmetric to a non-congested one, or is sub-optimal and therefore can be eliminated from the search space.

Lemma 1. *There is a solution to $\mathcal{P}$ that is not congested at any time step.*

Proof. Let $S = \langle h, A, \ell, b \rangle$ be a solution to $\mathcal{P}$. Assume that i is a time step at which S is congested. Now i cannot be equal to h, since otherwise we can improve the miscoverage by setting $A(c, i) = 0$ for $c \in C$. Moreover, if $A(c, i) = 1$, then $A(c, i+1) = 0$, otherwise we can improve the miscoverage of S by setting $A(c, i) = 0$, which contradicts the optimality of S. Also, if $A(c, i) = 1$, then $cnt_S(c, i + \rho(c)) = 1$, otherwise we can improve the miscoverage by setting $A(c, i) = 0$ and $A(c, i+1) = 1$. Considering these properties, we transform S to a schedule $S' = \langle h, A', \ell, b \rangle$ by setting $A'(c, i) = 0$ for all $c \in C$, $A'(c, i+1) = 1$ for all $c \in C$ such that $A(c, i) = 1$ or $A(c, i+1) = 1$, and $A'(c, t) = A(c, t)$ for all $c \in C$ and $1 \leq t \leq b$ such that $t \neq i$ and $t \neq i+1$. This transformation moves all service actions at time step i to the time step $i+1$ and we have $mc(S', \mathcal{M}) = mc(S, \mathcal{M})$. Note that the number of maintenance breaks does not increase by this transformation. Therefore, S' is also a solution to $\mathcal{P}$. Furthermore, we have $\|S\|_{\mathcal{M}} < \|S'\|_{\mathcal{M}}$. Since $\|\cdot\|_{\mathcal{M}}$ is bounded from above, by repetition of this transformation one can produce a solution to $\mathcal{P}$ that is not congested at any time step. □

We now show that over-tight schedules are also redundant, because they are either symmetric to schedules that are not over-tight, or can be pruned as sub-optimal. The main observation here is that one can delay service actions in over-tight schedules without increasing miscoverage.

Lemma 2. *There is a solution to $\mathcal{P}$ that is not over-tight for any component at any time step.*

Proof. Let $S = \langle h, A, \ell, b \rangle$ be a solution to $\mathcal{P}$. Assume that i is the smallest number such that S is over-tight for component $c \in C$ at time step i. Then, there must exist time step $j < i$ such that $A(c, j) = 1$ and $A(c, k) = 0$ for $j < k < i$. We have $cnt_S(c, k) = 2$ for $j \leq k < i$. Moreover, we have $cnt_S(c, \rho(c) + k) = 1$ for $j < k < i$, otherwise we can improve miscoverage by setting $A(c, j) = 0$ and $A(c, i) = 1$, which contradicts the optimality of S. Moreover, $A(c, i) \neq 1$, otherwise we can decrease miscoverage by setting $A(c, j) = 0$. We transform S to a schedule $S' = \langle h, A', \ell, b \rangle$, by setting $A'(c, j) = 0$ and $A'(c, i) = 1$, and $A'(c', t) = A(c', t)$ for $\langle c', t \rangle \neq \langle c, i \rangle$. Then S' is not over-tight for c at any time step $j \leq i$, and we have $mc(S', \mathcal{M}) = mc(S, \mathcal{M})$. By repetitive application of the mentioned transformation, one can produce a solution to $\mathcal{P}$ that is not over-tight for any component at any time step. □

Lemma 3 shows that elimination processes of congested and over-tight schedules explained in the proofs of Lemmas 1 and 2 do not conflict with each other.

Lemma 3. *There is a solution to $\mathcal{P}$ that (i) is not congested at any time step; and (ii) has no component making the solution over-tight at any time step.*

Proof. The transformations used in the proofs of Lemmas 1 and 2 increase $\|\cdot\|_{\mathcal{M}}$. The proof is complete by considering that $\|\cdot\|_{\mathcal{M}}$ is bounded from above. □

In analogy to congested and over-tight schedules, lagging and under-tight schedules can either be eliminated as symmetric to other schedules or pruned as sub-optimal by preceding the service actions without increasing miscoverage. We show by the following two lemmas that this is possible without reproducing congested or over-tight schedules.

Lemma 4. *There is a solution to $\mathcal{P}$ that is not lagging nor congested at any time step, nor is over-tight for any component at any time step.*

Proof. According to Lemma 3, there is a solution $S = \langle h, A, \ell, b \rangle$ to $\mathcal{P}$ that is not congested at any time step, nor is over-tight for any component at any time step.

Assume that i is a time step at which S is lagging. By definition, i cannot be equal to 1. Also, if $A(c,i) = 1$, then $i+\rho(c)-1 \leq h$ and $cnt_S(c, i+\rho(c)-1) = 1$, otherwise we can improve miscoverage by setting $A(c,i) = 0$ and $A(c,i-1) = 1$. Considering these properties, we transform S to a schedule $S' = \langle h, A', \ell, b\rangle$, by setting $A'(c,i) = 0$ for every $c \in C$, $A'(c,i-1) = 1$ for every $c \in C$ such that $A(c,i) = 1$, and $A'(c,t) = A(c,t)$ for every $c \in C$ and $1 \leq t \leq b$ such that $t \neq i-1$ and $t \neq i$. We have $mc(S', \mathcal{M}) = mc(S, \mathcal{M})$. This transformation moves all service actions at time step i to time step $i-1$ in the case that no component is being serviced at time step $i-1$. Therefore, the number of maintenance breaks does not increase, and also, for each $c \in C$ and $1 \leq t \leq h$, $cnt_{S'}(c,t) = 2$ only if $cnt_S(c,t) = 2$. We can conclude that S' is not congested at any time step, nor over-tight for any component at any time step. Furthermore, we have $\|S\|_{\mathcal{M}} > \|S'\|_{\mathcal{M}}$. Since $\|\cdot\|_{\mathcal{M}}$ is bounded from below, by repetition of the this transformation one can produce a solution to $\mathcal{P}$ that is not lagging nor congested at any time step, nor is over-tight for any component at any time step. □

Lemma 5. *There is a solution to $\mathcal{P}$ that is not congested at any time step, nor is under-tight nor over-tight for any component at any time step.*

Proof. According to Lemma 3, there is a solution $S = \langle h, A, \ell, b\rangle$ to $\mathcal{P}$ that is not congested at any time step, nor is over-tight for any component at any time step. Assume that i is the smallest number such that S is under-tight for component $c \in C$ at time step i. Then, there must exist time step $j > i$ such that $A(c,j) = 1$ and $cnt_S(c,k) = 0$ for $i \leq k < j$, otherwise we can improve miscoverage by setting $A(c,i) = 1$, which contradicts the optimality of S. Moreover, by definition, i cannot be equal to 1. We transform S to a schedule $S' = \langle h, A', \ell, b\rangle$, by setting $A'(c,j) = 0$ and $A'(c,i) = 1$, and $A'(c',t) = A(c',t)$ for any $\langle c', t\rangle \neq \langle c, i\rangle$. The schedule S' is not under-tight for c at any time step $j \leq i$, and we have $mc(S', \mathcal{M}) = mc(S, \mathcal{M})$. This transformation moves the service action of c from time step j to time step i where c is not being serviced at any time step between i and j. Therefore, for any time step $1 \leq t \leq b$, $cnt_{S'}(c,t) = 2$ only if $cnt_S(c,t) = 2$. Thus, S' is not congested at any time step, nor over-tight for any component at any time step. By iterating this transformation, one can produce a solution to $\mathcal{P}$ that gets never congested, nor is under-tight nor over-tight for any component at any time step. □

We now show that if a schedule is over-serving for some component, then a service action for that component can be removed without increasing miscoverage, and this is possible without reproducing congested or over-tight schedules.

Lemma 6. *There is a solution to $\mathcal{P}$ that is not congested at any time step, nor is over-tight nor over-serving for any component at any time step.*

Proof. According to Lemma 3, there is a solution $S = \langle h, A, \ell, b\rangle$ to $\mathcal{P}$ that is not congested at any time step, nor is over-tight for any component at any time step. If S is over-serving for a component $c_i \in C$ at time step t_1, we can transform it to the schedule $S' = \langle h, A', \ell, b\rangle$ by setting $A'(c,t) = 0$, for any $\langle c, t\rangle = \langle c_i, t_1\rangle$,

Listing 3. Pruning constraints

```
:- comp(C), break(T), scnt(C,T-1,2).
:- comp(C), break(T), scnt(C,T,0).

:- comp(C), serv(C,T1),
   0 >= #sum{ -1,T2: scnt(C,T2,2), time(T2), T2>=T1, T2<T1+R;
               1,T2: scnt(C,T2,1), time(T2), T2>=T1, T2<T1+R }.
:- comp(C), not serv(C,T1), break(T1),
   0 > #sum{ -1,T2: not scnt(C,T2,1), time(T2), T2>=T1, T2<T1+R;
              1,T2: scnt(C,T2,1),     time(T2), T2>=T1, T2<T1+R }.

:- break(T), scnt(C,T,2): comp(C).
:- T>1, break(T), not scnt(C,T-1,1): comp(C).
```

and $A'(c,t) = A(c,t)$, otherwise. Then we have $mc(S,\mathcal{M}) - mc(S',\mathcal{M}) = \sigma(S,c_i,t_1,t_2,2) - \sigma(S,c_i,t_1,t_2,1) \geq 0$, where $t_2 = min(t_1+\rho(c_i),h)$. This transformation removes the preventive maintenance action used to service component c_i at time step t_1, and therefore does not cause S' to be congested at any time step, nor be over-tight for any component at any time step. By repetitive application of this transformation, one can produce a solution to $\mathcal{P}$ that is not congested at any time step, nor is over-tight nor over-serving for any component at any time step. □

Note that according to the proof of Lemma 6, S is symmetric to S' if $\sigma(S,c_i,t_1,t_2,2) = \sigma(S,c_i,t_1,t_2,1)$. Also, S is sub-optimal if $\sigma(S,c,c_i,t_1,t_2,2) > \sigma(S,c_i,t_1,t_2,1)$. In both cases, S can safely be eliminated from the search space.

Similarly, we show that if a schedule is under-serving for some component, then a service action for that component can be added to the schedule, which decreases miscoverage. In other words, under-serving schedules can be safely pruned as sub-optimal ones.

Lemma 7. *No solution to $\mathcal{P}$ is under-serving for any component at any time step.*

Proof. Assume the contrary, i.e., let $S = \langle h, A, \ell, b\rangle$ be a solution to $\mathcal{P}$ such that for some $c_i \in C$, we have $A(c_i,t_1) = 0$ and $\sigma(S,c_i,t_1,t_2,0) > \sigma(S,c_i,t_1,t_2,1)$, where $t_2 = min(t_1+\rho(c),h)$. Let $S' = \langle h, A', \ell, b\rangle$, where $A'(c,t) = 1$ for $\langle c,t\rangle = \langle c_i,t_1\rangle$, and $A'(c,t) = A(c,t)$, otherwise. We have $mc(S,\mathcal{M}) - mc(S',\mathcal{M}) = \sigma(S,c_i,t_1,t_2,0) - \sigma(S,c_i,t_1,t_2,1) > 0$, contradicting the optimality of S. □

We are now ready to present the main theoretical result of this paper. Theorem 1 shows that at least one optimal schedule will exist after pruning all schedules that have any of the properties introduced by Definition 5.

Theorem 1. *There is a solution to $\mathcal{P}$ that is not congested nor lagging at any time step, nor is under-tight, over-tight, over-serving, nor under-serving for any component at any time step.*

Proof. According to Lemma 3, there is a solution $S = \langle h, A, \ell, b\rangle$ to $\mathcal{P}$ that is not congested at any time step, nor is over-tight for any component at any time step.

By Lemma 7, S is not under-serving for any component at any time step. The transformations used in the proofs of Lemmas 4, 5, and 6 decrease the measure $\|\cdot\|_{\mathcal{M}}$. The proof is complete by noting that $\|\cdot\|_{\mathcal{M}}$ is bounded from below. □

Listing 3 shows the ASP constraints for encoding our pruning constraints when the Elevator encoding explained in Sect. 3 has been used as the baseline encoding. In the case of a different baseline encoding, minor modifications might be necessary. Lines 1 and 2 eliminate answer sets representing over-tight and under-tight schedules from the search space, respectively. Lines 4–6 and 7–9 guarantee that no answer set represents an over-serving nor an under-serving schedule, respectively. Lines 11 and 12 forbid congested and lagging schedules to be represented by any answer set, respectively. Thus, adding these constraints to the baseline ASP encoding is safe, as it does not cause all optimal answer sets to be pruned.

5 Experiments

Our experiments were conducted on a SLURM-based cluster containing 140 nodes with on Intel® Xeon® CPUs (E5-2620 v3 at 2.40 GHz) and over 3000+ cores. The search for optimal schedules was implemented on this cluster using the version 3.3.8 of the CLASP solver [8] and WASP 2.0 [1]. The specific goal of the experiments was to quantify the execution time differences between baseline encodings and their variants modified to include pruning constraints. For the scaling experiments, we generated ten random machines for each machine size from 1 to 16 components, thus giving rise to 160 test instances in total. For each of these, the *empty schedule* (with $B = \emptyset$) is one of the feasible solutions but does not usually reach the optimal miscoverage for the machine. In addition, ten random 8-component machines were generated and used to check the effect of a large timelines on the execution time. All experiments respected the timeout at 2^{15} s (roughly 9 h).

The first experiment measured the runtimes of ten different ASO encodings and two optimization strategies of CLASP. The results of the experiment are shown in the first cactus plot in Fig. 1. The line colors denote five baseline encodings: 1-Level, 2-Level, Mixed, Compact 2-Level, and Elevator. Four solving methods are indicated with four line shapes, showing optimization based on both branch-and-bound (BB) and unsatisfiable core (USC) strategies of CLASP, and the contrast between the unmodified encoding and the one modified with pruning constraints. We observe some clearly notable differences between the runtimes. The first observation highlights the strategy: the BB strategy handles 95–111 instances with various baseline encodings, and 137–153 instances with extended encodings. On the contrary, the USC strategy handles 150–158 instances with baseline encodings and 159–160 instances with extended encodings. Thus, regardless of the encoding used, the USC strategy of CLASP handles more instances than the BB strategy. Therefore, we decided to use the USC strategy in the later experiments only.

The second observation is that pruning constraints help CLASP especially when the BB strategy is used. In addition, they also help to treat more difficult instances when the USC strategy is used. The improvement with the USC

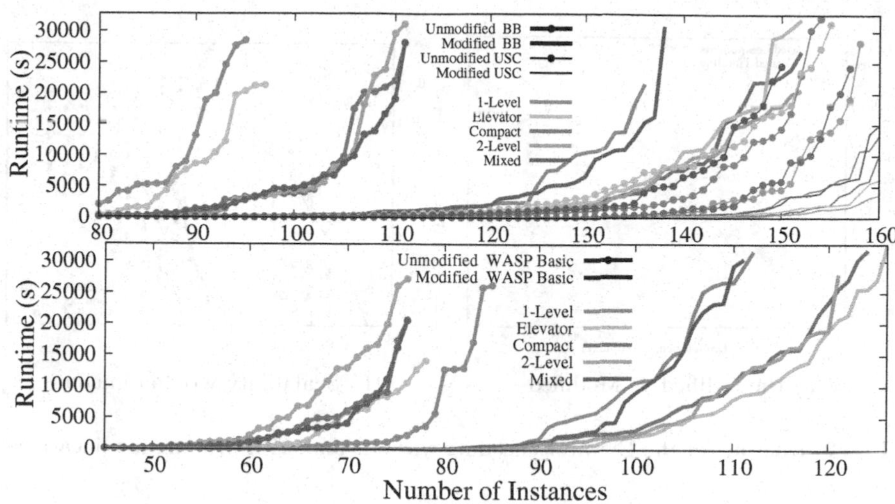

Fig. 1. Cactus plots of the execution times of different encodings showing the effect of modification and the optimization strategy on two solvers: CLASP (above) and WASP (below).

strategy seems to be slightly greater than the effect of pruning alone, but the fastest method is obtained, as expected, when we combine the USC strategy with pruning. The observation about the effectiveness of the pruning constraints is also replicated on WASP in the second plot of Fig. 1 showing a similar contrast between the unmodified and modified encodings when using the basic optimization strategy of the solver.

The remaining observations concern the effect of pruning constraints when added to the baseline encodings on the CLASP solver. First, the baseline Elevator is among the slowest encodings under both BB and USC strategies, but its modified variant runs up to the proximity of 2-Level encoding. Second, 2-Level and Compact 2-Level encodings are fast and behave pretty much alike under all four solving methods. Yet the modified 2-level encoding seems to tackle difficult instances in less time than the Compact encoding. Third, we observe that the USC optimization strategy and pruning constraints are less effective with the Mixed encoding than with the other encodings: it is among the fastest baseline encodings under the BB strategy, while others benefit more from pruning constraints and the USC strategy.

A different kind of experiment on CLASP focused on the Elevator encoding. The scatter plot in Fig. 2a summarizes two such experiments where the first one tested the versions of the encoding with a fixed timeline $h = 32$ and random machines whose component sizes varied in the range from 1 to 16, and the second one tested the versions of the encoding with ten random 8-component machines and timelines ranging from 16 to 64 time steps. In these experiments, each parameter value was tested with 10 random machines. The diagonal lines indicate that the runtimes of the modified encoding are an order of magnitude

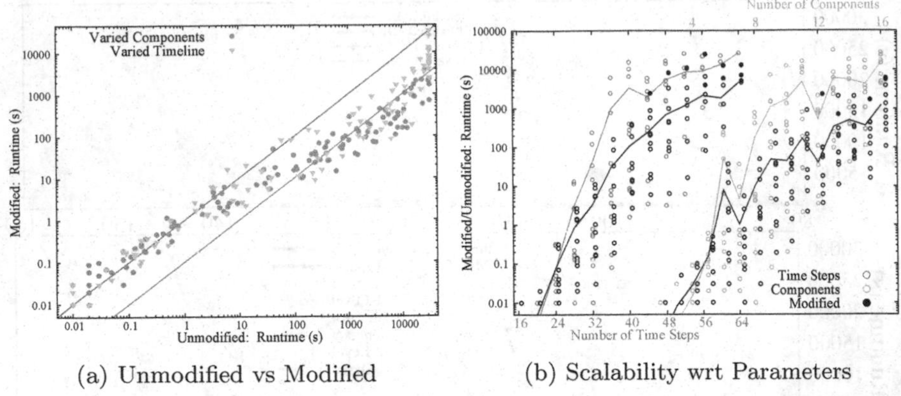

(a) Unmodified vs Modified

(b) Scalability wrt Parameters

Fig. 2. Runtime of the Elevator encoding without and with pruning modifications

better than the baseline in both experiments. The same experiments have been viewed from another perspective in Fig. 2b. This scatter plot and the average lines show how the parameter of each experiment correlates with the running time. By them, the running time grows exponentially according to both the timeline size and the machine size. The effect of increasing the number of time steps seems to be moderated, probably because of the small number of breaks with which the over-coverage can be fully avoided. Furthermore, we observe that the baseline encodings modified by adding the pruning constraints improve the runtimes roughly by a factor of 10, and a few solid circles indicate that the modified encoding can solve more instances than the unmodified encoding.

Based on our observations, we recap that the experiments support a significant advantage when extending the baseline encodings with pruning constraints. The effect of this modification seemed to be largely independent of the choice of the optimization strategy and the combination of both techniques gives the best result. Moreover, these improvements turned out to depend on the base encoding used, but similar for all. Regardless of the choice of these techniques, 2-Level and Compact 2-Level encodings kept their leading positions consistently. In our tests, scalability with respect to both components and timeline appear very similar. As a further reflection, we note that the present experiments lancer more rigorous practice in comparison to the earlier work [17]. All tests were carried out with random machines, while allowing test reiteration, and averages were replaced with matched scatter plots. We also extended the scalability tests to the effect of timeline, an important dimension of the problem instances. However, larger and new kinds of experiments to separate scalability with respect to components and timelines seem to be necessary in the future.

6 Conclusion

In this article, we continue the previous research [17] on an intractable optimization problem, viz. the minimization of miscoverage in schedules devised for the preventive maintenance of a multi-component machine. In this follow-up study, we design additional constraints for pruning the search space and evaluate their effect on the search for optimal solutions. The original preventive maintenance scheduling (PMS) problem and the novel pruning constraints are both encoded as logic programs in the framework of answer set programming (ASP) and optimization (ASO). In this way, we follow the overall methodology set up in [17]. As regards technical results, we develop a principled approach to motivate pruning based on the problem structure. In particular, we establish seven lemmas about the properties of *pruning constraints* and their correctness and draw these lemmas together in a cap-stone theorem, Theorem 1.

To quantify the effect of the pruning constraints on optimization time, we carried out several experiments with random multi-component machines and ranges of parameters that included machine size (1–16 components) and timeline length (16–64 time steps) as the basis for scalability analysis. These machines are used to instantiate five different (baseline) encodings of the PMS problem, subsequently solved through two different optimization strategies, viz. branch and bound (BB) and unsatisfiable core (USC), as implemented in the CLINGO system. According to our experiments in general, the pruning constraints often improve the efficiency of all encodings by a factor of 10, except when the problem instance gets so small that the overhead of introducing the pruning constraints dominates the time taken by optimization. The experiments also extend to WASP with very similar results on the effectiveness of the pruning constraints.

It is worth emphasizing that pruning constraints designed in this work retain at least one optimal schedule in the search space, i.e., there is no loss in the quality of the schedules in this respect. This is possible since the solutions to the optimization problem are not further constrained by other constraints. However, if the problem is extended by *resource constraints*, e.g., concerning the availability and expertise of service personnel, pruning constraints may have to be treated as soft constraints or as secondary components in the objective function. Although this might cancel some of the speed-up perceived in the experiments reported in this work, pruning constraints nevertheless demonstrate that a tightened representation of the search space can push the practical solvability limits of the PMS optimization problem forward.

In the future, the interaction between pruning constraints and supplementary resource constraints in the problem formulation call for further attention. Yet another dimension is provided by the stochastic aspects of component failures which can be used to enrich the models of preventive maintenance.

Acknowledgments. The support from the Academy of Finland within the projects AI-ROT (#335718) and XAILOG (#345633) is gratefully acknowledged.

References

1. Alviano, M., Dodaro, C., Leone, N., Ricca, F.: Advances in WASP. In: Calimeri, F., Ianni, G., Truszczynski, M. (eds.) LPNMR 2015. LNCS (LNAI), vol. 9345, pp. 40–54. Springer, Cham (2015). https://doi.org/10.1007/978-3-319-23264-5_5

2. Banbara, M., et al.: teaspoon: solving the curriculum-based course timetabling problems with answer set programming. Ann. Oper. Res. **275**, 3–37 (2019). https://doi.org/10.1007/s10479-018-2757-7
3. Brewka, G., Eiter, T., Truszczynski, M.: Answer set programming at a glance. Commun. ACM **54**(12), 92–103 (2011). https://doi.org/10.1145/2043174.2043195
4. Do, P., Vu, H.C., Barros, A., Bérenguer, C.: Maintenance grouping for multi-component systems with availability constraints and limited maintenance teams. Reliab. Eng. Syst. Saf. **142**, 56–67 (2015). https://doi.org/10.1016/j.ress.2015.04.022
5. Dodaro, C., Maratea, M.: Nurse scheduling via answer set programming. In: Balduccini, M., Janhunen, T. (eds.) LPNMR 2017. LNCS (LNAI), vol. 10377, pp. 301–307. Springer, Cham (2017). https://doi.org/10.1007/978-3-319-61660-5_27
6. Drescher, C., Tifrea, O., Walsh, T.: Symmetry-breaking answer set solving. AI Commun. **24**(2), 177–194 (2011)
7. Eiter, T., Geibinger, T., Musliu, N., Oetsch, J., Skočovský, P., Stepanova, D.: Answer-set programming for lexicographical makespan optimisation in parallel machine scheduling. In: Proceedings of the 18th International Conference on Principles of Knowledge Representation and Reasoning (KR 2021), pp. 280–290. IJCAI Organization (2021). https://doi.org/10.24963/kr.2021/27
8. Gebser, M., Kaminski, R., Kaufmann, B., Romero, J., Schaub, T.: Progress in *clasp* series 3. In: Calimeri, F., Ianni, G., Truszczynski, M. (eds.) LPNMR 2015. LNCS (LNAI), vol. 9345, pp. 368–383. Springer, Cham (2015). https://doi.org/10.1007/978-3-319-23264-5_31
9. Gebser, M., Kaminski, R., Ostrowski, M., Schaub, T., Thiele, S.: On the input language of ASP grounder *Gringo*. In: Erdem, E., Lin, F., Schaub, T. (eds.) LPNMR 2009. LNCS (LNAI), vol. 5753, pp. 502–508. Springer, Heidelberg (2009). https://doi.org/10.1007/978-3-642-04238-6_49
10. Heule, M., Walsh, T.: Symmetry in solutions. In: AAAI 2010, pp. 77–82 (2010). https://doi.org/10.1609/aaai.v24i1.7549
11. Li, C.M., Manyà, F.: MaxSAT, hard and soft constraints. In: Handbook of Satisfiability, Frontiers in Artificial Intelligence and Applications, vol. 185, pp. 613–631. IOS Press (2009)
12. Luukkala, V., Niemelä, I.: Enhancing a smart space with answer set programming. In: Dean, M., Hall, J., Rotolo, A., Tabet, S. (eds.) RuleML 2010. LNCS, vol. 6403, pp. 89–103. Springer, Heidelberg (2010). https://doi.org/10.1007/978-3-642-16289-3_9
13. Rossi, F., van Beek, P., Walsh, T.: Constraint programming. In: Handbook of Knowledge Representation, Foundations of Artificial Intelligence, vol. 3, pp. 181–211. Elsevier (2008). https://doi.org/10.1016/S1574-6526(07)03004-0
14. Simons, P., Niemelä, I., Soininen, T.: Extending and implementing the stable model semantics. Artif. Intell. **138**(1–2), 181–234 (2002). https://doi.org/10.1016/S0004-3702(02)00187-X
15. Tarzariol, A.: A model-oriented approach for lifting symmetry-breaking constraints in answer set programming. In: IJCAI 2022, pp. 5875–5876 (2022). https://doi.org/10.24963/ijcai.2022/840
16. Tarzariol, A., Schekotihin, K., Gebser, M., Law, M.: Efficient lifting of symmetry breaking constraints for complex combinatorial problems. Theory Pract. Logic Program. **22**(4), 606–622 (2022). https://doi.org/10.1017/S1471068422000151
17. Yli-Jyrä, A., Janhunen, T.: Applying answer set optimization to preventive maintenance scheduling for rotating machinery. In: Governatori, G., Turhan, A.Y. (eds.) Rules and Reasoning. LNCS, vol. 13752, pp. 3–19. Springer, Cham (2022). https://doi.org/10.1007/978-3-031-21541-4_1

Automatic Rollback Suggestions for Incremental Datalog Evaluation

David Zhao[1(✉)], Pavle Subotić[2], Mukund Raghothaman[3], and Bernhard Scholz[1]

[1] University of Sydney, Sydney, Australia
{david.zhao,bernhard.scholz}@sydney.edu.au
[2] Microsoft, Redmond, USA
pavlesubotic@microsoft.com
[3] University Southern California, Los Angeles, USA
raghotha@usc.edu

Abstract. Advances in incremental Datalog evaluation strategies have made Datalog popular among use cases with constantly evolving inputs such as static analysis in continuous integration and deployment pipelines. As a result, new logic programming debugging techniques are needed to support these emerging use cases.

This paper introduces an incremental debugging technique for Datalog, which determines the failing changes for a *rollback* in an incremental setup. Our debugging technique leverages a novel incremental provenance method. We have implemented our technique using an incremental version of the Soufflé Datalog engine and evaluated its effectiveness on the DaCapo Java program benchmarks analyzed by the Doop static analysis library. Compared to state-of-the-art techniques, we can localize faults and suggest rollbacks with an overall speedup of over 26.9× while providing higher quality results.

1 Introduction

Datalog has achieved widespread adoption in recent years, particularly in static analysis use cases [2,4,8,18,19,22,41] that can benefit from incremental evaluation. In an industrial setting, static analysis tools are deployed in continuous integration and deployment setups to perform checks and validations after changes are made to a code base [1,12]. Assuming that changes between analysis runs (aka. epochs) are small enough, a static analyzer written in Datalog can be effectively processed by incremental evaluation strategies [27,28,30,39] which recycle computations of previous runs. When a fault appears from a change in the program, users commonly need to (1) localize which changes caused the fault and (2) partially roll back the changes so that the faults no longer appear. However, manually performing this bug localization and the subsequent rollback is impractical, and users typically perform a full rollback while investigating the fault's actual cause [35,36]. The correct change is re-introduced when the fault is

M. Hanus and D. Inclezan (Eds.): PADL 2023, LNCS 13880, pp. 295–312, 2023.
https://doi.org/10.1007/978-3-031-24841-2_19

found and addressed, and the program is re-analyzed. This entire debugging process can take significant time. Thus, an automated approach for detecting and performing partial rollbacks can significantly enhance developer productivity.

Existing state-of-the-art Datalog debugging techniques that are available employ data provenance [25,40] or algorithmic debugging [10] to provide explanations. However, these techniques require a deep understanding of the tool's implementation and target the ruleset, not the input. Therefore, such approaches are difficult to apply to automate input localization and rollback. The most natural candidate for this task is *delta debugging* [37,38], a debugging framework for generalizing and simplifying a failing test case. This technique has recently been shown to scale well when integrated with state-of-the-art Datalog synthesizers [29] to obtain better synthesis constraints. Delta debugging uses a divide-and-conquer approach to localize the faults when changes are made to a program, thus providing a concise witness for the fault. However, the standard delta debugging approach is programming language agnostic and requires programs to be re-run, which may require significant time.

In this paper, we introduce a novel approach to automating localize-rollback debugging. Our approach comprises a novel incremental provenance technique and two intertwined algorithms that diagnose and compute a rollback suggestion for a set of faults (missing and unwanted tuples). The first algorithm is a *fault localization* algorithm that reproduces a set of faults, aiding the user in diagnosis. Fault localization traverses the incremental proof tree provided by our provenance technique, producing the subset of an incremental update that causes the faults to appear in the current epoch. The second algorithm performs an *input repair* to provide a local *rollback suggestion* to the user. A rollback suggestion is a subset of an incremental update, such that the faults are fixed when it is rolled back.

We have implemented our technique using an extended incremental version of the Soufflé [23,39] Datalog engine and evaluated its effectiveness on DaCapo [6] Java program benchmarks analyzed by the Doop [8] static analysis tool. Compared to delta debugging, we can localize and fix faults with a speedup of over 26.9× while providing smaller repairs in 27% of the benchmarks. To the best of our knowledge, we are the first to offer such a debugging feature in a Datalog engine, particularly for large workloads within a practical amount of time. We summarize our contributions as follows:

- We propose a novel debugging technique for incremental changing input. We employ localization and rollback techniques for Datalog that scale to real-world program analysis problems.
- We propose a novel incremental provenance mechanism for Datalog engines. Our provenance technique leverages incremental information to construct succinct proof trees.
- We implement our technique in the state-of-the-art Datalog engine Soufflé, including extending incremental evaluation to compute provenance.

- We evaluate our technique with Doop static analysis for large Java programs and compare it to a delta-debugging approach adapted for the localization and rollback problem.

2 Overview

```
admin = new Admin();
sec = new AdminSession();
ins = new InsecureSession();
admin.session = ins;
if (admin.isAdmin && admin.isAuth)
    admin.session = sec;
else
    userSession = ins;
```

(a) Input Program

```
new(admin,L1).
new(sec,L2).
new(ins,L3).
store(admin,session,ins).

store(admin,session,sec).

assign(userSession,ins).
```

(b) EDB Tuples

```
// r1: var = new Obj()
vpt(Var, Obj) :- new(Var, Obj).
// r2: var = var2
vpt(Var, Obj) :- assign(Var, Var2), vpt(Var2, Obj).
// r3: v = i.f; i2.f = v2 where i, i2 point to same obj
vpt(Var, Obj) :- load(Var, Inter, F), store(Inter2, F, Var2),
                 vpt(Inter, InterObj), vpt(Inter2, InterObj),
                 vpt(Var2, Obj).
// r4: v1, v2 point to same obj
alias(V1, V2) :- vpt(V1, Obj), vpt(V2, Obj), V1 != V2.
```

(c) Datalog Points-to Analysis

Fig. 1. Program analysis datalog setup

2.1 Motivating Example

Consider a Datalog pointer analysis in Fig. 1. Here, we show an input program to analyze (Fig. 1a), which is encoded as a set of tuples (Fig. 1b) by an *extractor*, which maintains a mapping between tuples and source code [23,31,33]. We have relations `new`, `assign`, `load`, and `assign` capturing the semantics of the input program to analyze. These relations are also known as the *Extensional Database* (EDB), representing the analyzer's input. The analyzer written in Datalog computes relations `vpt` (*Variable Points To*) and `alias` as the output, which is also known as the *Intensional Database* (IDB). For the points-to analysis, Fig. 1c has four rules. A rule is of the form: $R_h(X_h)$:- $R_1(X_1), \ldots, R_k(X_k)$. Each $R(X)$ is a *predicate*, with R being a *relation* name and X being a vector of *variables* and *constants* of appropriate arity. The predicate to the left of :- is the *head* and the sequence of predicates to the right is the *body*. A Datalog rule can be read from right to left: "for all rule instantiations, if every tuple in the body is derivable, then the corresponding tuple for the head is also derivable".

For example, r_2 is `vpt(Var,Obj) :- assign(Var,Var2), vpt(Var2,Obj)`, which can be interpreted as "if there is an assignment from `Var` to `Var2`, and if `Var2` may point to `Obj`, then also `Var` may point to `Obj`". In combination, the four rules represent a *flow-insensitive* but *field-sensitive* points-to analysis. The IDB relations `vpt` and `alias` represent the analysis result: variables may point to objects and pairs of variables that may be an alias with each other.

Suppose the input program in Fig. 1a changes by adding a method to upgrade a user session to an admin session with the code:

```
upgradedSession = userSession;
userSession = admin.session;
```

The result of the points-to analysis can be incrementally updated by inserting the tuples `assign(upgradedSession, userSession)` and `load(userSession, admin, session)`. After computing the incremental update, we would observe that `alias(userSession, sec)` is now contained in the output. However, we may wish to maintain that `userSession` *should not* alias with the secure session `sec`. Consequently, the incremental update has introduced a *fault*, which we wish to localize and initiate a rollback.

A fault localization provides a subset of the incremental update that is sufficient to reproduce the fault, while a rollback suggestion is a subset of the update which fixes the faults. In this particular situation, the fault localization and rollback suggestion are identical, containing only the insertion of the second tuple, `load(userSession, admin, session)`. Notably, the other tuple in the update, `assign(upgradedSession, userSession)`, is irrelevant for reproducing or fixing the fault and thus is not included in the fault localization/rollback.

In general, an incremental update may contain thousands of inserted and deleted tuples, and a set of faults may contain multiple tuples that are changed in the incremental update. Moreover, the fault tuples may have multiple alternative derivations, meaning that the localization and rollback results are different. In these situations, automatically localizing and rolling back the faults to find a small relevant subset of the incremental update is essential to provide a concise explanation of the faults to the user.

The scenario presented above is common during software development, where making changes to a program causes faults to appear. While our example concerns a points-to analysis computed for a source program, our fault localization and repair techniques are, in principle, applicable to any Datalog program.

Problem Statement: Given an incremental update with its resulting faults, automatically find a fault localization and rollback suggestion.

2.2 Approach Overview

An overview of our approach is shown in Fig. 2. The first portion of the system is the incremental Datalog evaluation. Here, the incremental evaluation takes an

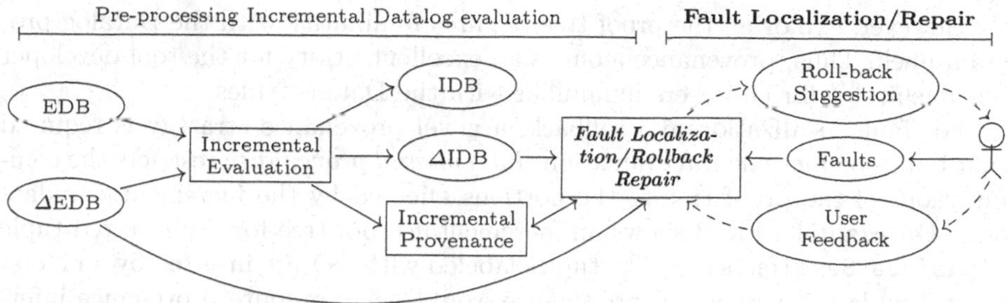

Fig. 2. Fault localization and repair system

EDB and an incremental update containing tuples inserted or deleted from the EDB, denoted ΔEDB. The result of the incremental evaluation is the output IDB, along with the set of IDB insertions and deletions from the incremental update, denoted ΔIDB. The evaluation also enables incremental provenance, producing a proof tree for a given query tuple.

The second portion of the system is the fault localization/rollback repair. This process takes a set of faults provided by the user, which is a subset of ΔIDB where each tuple is either unwanted and inserted in ΔIDB or is desirable but deleted in ΔIDB. Then, the fault localization and rollback repair algorithms use the full ΔIDB and ΔEDB, along with incremental provenance, to produce a localization or rollback suggestion.

The main fault localization and rollback algorithms work in tandem to provide localizations or rollback suggestions to the user. The key idea of these algorithms is to compute proof trees for fault tuples using the provenance utility provided by the incremental Datalog engine. These proof trees directly provide localization for the faults. For fault rollback, the algorithms create an Integer Linear Programming (ILP) instance that encodes the proof trees, with the goal of *disabling* all proof trees to prevent the fault tuples from appearing.

The result is a localization or rollback suggestion, which is a subset of ΔEDB. For localization, the subset $S \subseteq \Delta\text{EDB}$ is such that if we were to apply S to EDB as the diff, the set of faults would be reproduced. For a rollback suggestion, the subset $S \subseteq \Delta\text{EDB}$ is such that if we were to remove S from ΔEDB, then the resulting diff would *not* produce the faults.

3 Incremental Provenance

Provenance [10,29,40] provides machinery to explain the existence of a tuple. For example, the tuple `vpt(userSession, L3)` could be explained in our running example by the following proof tree:

$$\dfrac{\texttt{assign(userSession, ins)} \qquad \dfrac{\texttt{new(ins, L3)}}{\texttt{vpt(ins, L3)}}\, r_1}{\texttt{vpt(userSession, L3)}}\, r_2$$

However, exploring the proof tree requires familiarity with the Datalog program itself. Thus, provenance alone is an excellent utility for the tool developer but unsuitable for end-users unfamiliar with the Datalog rules.

For fault localization and rollback, a novel provenance strategy is required that builds on incremental evaluation. *Incremental provenance* restricts the computations of the proof tree to the portions affected by the incremental update only. For example, Fig. 3 shows an incremental proof tree for the inserted tuple `alias(userSession,sec)`. The tuples labeled with `(+)` are inserted by an incremental update. Incremental provenance would only compute provenance information for these newly inserted tuples and would not explore the tuples in red already established in a previous epoch.

```
                                           new(a,L1)   new(a,L1)   new(s,L2)
load(u,a,s)(+)   store(a,s,s)              vpt(a,L1)   vpt(a,L1)   vpt(s,L2)   new(s,L2)
                        vpt(userSession,L2)(+)                                 vpt(s,L2)
                                      alias(userSession,sec)(+)
```

Fig. 3. The proof tree for `alias(userSession,sec)`. `(+)` denotes tuples that are inserted as a result of the incremental update, red denotes tuples that were not affected by the incremental update.

To formalize incremental provenance, we define inc-prov as follows. Given an incremental update ΔE, inc-prov$(t, \Delta E)$ should consist of tuples that were updated due to the incremental update.

Definition 1. *The set* inc-prov$(t, \Delta E)$ *is the set of tuples that appear in the proof tree for* t*, that are also inserted as a result of* ΔE*.*

A two-phase approach for provenance was introduced in [40]. In the first phase, tuples are annotated with *provenance annotations* while computing the IDB. In the second phase, the user can query a tuple's proof tree using the annotations from the first phase. For each tuple, the important annotation is its minimal proof tree height. For instance, our running example produces the tuple `vpt(userSession, L3)`. This tuple would be annotated with its minimal proof tree height of 2. Formally, the height annotation is 0 for an EDB tuple, or $h(t) = \max\{h(t_1), \ldots, h(t_k)\} + 1$ if t is computed by a rule instantiation t :- $t_1, \ldots, t_k$. The provenance annotations are used in a provenance construction stage, where the annotations form constraints to guide the proof tree construction.

For incremental evaluation, the standard strategies [27,28,30,39] use *incremental annotations* to keep track of when tuples are computed during the execution. In particular, for each tuple, [39] stores the iteration number of the fixpoint computation and a count for the number of derivations. To compute provenance information in an incremental evaluation setting, we observe a correspondence between the iteration number and provenance annotations. A tuple is produced in some iteration if at least one of the body tuples was produced in the previous iteration. Therefore, the iteration number I for a tuple produced in a fixpoint is equivalent to $I(t) = \max\{I(t_1), \ldots, I(t_k)\} + 1$ if t is computed by rule instantiation t :- $t_1, \ldots, t_k$. This definition of iteration number corresponds closely to the height annotation in provenance. Therefore, the iteration number is suitable for constructing proof trees similar to provenance annotations.

For fault localization and rollback, it is also important that the Datalog engine produces only provenance information that is *relevant* for faults that appear after an incremental update. Therefore, the provenance information produced by the Datalog engine should be restricted to tuples inserted or deleted by the incremental update. Thus, we adapt the user-driven proof tree exploration process in [40] to use an automated procedure that enumerates exactly the portions of the proof tree that have been affected by the incremental update.

As a result, our approach for incremental provenance produces proof trees containing only tuples inserted or deleted due to an update. For fault localization and rollback, this property is crucial for minimizing the search space when computing localizations and rollback suggestions.

4 Fault Localization and Rollback Repair

This section describes our approach and algorithms for both the fault localization and rollback problems. We begin by formalizing the problem and then presenting basic versions of both problems. Finally, we extend the algorithms to handle missing faults and negation.

4.1 Preliminaries

We first define a fault to formalize the fault localization and rollback problems. For a Datalog program, a fault may manifest as either (1) an undesirable tuple that appears or (2) a desirable tuple that disappears. In other words, a fault is a tuple that does not match the *intended semantics* of a program.

Definition 2 (Intended Semantics). *The* intended semantics *of a Datalog program P is a pair of sets (I_+, I_-) where I_+ and I_- are desirable and undesirable tuple sets, respectively. An input set E is* correct *w.r.t P and (I_+, I_-) if $I_+ \subseteq P(E)$ and $I_- \cap P(E) = \emptyset$.*

Given an intended semantics for a program, a *fault* can be defined as follows:

Definition 3 (Fault). *Let P be a Datalog program, with input set E and intended semantics (I_+, I_-). Assume that E is incorrect w.r.t. P with (I_+, I_-). Then, a* fault *of E is a tuple t such that either t is desirable but missing, i.e., $t \in I_+ \backslash P(E)$ or t is undesirable but produced, i.e., $t \in P(E) \cap I_-$.*

We can formalize the situation where an incremental update for a Datalog program introduces a fault. Let P be a Datalog program with intended semantics $I_\checkmark = (I_+, I_-)$ and let E_1 be an input EDB. Then, let $\Delta E_{1\to2}$ be an incremental update, such that $E_1 \uplus \Delta E_{1\to2}$ results in another input EDB, E_2. Then, assume that E_1 is correct w.r.t $I_\checkmark$, but E_2 is incorrect.

Fault Localization. The fault localization problem allows the user to pinpoint the sources of faults. This is achieved by providing a minimal subset of the incremental update that can still reproduce the fault.

Definition 4 (Fault Localization). *A* fault localization *is a subset* $\delta E \subseteq \Delta E_{1\to2}$ *such that* $P(E_1 \uplus \delta E)$ *exhibits* all *faults of* E_2.

Rollback Suggestion. A rollback suggestion provides a subset of the diff, such that its removal from the diff would fix all faults.

Definition 5 (Rollback Suggestion). *A* rollback suggestion *is a subset* $\delta E_\times \subseteq \Delta E_{1\to2}$ *such that* $P(E_1 \uplus (\Delta E_{1\to2} \backslash \delta E_\times))$ *does not produce any faults of* E_2.

Ideally, fault localizations and rollback suggestions should be of minimal size.

4.2 Fault Localization

In the context of incremental Datalog, the *fault localization problem* provides a small subset of the incremental changes that allow the fault to be reproduced.

We begin by considering a basic version of the fault localization problem. In this basic version, we have a positive Datalog program (i.e., with no negation), and we localize a set of faults that are undesirable but appear (i.e., $P(E) \cap I_-$). The main idea of the fault localization algorithm is to compute a proof tree for each fault tuple. The tuples forming these proof trees are sufficient to localize the faults since these tuples allow the proof trees to be valid and, thus, the fault tuples to be reproduced.

Algorithm 1. Localize-Faults(P, E_2, $\Delta E_{1\to2}$, F): Given a diff $\Delta E_{1\to2}$ and a set of fault tuples F, returns $\delta E \subseteq \Delta E_{1\to2}$ such that $E_1 \uplus \delta E$ produces all $t \in F$

1: **for** tuple $t \in F$ **do**
2: Let inc-prov$(t, \Delta E_{1\to2})$ be an incremental proof tree of t w.r.t P and E_2, containing tuples that were inserted due to $\Delta E_{1\to2}$
3: **return** $\bigcup_{t\in F}(\text{inc-prov}(t, \Delta E_{1\to2}) \cap \Delta E_{1\to2})$

The basic fault localization is presented in Algorithm 1. For each fault tuple $t \in F$, the algorithm computes one incremental proof tree inc-prov$(t, \Delta E_{1\to2})$. These proof trees contain the set of tuples that were inserted due to the incremental update $\Delta E_{1\to2}$ and cause the existence of each fault tuple t. Therefore, by returning the union $\bigcup_{t\in F}(\text{inc-prov}(t, \Delta E_{1\to2}) \cap \Delta E_{1\to2})$, the algorithm produces a subset of $\Delta E_{1\to2}$ that reproduces the faults.

4.3 Rollback Repair

The rollback repair algorithm produces a rollback suggestion. As with fault localization, we begin with a basic version of the rollback problem, where we have only a positive Datalog program and wish to roll back a set of unwanted fault tuples. The basic rollback repair algorithm involves computing *all* non-cyclic proof trees for each fault tuple and 'disabling' each of those proof trees, as shown in Algorithm 2. If all proof trees are invalid, the fault tuple will no longer be computed by the resulting EDB.

Algorithm 2. Rollback-Repair(P, E_2, $\Delta E_{1\to 2}$, F): Given a diff $\Delta E_{1\to 2}$ and a set of fault tuples F, return a subset $\delta E \subseteq \Delta E_{1\to 2}$ such that $E_1 \uplus (\Delta E_{1\to 2} \backslash \delta E)$ does not produce t_r

---

Let all-inc-prov$(t, \Delta E_{1\to 2}) = \{T_1, \ldots, T_n\}$ be the total incremental provenance for a tuple t w.r.t P and E_2, where each T_i is a non-cyclic proof tree containing tuples inserted due to $\Delta E_{1\to 2}$.
Construct an integer linear program instance as follows:
Create a 0/1 integer variable x_{t_k} for each tuple t_k that occurs in the proof trees in all-inc-prov$(t, \Delta E_{1\to 2})$ for each fault tuple $t \in F$
for each tuple $t_f \in F$ **do**
    for each proof tree $T_i \in$ all-inc-prov$(t, \Delta E_{1\to 2})$ **do**
        for each line $t_h \leftarrow t_1 \wedge \ldots \wedge t_k$ in T_i **do**
            Add a constraint $x_{t_1} + \ldots + x_{t_k} - x_{t_h} \le k - 1$
    Add a constraint $x_{t_f} = 0$
Add the objective function maximize $\sum_{t_e \in EDB} x_{t_e}$
Solve the ILP
Return $\{t \in \Delta E_{1\to 2} \mid x_t = 0\}$

Algorithm 2 computes a minimum subset of the diff $\Delta E_{1\to 2}$, which would prevent the production of each $t \in F$ when excluded from the diff. The key idea is to use integer linear programming (ILP) [32] as a vehicle to disable EDB tuples so that the fault tuples vanish in the IDB. We phrase the proof trees as a pseudo-Boolean formula [21] whose propositions represent the EDB and IDB tuples. In the ILP, the faulty tuples are constrained to be false, and the EDB tuples assuming the true value are to be maximized, i.e., we wish to eliminate the least number of tuples in the EDB for repair. The ILP created in Algorithm 2 introduces a variable for each tuple (either IDB or EDB) that appears in *all* incremental proof trees for the fault tuples. For the ILP model, we have three types of constraints: (1) to encode a single-step proof, (2) to enforce that fault tuples are false, and (3) to ensure that variables are in the 0-1 domain. The constraints encoding proof trees correspond to each one-step derivation which can be expressed as a Boolean constraint $t_1 \wedge \ldots \wedge t_k \implies t_h$ for the rule application, where $t_1, \ldots, t_k$ and t_h are Boolean variables. Using propositional logic rules, this is equivalent to $\overline{t_1} \vee \ldots \vee \overline{t_k} \vee t_h$. This formula is then transformed into a pseudo (linear) Boolean formula where φ maps a Boolean function to the 0–1 domain, and x_t corresponds to the 0–1 variable of proposition t in the ILP.

$$\begin{aligned} \varphi\left(\overline{t_1} \vee \ldots \vee \overline{t_k} \vee t_h\right) &\equiv (1 - x_{t_1}) + \ldots + (1 - x_{t_k}) + t_h > 0 \\ &\equiv x_{t_1} + \ldots + x_{t_k} - x_{t_h} \le k - 1 \end{aligned}$$

The constraints assuming false values for fault tuples $t_f \in F$ are simple equalities, i.e., $x_{t_f} = 0$. The objective function for the ILP is to maximize the number of inserted tuples that are kept, which is equivalent to minimizing the number of tuples in $\Delta E_{1\to2}$ that are disabled by the repair. In ILP form, this is expressed as maximizing $\sum_{t\in\Delta E_{1\to2}} t$.

$$\begin{array}{lll} \text{max.} & \sum_{t\in\Delta E_{1\to2}} x_t & \\ \text{s.t.} & x_{t_1} + \ldots x_{t_k} - x_{t_h} \leq k-1 & (\forall t_h \Leftarrow t_1 \wedge \ldots \wedge t_k \in T_i) \\ & x_{t_f} = 0 & (\forall t_f \in F) \\ & x_t \in \{0,1\} & (\forall \text{tuples } t) \end{array}$$

The solution of the ILP permits us to determine the EDB tuples for repair:

$$\delta E = \{t \in \Delta E_{1\to2} \mid x_t = 0\}$$

This is a minimal set of inserted tuples that must be removed from $\Delta E_{1\to2}$ so that the fault tuples disappear.

This ILP formulation encodes the problem of disabling all proof trees for all fault tuples while maximizing the number of inserted tuples kept in the result. If there are multiple fault tuples, the algorithm computes proof trees for each fault tuple and combines all proof trees in the ILP encoding. The result is a set of tuples that is minimal but sufficient to prevent the fault tuples from being produced.

4.4 Extensions

Missing Tuples. The basic versions of the fault localization and rollback repair problem only handle a tuple which is undesirable but appears. The opposite kind of fault, i.e., a tuple which is desirable but missing, can be localized or repaired by considering a *dual* version of the problem. For example, consider a tuple t that disappears after applying a diff $\Delta E_{1\to2}$, and appears in the update in the opposite direction, $\Delta E_{2\to1}$. Then, the dual problem of localizing the disappearance of t is to *rollback* the appearance of t after applying the opposite diff, $\Delta E_{2\to1}$.

To localize a disappearing tuple t, we want to provide a small subset δE of $\Delta E_{1\to2}$ such that t is still not computable after applying δE to E_1. To achieve this, *all* ways to derive t must be invalid after applying δE. Considering the dual problem, rolling back the appearance of t in $\Delta E_{2\to1}$ results in a subset δE such that $E_2 \uplus (\Delta E_{2\to1} \backslash \delta E)$ does not produce t. Since $E_1 = E_2 \uplus \Delta E_{2\to1}$, if we were to apply the reverse of δE (i.e., insertions become deletions and vice versa), we would arrive at the same EDB set as $E_2 \uplus (\Delta E_{2\to1} \backslash \delta E)$. Therefore, the reverse of δE is the desired minimal subset that localizes the disappearance of t.

Similarly, to roll back a disappearing tuple t, we apply the dual problem of *localizing* the appearance of t after applying the opposite diff $\Delta E_{2\to1}$. Here, to roll back a disappearing tuple, we introduce *one* way of deriving t. Therefore, localizing the appearance of t in the opposite diff provides one derivation for t and thus is the desired solution. In summary, to localize or rollback a tuple t that is missing after applying $\Delta E_{1\to 2}$, we compute a solution for the dual problem. The dual problem for localization is to roll back the appearance of t after applying $\Delta E_{2\to 1}$, and similarly, the dual problem for rollback is localization.

Stratified Negation. Stratified negation is a common extension for Datalog. With stratified negation, atoms in the body of a Datalog rule may appear negated, e.g., $R_h(X_h)$:- $R_1(X_1), \ldots, !R_k(X_k), \ldots, R_n(X_n)$. The negated atoms are denoted with !, and any variables contained in negated atoms must also exist in some positive atom in the body of the rule (a property called *groundedness*). Semantically, negations are satisfied if the instantiated tuple *does not* exist in the corresponding relation. The 'stratified' in 'stratified negation' refers to the property that no cyclic negations can exist. For example, the rule $A(x)$:- $B(x,y), !A(y)$ causes a dependency cycle where A depends on the negation of A and thus is not allowed under stratified negation.

Consider the problem of localizing the appearance of an unwanted tuple t. If the Datalog program contains stratified negation, then the appearance of t can be caused by two possible situations. Either (1) there is a positive tuple in the proof tree of t that appears, or (2) there is a negated tuple in the proof tree of t that disappears. The first case is the standard case, but in the second case, if a negated tuple disappears, then its disappearance can be localized or rolled back by computing the dual problem as in the missing tuple strategy presented above. We may encounter further negated tuples in executing the dual version of the problem. For example, consider the set of Datalog rules $A(x)$:- $B(x), !C(x)$ and $C(x)$:- $D(x), !E(x)$. If we wish to localize an appearing (unwanted) tuple `A(x)`, we may encounter a disappearing tuple `C(x)`. Then, executing the dual problem, we may encounter an appearing tuple `E(x)`. We can generally continue flipping between the dual problems to solve the localization or repair problem. This process is guaranteed to terminate due to the stratification of negations. Each time the algorithm encounters a negated tuple, it must appear in an earlier stratum than the previous negation. Therefore, eventually, the negations will reach the input EDB, and the process terminates.

4.5 Full Algorithm

The full rollback repair algorithm presented in Algorithm 3 incorporates the basic version of the problem and all of the extensions presented above. The result of the algorithm is a rollback suggestion, which fixes all faults. Algorithm 3 begins by initializing the EDB after applying the diff (line 1) and separate sets of unwanted faults (lines 2) and missing faults (3). The set of candidate tuples forming the repair is initialized to be empty (line 4).

The main part of the algorithm is a worklist loop (lines 5 to 13). In this loop, the algorithm first processes all unwanted but appearing faults (F^+, line 6) by computing the repair of F^+. The result is a subset of tuples in the diff such that the faults F^+ no longer appear when the subset is excluded from the diff. In the provenance system, negations are treated as EDB tuples, and thus the resulting repair may contain negated tuples. These negated tuples are added to

Algorithm 3 . Full-Rollback-Repair(P, E_1, $\Delta E_{1\to 2}$, (I_+, I_-)): Given a diff $\Delta E_{1\to 2}$ and an intended semantics $(I_+, I_,)$, compute a subset $\delta E \subseteq \Delta E_{1\to 2}$ such that $\Delta E_{1\to 2} \backslash \delta E$ satisfies the intended semantics

```
1: Let E_2 be the EDB after applying the diff: E_1 ⊎ ΔE_{1→2}
2: Let F^+ be appearing unwanted faults: {I_- ∩ P(E_2)}
3: Let F^- be missing desirable faults: {I_+ \ P(E_2)}
4: Let L be the set of repair tuples, initialized to ∅
5: while both F^+ and F^- are non-empty do
6:     Add Rollback-Repair(P, E_2, ΔE_{1→2}, F^+) to L
7:     for negated tuples !t ∈ L do
8:         Add t to F^-
9:     Clear F^+
10:    Add Localize-Faults(P, E_1, ΔE_{2→1}, F^-) to L
11:    for negated tuples !t ∈ L do
12:        Add t to F^+
13:    Clear F^-
14: return L
```

F^- (line 7) since a tuple appearing in F^+ may be caused by a negated tuple disappearing. The algorithm then repairs the tuples in F^- by computing the dual problem, i.e., localizing F^- with respect to $\Delta E_{2\to 1}$. Again, this process may result in negated tuples, which are added to F^+, and the loop begins again. This worklist loop must terminate, due to the semantics of stratified negation, as discussed above. At the end of the worklist loop, L contains a candidate repair.

While Algorithm 3 presents a full algorithm for rollback, the fault localization problem can be solved similarly. Since rollback and localization are dual problems, the full fault localization algorithm swaps Rollback-Repair in line 6 and Localize-Faults in line 10.

Example. We demonstrate how our algorithms work by using our running example. Recall that we introduce an incremental update consisting of inserting two tuples `assign(upgradedSession, userSession)` and `load(userSession, admin, session)`. As a result, the system computes the unwanted fault tuple `alias(userSession, sec)`. To rollback the appearance of the fault tuple, the algorithms start by computing its provenance, as shown in Fig. 3. The algorithm then creates a set of ILP constraints, where each tuple (with shortened variables) represents an ILP variable:

$$\text{maximize} \sum \texttt{load(u, a, s)} \text{ such that}$$

$$\texttt{load(u, a, s)} - \texttt{vpt(u, L2)} \leq 0,\ \texttt{vpt(u, L2)} - \texttt{alias(u, s)} \leq 0,\ \texttt{alias(u, s)} = 0$$

For this simple ILP, the result indicates that the insertion of `load(userSession, admin, session)` should be rolled back to fix the fault.

5 Experiments

This section evaluates our technique on real-world benchmarks to determine its effectiveness and applicability. We consider the following research questions:

- RQ1: Is the new technique faster than a delta-debugging strategy?
- RQ2: Does the new technique produce more precise localization/repair candidates than delta debugging?

Experimental Setup:[1] We implemented the fault localization and repair algorithms using Python[2]. The Python code calls out to an incremental version of the Soufflé Datalog engine [39] extended with incremental provenance. Our implementation of incremental provenance uses the default metric of minimizing proof tree height, as it provides near-optimal repairs with slight runtime improvements. For solving integer linear programs, we use the GLPK library.

Our main point of comparison in our experimental evaluation is the delta debugging approach, such as that used in the ProSynth Datalog synthesis framework [29]. We adapted the implementation of delta debugging used in ProSynth to support input tuple updates. Like our fault repair implementation, the delta debugging algorithm was implemented in Python; however, it calls out to the standard Soufflé engine since that provides a lower overhead than the incremental or provenance versions.

For our benchmarks, we use the Doop program analysis framework [8] with the DaCapo set of Java benchmarks [7]. The analysis contains approx. 300 relations, 850 rules, and generates approx. 25 million tuples from an input size of 4–9 million tuples per DaCapo benchmark. For each of the DaCapo benchmarks, we selected an incremental update containing 50 tuples to insert and 50 tuples to delete, which is representative of the size of a typical commit in the underlying source code. From the resulting IDB changes, we selected four different arbitrary fault sets for each benchmark, which may represent an analysis error.

Performance: The results of our experiments are shown in Table 1. Our fault repair technique performs much better overall compared to the delta debugging technique. We observe a geometric mean speedup of over 26.9×[3] compared to delta debugging. For delta debugging, the main cause of performance slowdown is that it is a black-box search technique, and it requires multiple iterations of invoking Soufflé (between 6 and 19 invocations for the presented benchmarks) to direct the search. This also means that any intermediate results generated in a previous Soufflé run are discarded since no state is kept to allow the reuse of results. Each invocation of Soufflé takes between 30–50 s, depending on the benchmark and EDB. Thus, the overall runtime for delta debugging is in the hundreds of seconds at a minimum. Indeed, we observe that delta debugging takes between 370 and 6135 s on our benchmarks, with one instance timing out after two hours (7200 s).

[1] We use an Intel Xeon Gold 6130 with 192 GB RAM, GCC 10.3.1, and Python 3.8.10.

[2] Available at github.com/davidwzhao/souffle-fault-localization.

[3] We say "over" because we bound timeouts to 7200 s.

Table 1. Repair size and runtime of our technique compared to delta debugging

Program	No.	Rollback repair				Delta debugging		Speedup
		Size	Overall (s)	Local(s)	Repair(s)	Size	Runtime (s)	
antlr	1	2	73.6	0.51	73.1	3	3057.8	41.5
	2	1	79.4	0.00	79.4	1	596.5	7.5
	3	1	0.95	0.95	–	1	530.8	558.7
	4	2	77.8	1.89	75.9	3	3017.6	38.8
bloat	1	2	3309.5	0.02	3294.1	2	2858.6	0.9
	2	1	356.3	0.00	355.4	1	513.6	1.4
	3	1	0.33	0.33	–	1	557.7	1690.0
	4	3	3870.6	0.10	3854.7	2	2808.3	0.7
chart	1	1	192.6	0.00	192.6	1	685.0	3.6
	2	1	3.01	3.01	–	1	675.3	224.4
	3	1	78.8	0.00	78.8	1	667.6	8.5
	4	2	79.9	3.24	76.7	3	3001.1	37.6
eclipse	1	2	177.3	0.04	177.2	3	2591.2	14.6
	2	1	79.2	0.00	79.1	1	416.1	5.3
	3	1	0.12	0.12	–	1	506.3	4219.2
	4	3	91.9	0.09	91.8	3	2424.4	26.4
fop	1	2	83.8	0.05	83.8	2	3446.6	41.1
	2	1	76.9	0.00	76.9	1	670.7	8.7
	3	1	0.66	0.66	–	1	721.8	1093.6
	4	6	74.8	0.50	74.3	Timeout (7200)		96.3+
hsqldb	1	2	83.3	0.04	83.3	2	2979.2	35.8
	2	1	79.4	0.00	79.4	1	433.8	5.5
	3	1	74.0	0.00	74.0	1	663.1	9.0
	4	3	75.5	0.04	75.5	5	6134.8	81.3
jython	1	1	83.3	0.00	83.3	1	609.4	7.3
	2	1	78.2	0.00	78.2	1	590.4	7.5
	3	1	76.6	0.00	76.6	1	596.2	7.8
	4	1	75.8	0.00	75.8	1	587.6	7.8
luindex	1	2	81.3	0.07	81.2	3	2392.1	29.4
	2	1	79.8	0.00	79.8	1	511.0	6.4
	3	1	0.10	0.10	–	1	464.8	4648.0
	4	4	77.9	0.12	77.8	5	4570.4	58.7
lusearch	1	2	110.2	0.06	110.0	3	2558.8	23.2
	2	1	1062.1	0.00	1057.4	1	370.4	0.3
	3	1	0.12	0.12	–	1	369.6	3080.0
	4	2	294.2	0.06	293.2	3	2420.9	8.2
pmd	1	2	78.1	0.02	78.1	3	3069.8	39.3
	2	1	77.0	0.00	77.0	1	600.2	7.8
	3	1	0.08	0.08	–	1	717.8	8972.5
	4	3	74.3	0.08	74.2	3	2828.3	38.1
xalan	1	1	84.9	0.00	84.9	1	745.3	8.8
	2	1	82.2	0.00	82.2	1	728.9	8.9
	3	1	100.1	0.00	100.1	1	1243.7	12.4
	4	1	521.6	0.00	518.3	1	712.5	1.4

On the other hand, our rollback repair technique calls for provenance information from an already initialized instance of incremental Soufflé. This incrementality allows our technique to reuse the already computed IDB for each provenance query. For eight of the benchmarks, the faults only contained missing tuples. Therefore, only the Localize-Faults method was called, which only computes one proof tree for each fault tuple and does not require any ILP solving. The remainder of the benchmarks called the Rollback-Repair method, where the main bottleneck is for constructing and solving the ILP instance. For three of the benchmarks, `bloat-1`, `bloat-4`, and `lusearch-2`, the runtime was slower than delta debugging. This result is due to the fault tuples in these benchmarks having many different proof trees, which took longer to compute. In addition, this multitude of proof trees causes a larger ILP instance to be constructed, which took longer to solve.

Quality: While the delta debugging technique produces 1-minimal results, we observe that despite no overall optimality guarantees, our approach was able to produce more minimal repairs in 27% of the benchmarks. Moreover, our rollback repair technique produced a larger repair in only one of the benchmarks. This difference in quality is due to the choices made during delta debugging. Since delta debugging has no view of the internals of Datalog execution, it can only partition the EDB tuples randomly. Then, the choices made by delta debugging may lead to a locally minimal result that is not globally optimal. For our fault localization technique, most of the benchmarks computed one iteration of repair and did not encounter any negations. Therefore, due to the ILP formulation, the results were optimal in these situations. Despite our technique overall not necessarily being optimal, it still produces high-quality results in practice.

6 Related Work

Logic Programming Input Repair. A plethora of logic programming paradigms exist that can express diagnosis and repair by EDB regeneration [13,16,17,24]. Unlike these logic programming paradigms, our technique is designed to be embedded in high-performance modern Datalog engines. Moreover, our approach can previous computations (proof trees and incremental updates) to localize and repair only needed tuples. This bounds the set of repair candidates and results in apparent speedups. Other approaches, such as the ABC Repair System [26], use a combination of provenance-like structures and user-guided search to localize and repair faults. However, that approach is targeted at the level of the Datalog specification and does not always produce effective repairs. Techniques such as delta debugging have recently been used to perform state-of-the-art synthesis of Datalog programs efficiently [29]. Our delta debugging implementation adapts this method, given it produces very competitive synthesis performance and can be easily re-targeted to diagnose and repair inputs.

Database Repair. Repairing inconsistent databases with respect to integrity constraints has been extensively investigated in the database community [3,9,15,16].

Unlike our approach, integrity constraints are much less expressive than Datalog; in particular, they do not allow fixpoints in their logic. The technique in [16] shares another similarity in that it also presents repair for incremental SQL evaluation. However, this is limited to relational algebra, i.e., SQL and Constrained Functional Dependencies (CFDs) that are less expressive than Datalog. A more related variant of database repair is consistent query answering (CQA) [3,9]. These techniques repair query answers given a database, integrity constraints and an SQL query. Similarly, these approaches do not support recursive queries, as can be expressed by Datalog rules.

Program Slicing. Program slicing [5,14,20,34] encompasses several techniques that aim to compute portions (or *slices*) of a program that contribute to a particular output result. For fault localization and debugging, program slicing can be used to localize slices of programs that lead to a fault or error. The two main approaches are *static* program slicing, which operates on a static control flow graph, and *dynamic* program slicing, which considers the values of variables or execution flow of a particular execution. As highlighted by [11], data provenance is closely related to slicing. Therefore, our technique can be seen as a form of static slicing of the Datalog EDB with an additional rollback repair stage.

7 Conclusion

We have presented a new debugging technique that localizes faults and provides rollback suggestions for Datalog program inputs. Unlike previous approaches, our technique does not entirely rely on a black-box solver to perform the underlying repair. Instead, we utilize incremental provenance information. As a result, our technique exhibits speedups of 26.9× compared to delta debugging and finds more minimal repairs 27% of the time.

There are also several potential future directions for this research. One is to adapt our technique to repair changes in Datalog *rules*, as well as changes in input tuples. Another direction is to adopt these techniques for different domain areas outside the use cases of program analyses.

Acknowledgments. M.R. was funded by U.S. NSF grants CCF-2146518, CCF-2124431, and CCF-2107261.

References

1. GitHub CodeQL (2021). https://codeql.github.com/. Accessed 19 Oct 2021
2. Allen, N., Scholz, B., Krishnan, P.: Staged points-to analysis for large code bases. In: Franke, B. (ed.) CC 2015. LNCS, vol. 9031, pp. 131–150. Springer, Heidelberg (2015). https://doi.org/10.1007/978-3-662-46663-6_7
3. Arenas, M., Bertossi, L.E., Chomicki, J.: Answer sets for consistent query answering in inconsistent databases. Theory Pract. Log. Program. **3**(4–5), 393–424 (2003)

4. Backes, J., et al.: Reachability analysis for AWS-based networks. In: Dillig, I., Tasiran, S. (eds.) CAV 2019. LNCS, vol. 11562, pp. 231–241. Springer, Cham (2019). https://doi.org/10.1007/978-3-030-25543-5_14
5. Binkley, D.W., Gallagher, K.B.: Program slicing. Adv. Comput. **43**, 1–50 (1996)
6. Blackburn, S.M., et al.: The DaCapo benchmarks: Java benchmarking development and analysis. In: OOPSLA 2006: Proceedings of the 21st annual ACM SIGPLAN conference on Object-Oriented Programing, Systems, Languages, and Applications, pp. 169–190. ACM Press, New York (2006). http://doi.acm.org/10.1145/1167473.1167488
7. Blackburn, S.M., et al.: The dacapo benchmarks: Java benchmarking development and analysis. In: Proceedings of the 21st Annual ACM SIGPLAN Conference on Object-Oriented Programming Systems, Languages, and Applications, pp. 169–190 (2006)
8. Bravenboer, M., Smaragdakis, Y.: Strictly declarative specification of sophisticated points-to analyses. SIGPLAN Not. **44**(10), 243–262 (2009)
9. Bravo, L., Bertossi, L.E.: Consistent query answering under inclusion dependencies. In: Lutfiyya, H., Singer, J., Stewart, D.A. (eds.) Proceedings of the 2004 conference of the Centre for Advanced Studies on Collaborative research, 5–7 October 2004, Markham, Ontario, Canada, pp. 202–216. IBM (2004)
10. Caballero, R., Riesco, A., Silva, J.: A survey of algorithmic debugging. ACM Comput. Surv. (CSUR) **50**(4), 60 (2017)
11. Cheney, J.: Program slicing and data provenance. IEEE Data Eng. Bull. **30**(4), 22–28 (2007)
12. Distefano, D., Fähndrich, M., Logozzo, F., O'Hearn, P.W.: Scaling static analyses at Facebook. Commun. ACM **62**(8), 62–70 (2019)
13. El-Hassany, A., Tsankov, P., Vanbever, L., Vechev, M.: Network-wide configuration synthesis. In: Majumdar, R., Kunčak, V. (eds.) CAV 2017. LNCS, vol. 10427, pp. 261–281. Springer, Cham (2017). https://doi.org/10.1007/978-3-319-63390-9_14
14. Ezekiel, S., Lukas, K., Marcel, B., Zeller, A.: Locating faults with program slicing: an empirical analysis. Empir. Softw. Eng. **26**(3), 1–45 (2021)
15. Fan, W.: Constraint-driven database repair. In: Liu, L., Özsu, M.T. (eds.) Encyclopedia of Database Systems, pp. 458–463. Springer, Boston (2009). https://doi.org/10.1007/978-0-387-39940-9_599
16. Fan, W., Geerts, F., Jia, X.: Semandaq: a data quality system based on conditional functional dependencies. Proc. VLDB Endow. **1**(2), 1460–1463 (2008). https://doi.org/10.14778/1454159.1454200
17. Gelfond, M., Lifschitz, V.: The Stable Model Semantics for Logic Programming, pp. 1070–1080. MIT Press (1988)
18. Grech, N., Brent, L., Scholz, B., Smaragdakis, Y.: Gigahorse: thorough, declarative decompilation of smart contracts. In: Proceedings of the 41th International Conference on Software Engineering, ICSE 2019, p. (to appear). ACM, Montreal (2019)
19. Grech, N., Kong, M., Jurisevic, A., Brent, L., Scholz, B., Smaragdakis, Y.: MadMax: surviving out-of-gas conditions in ethereum smart contracts. In: SPLASH 2018 OOPSLA (2018)
20. Harman, M., Hierons, R.: An overview of program slicing. Softw. Focus **2**(3), 85–92 (2001)
21. Hooker, J.: Generalized resolution for 0–1 linear inequalities. Ann. Math. Artif. Intell. **6**, 271–286 (1992). https://doi.org/10.1007/BF01531033

22. Huang, S.S., Green, T.J., Loo, B.T.: Datalog and emerging applications: an interactive tutorial. In: Proceedings of the 2011 ACM SIGMOD International Conference on Management of Data, SIGMOD 2011, pp. 1213–1216. ACM (2011). https://doi.org/10.1145/1989323.1989456, https://doi.acm.org/10.1145/1989323.1989456
23. Jordan, H., Scholz, B., Subotić, P.: SOUFFLÉ: on synthesis of program analyzers. In: Chaudhuri, S., Farzan, A. (eds.) CAV 2016. LNCS, vol. 9780, pp. 422–430. Springer, Cham (2016). https://doi.org/10.1007/978-3-319-41540-6_23
24. Kakas, A.C., Kowalski, R.A., Toni, F.: Abductive logic programming (1993)
25. Karvounarakis, G., Ives, Z.G., Tannen, V.: Querying data provenance. In: SIGMOD 2010, p. 951–962. Association for Computing Machinery, New York (2010). https://doi.org/10.1145/1807167.1807269
26. Li, X., Bundy, A., Smaill, A.: ABC repair system for datalog-like theories. In: KEOD, pp. 333–340 (2018)
27. McSherry, F., Murray, D.G., Isaacs, R., Isard, M.: Differential dataflow. In: CIDR (2013)
28. Motik, B., Nenov, Y., Piro, R., Horrocks, I.: Maintenance of datalog materialisations revisited. Artif. Intell. **269**, 76–136 (2019)
29. Raghothaman, M., Mendelson, J., Zhao, D., Naik, M., Scholz, B.: Provenance-guided synthesis of datalog programs. Proc. ACM Program. Lang. **4**(POPL), 1–27 (2019)
30. Ryzhyk, L., Budiu, M.: Differential datalog. Datalog **2**, 4–5 (2019)
31. Schäfer, M., Avgustinov, P., de Moor, O.: Algebraic data types for object-oriented datalog (2017)
32. Schrijver, A.: Theory of Linear and Integer Programming. Wiley, USA (1986)
33. Vallée-Rai, R. Co, P., Gagnon, E., Hendren, L., Lam, P., Sundaresan, V.: Soot: a Java bytecode optimization framework. In: CASCON First Decade High Impact Papers, pp. 214–224 (2010)
34. Weiser, M.: Program slicing. IEEE Trans. Software Eng. **4**, 352–357 (1984)
35. Yan, M., Xia, X., Lo, D., Hassan, A.E., Li, S.: Characterizing and identifying reverted commits. Empir. Softw. Eng. **24**(4), 2171–2208 (2019). https://doi.org/10.1007/s10664-019-09688-8
36. Yoon, Y., Myers, B.A.: An exploratory study of backtracking strategies used by developers. In: Proceedings of the 5th International Workshop on Co-operative and Human Aspects of Software Engineering, CHASE 2012, pp. 138–144. IEEE Press (2012)
37. Zeller, A.: Yesterday, my program worked. Today, it does not. Why? ACM SIGSOFT Softw. Eng. Notes **24**(6), 253–267 (1999)
38. Zeller, A., Hildebrandt, R.: Simplifying and isolating failure-inducing input. IEEE Trans. Software Eng. **28**(2), 183–200 (2002)
39. Zhao, D., Subotic, P., Raghothaman, M., Scholz, B.: Towards elastic incrementalization for datalog. In: 23rd International Symposium on Principles and Practice of Declarative Programming, pp. 1–16 (2021)
40. Zhao, D., Subotić, P., Scholz, B.: Debugging large-scale datalog: a scalable provenance evaluation strategy. ACM Trans. Program. Lang. Syst. (TOPLAS) **42**(2), 1–35 (2020)
41. Zhou, W., Sherr, M., Tao, T., Li, X., Loo, B.T., Mao, Y.: Efficient querying and maintenance of network provenance at internet-scale. In: Proceedings of the 2010 ACM SIGMOD International Conference on Management of Data, pp. 615–626 (2010)

Author Index

Printed in the United States
by Baker & Taylor Publisher Services